GREEN
NANOTECHNOLOGY
Solutions for Sustainability and Energy in the Built Environment

GREEN
NANOTECHNOLOGY
Solutions for Sustainability and Energy in the Built Environment

Geoffrey B. Smith
Claes G. Granqvist

CRC Press
Taylor & Francis Group
Boca Raton   London   New York

CRC Press is an imprint of the
Taylor & Francis Group, an **informa** business

CRC Press
Taylor & Francis Group
6000 Broken Sound Parkway NW, Suite 300
Boca Raton, FL 33487-2742

Library of Congress Cataloging-in-Publication Data

Smith, Geoffrey B. (Geoffrey Burton)
  Green nanotechnology : solutions for sustainability and energy in the built environment / authors, Geoffrey B. Smith and Claes-Goran S. Granqvist.
    p. cm.
  "A CRC title."
  Includes bibliographical references and index.
  ISBN 978-1-4200-8532-7 (alk. paper)
  1. Sustainable buildings--Design and construction. 2. Nanotechnology. 3. Buildings--Electric equipment. 4. Buildings--Energy conservation. 5. Sustainable construction. 6. Electronic apparatus and appliances. I. Granqvist, Claes G. II. Title.

TH880.S65 2010
690.028'6--dc22   2010007968

Visit the Taylor & Francis Web site at
http://www.taylorandfrancis.com

and the CRC Press Web site at
http://www.crcpress.com

Contents

Preface

This book is about the science and technology of tiny structures that are able to improve the quality of life while simultaneously achieving huge reductions in the use of fossil fuels. The ultimate goal of this technology is to reduce carbon dioxide emission and other waste for the preservation and improvement of the world's ecosystems and for the well-being of humanity.

Our current ways of building and powering the modern world are damaging its atmosphere, oceans, lakes, forests, and fields. The threats to people, animals, and plants are manifold; Figure P1 illustrates some well-known aspects of these threats (based on selected references 1–6). Over the past 100 years or so, mankind has gradually and often painfully learned the importance of managing pollution and degradation of the environment due to deforestation, excessive irrigation, and too-intense farming. As a consequence, some products and practices have been modified or banned, and less-polluting new products have emerged. Better farming and land management are also gradually evolving. Unfortunately, most of the beneficial changes have followed only after serious environmental damage became obvious. We may be able to break the current cycle of accelerating damage from climate change, but climate is a hard nut to crack because it is influenced by so many factors and has both global and local aspects.

The whole earth–atmosphere ecosystem is under threat. This situation is an unprecedented challenge for humanity, and one that has implications for all aspects of our daily lives. Establishing better ways to provide and use energy is at the core of the solution and is a multifaceted task involving many products and most human activities.

And let us not forget that the world population is growing. Currently (2010), it is somewhat below 7 billion and, according to the United Nations, it is not expected to stabilize until it has exceeded 10 billion sometime in the coming 100–200 years. The enormity of this population growth cannot be understood without a historical perspective. Demographers estimate that there may have been 170–400 million people alive on earth in the year 1 A.D. It took more than one and a half

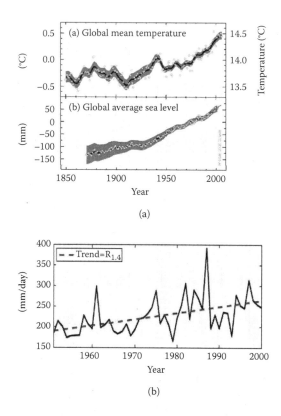

(a)

(b)

FIGURE P1 *A color version of this figure follows page 200.* Panorama over climate changes and some of their impacts. Panel (a) shows increasing global mean temperature and associated rise of global average sea level since 1850. (Reproduced with permission from Climate Change 2007—The Physical Science Basis. Working Group I Contribution to the Fourth Assessment Report of the IPCC [Intergovernmental Panel on Climate Change], Cambridge University Press, Cambridge, U.K., 2007; http://ipcc-wg1.ucar.edu/wg1/.) Panel (b) indicates the increasing trend of extreme rain events over India, specifically the mean rainfall during the four highest rain events every season during 1951–2000. (From B. N. Goswami et al., *Science* 314 (2006) 1442–1445. With permission.)

Continued

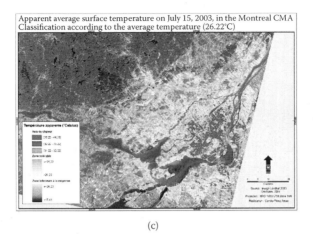

(c)

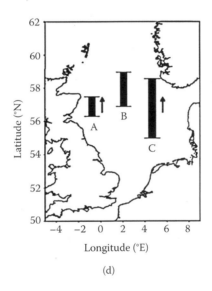

(d)

FIGURE P1 (*Continued*)   Panorama over climate changes and some of their impacts. Panel (c) depicts the urban heat island effect for the case of Montreal, Canada; yellow, red, and violet denote temperatures 5, 6, and more than 7°C above the mean temperature of 26.2°C as a result of the built environment. (From http://www.urbanheatislands.com. With permission.) Panel (d) indicates distribution shifts in marine fishes, specifically how North Sea fish distributions have moved to more northern latitudes in the case of cod (A), anglerfish (B), and snake blenny (C) during the period from 1977 to 2001. (From A. L. Perry et al., *Science* 308 (2005) 1912–1915. With permission.)   *Continued*

(e)

(f)

FIGURE P1 (*Continued*) Panorama over climate changes and some of their impacts. Panel (e) is a satellite image of wildfires outside San Diego, California. (From http://earthobservatory.nasa.gov/wildfires San Diego. With permission.) Panel (f) shows the break-up of ice sheets in the Arctic and the accompanying deterioration of the habitat for polar bears. (From J. Zillman, G. Pearman, Climate Change: Global and Local Implications, Presentation to the Australian Academy of Technological Sciences and Engineering, March 1, 2007; http://www.atse.org.au/uploads/ccsb010307. pdf. With permission.)

millennia, until 1650, for the figure to reach just 500 million. In the following centuries, however, the population began to grow more rapidly, so that by around 1800 there were 1 billion people in the world. By 1950, world population was over 2.5 billion, and in the following 50 years it more than doubled to 6 billion. As the population has exploded, so also has the need for more food, water, housing, transportation, consumer goods, social services, and energy.

The solutions to the energy problem lie in both technology and human behavior. The way we influence these is through new basic and applied science as well as global, national, and local community policies. Economists and politicians seek policies that balance between future economic losses as the environment degrades and the near-term costs and future economic benefits of various actions. Both sides of this equation involve huge numbers, amounting to trillions of dollars on a global scale.

Of all possible mechanisms for reducing energy use, the economically most viable one—even far ahead of requiring payment for carbon emissions—is the development and application of new low-cost technologies. These technologies must either drastically reduce emissions or act as super-efficient carbon sinks, capabilities that are made possible through green nanotechnology. This book is focused on the technologies to reduce emissions. The carbon sinks may be provided by new biotechnologies that are being advanced by nanoscience-based techniques enabling engineering and measurement at the level of the single cell. Developing both technologies will require large R&D resources, and the best scenario may include carbon trading in the near term to provide the incentives for the transition. If these new technologies eventually take over, future generations may look back and wonder why we ever put up with the dirty systems we have now. Taxing carbon emission will then be unnecessary. Indeed, we learned a long time ago that if one has the option of a clean water supply, one does not use water from a polluted stream. Once mankind experiences a new world based on clean systems, the parallel choice in energy will also be obvious.

All technologies ultimately use molecules or materials. Nanoscience is changing the world of materials, and hence its influence will be both broad and deep. We have observed at various forums around the world in the early 2000s that leading corporations and many governments and international bodies are very alert to opportunities in "green" nanotechnologies and keen to invest in them. It is also apparent to us that research efforts have tended to be too narrowly focused, with many scientists and business people following trends and glamour, just as in the "fashion" industry. This often means that media of various kinds—not excluding scientific journals—tend to emphasize costly and long-term solutions at the expense of functional, affordable, and near-term options. We hope this book can change this, and that it will open readers' eyes to a fascinating vision of how technology and nanoscience can merge and lead to commodity scale products to help preserve our planet. Commodities, by definition, are widely used and affordable, so most current solar technologies—including solar cells—do not yet qualify as such. These technologies have a long way to go, but we hope they will achieve commodity scale status with the help of nanotechnology.

A first step in imagining a clean and sustainable future is to think differently about everyday products—in particular, how they influence the amount of energy we utilize. Almost everything we use can be made much more energy efficient. Steel plate can be replaced with stronger lightweight porous panels; skyscrapers can be built in new ways; displays for computers and TVs can be made super energy efficient. The second step is to understand the role of nanotechnology in these changes, in the modification or replacement of materials and products. With these two steps, one can begin to appreciate the vast potential of nanotechnology for meeting the "green" challenge.

A valuable side effect of most nanotechnologies is that they inherently put us in closer touch with the natural world, which in itself has excellent psychological and health benefits. In more natural environments, school children learn faster, and workers enjoy their work more and are thus more productive. Unfortunately, many 20th-century technologies—such as fluorescent lamps and air conditioning—were sterile and depressing and separated people from the external environment. In contrast, the technologies we will cover in this book are dynamic, uplifting, and tuned to the environment. Figure P2 illustrates some of these new options (based on selected references 7–10).

Some words on terminology are in order. The title of this book includes "nanotechnology," and we have already used "nanoscience" several times above. Of course, science can be viewed as enabling technology, but in many cases the terms *nanotechnology* and *nanoscience* can be used interchangeably without loss of clarity. We see no need to be rigid on this matter. There is also a tendency today toward "nano-anything," and we add our own little bit in all innocence by using "nano-tortiglione" to describe some nanostructures. The interest in "nano" has soared tremendously during the last 10–15 years. However, the field is not as new as it may seem, and the apparent novelty is partly a result of terminology. Today's nanoparticles are often the same as the "ultrafine" particles in earlier literature.

This book is broadly organized as follows: Chapter 1 introduces "green nanotechnology" and explains what it means and why we are interested in it, as well as indicating what it can accomplish. This chapter is also intended as an invitation to those who want to enter studies of a new technology that has lots to offer if applied wisely. Chapter 2 then sets the scene for the bulk of the book by discussing the energy flows in nature—in particular, of solar and thermal radiation, and the light that we perceive with our eyes. This discussion will lead to a number of idealized properties of materials that make the most of what nature has to offer in order to accomplish energy efficiency and the utilization of solar energy and natural thermal flows, primarily for the built environment. We speak here of "materials in harmony with the environment." Chapter 3 takes us into the science of real materials and introduces their optical properties with a focus on nano-based features. The main point here is to demonstrate how real materials can mimic the idealized performance put forward in Chapter 2. The remaining Chapters 4–8 are topical and devoted to in-depth discussions of materials, and devices based on them, for a number of buildings-related applications. Chapter 4 treats windows capable of giving good visual indoors–outdoors contact, and Chapter 5 treats luminaires capable of illumination using electricity or daylight. Chapter 6 covers heat and electricity, and discusses nanomaterials-based solar collectors and solar cells. In Chapter 7 we then turn to coolness and how it is produced, which is an important topic that has been sadly neglected in the past. Our emphasis here is on surfaces with high albedo ("whiteness") and on materials and devices for using the clear sky as a heat sink. Chapter 8 contains brief discussions of a number of supporting nanotechnologies, specifically for air sensing using nanostructured oxides, air cleaning by photocatalysis, thermal insulation with nanomaterials, and novel devices incorporating nanomaterials for electrical storage. There are a few minor overlaps between the "basic" Chapters 2 and 3 and the "technical–topical" Chapters 4–8; these overlaps are intentional and serve, we hope, to facilitate the reading of the later chapters. Chapter 9 summarizes the main results and attempts to define the elements of a scenario in which we make good use of what green nanotechnology can offer to lessen the burdens caused by increasing global and local temperatures and waning natural resources of every kind. The book concludes with two appendices, one on thin-film deposition and the other on abbreviations, acronyms, and symbols.

The book can be read in different ways, and it does not have to be studied from the first page to the last. Chapter 2 is a foundation for most of what follows, though, so leaving out that part of the book is not a good idea. Chapter 3 is mainly of interest to materials scientists who want to understand how the optics of nanomaterials works from a fundamental perspective, and it can be omitted if the purpose of reading the book is to get an overview and not dig too deeply. The "topical–technical" Chapters 4–8 can be read in any order without serious loss of context. Throughout the chapters there are "boxes" devoted to discussions of topics in particular depth or for discussing matters that presuppose that the reader has a solid background in science.

This book represents the authors' best attempt to bring together a number of topics with a great deal of internal cohesion both for applications and for the significance of nanofeatures regarding the materials that are involved. Some aspects may be given a more cursory discussion than some readers would expect, whereas other aspects are covered in more detail than in earlier texts. For example, we do not discuss nanostructured solar cells in much detail, and there are two reasons for that. First, there are many excellent reviews already; second, large-scale applications of solar cells in the built environment may be problematic because they tend to heat up the air around the buildings and generate a need for increased air conditioning. On the other hand, we do discuss methods for preventing excessive temperatures and for passive cooling more elaborately than has been done in earlier buildings-related literature. Windows and luminaires are discussed in considerable detail. We cover air sensing and air cleaning by use of nanomaterials, although perhaps not to the extent these topics might warrant, since these technologies are in the early stages of R&D. The same could be said for nanomaterials-based thermal insulation.

FIGURE P2 (Facing Page) *A color version of this figure follows page 200.* Panorama over some aspects of green nanotechnologies. Panel (a) shows a low-energy commercial building in San Francisco with energy-efficient glazing and natural ventilation. (From PIERS, High Performance Commercial Building Systems, Lawrence Berkeley Laboratory, Berkeley, California, 2003; http://buildings.lbl.gov/hpcbs/pubs/HPCBS-FinalProgPPT10-28-03.pdf.) Panel (b) illustrates part of the roof of Melbourne's airport; it is coated with a paint that strongly reflects solar energy and also cools at night so that the air conditioning loads are much reduced. (From R. Lehmann, private communication (2009) and http://www.SkyCool.com.au.) Panel (c) depicts a daylit section of the "Soltag" experimental building at Velux A/S, Hørsholm, Denmark; photo by Adam Mørk. (From K. Valentin-Hansen, Velux A/S, Hørsholm, Denmark, private communication. With permission.) Panel (d) is an electrochromic smart window with electrically variable transparency; two rows of panes are fully dark, and one row is fully transparent. (From Sage Electrochromics, Inc. [2009]; http://www.sage-ec.com/pages/projgallery_comm.html. With permission.) Panels (e) and (f) show the authors of this book in action, specifically with GBS alongside a spectrophotometer set up to measure optical properties of small and large area nanostructured glazing and lighting panels, and CGG in front of an "advanced gas evaporator" for making nanoparticles of well-defined sizes. Equipment such as those in panels (e) and (f) underpin many aspects of green nanotechnology.

(a)

(b)

(c)

(d)

(e)

(f)

The scope of green nanotechnology is not limited to the one we focus on in this book, namely, solar energy and energy efficiency in buildings. Many human activities that we have not covered have large energy-related and environmental footprints. These include chemical and materials processing, water provision and quality, food production and processing, transport, biomedical technology, and manufacturing of construction materials. In these areas also, there is much scope for improvement through the application of nanoscience and nanotechnology. For example, metallic materials can have their properties enhanced by inclusions on the nanoscale. It is also noteworthy that the most used of all artificial materials, concrete, can be viewed as a nanocomposite, as can most ceramic materials.

We, the authors, have received much help and encouragement from our friends and colleagues across the globe. In particular we would like to acknowledge the assistance from the following persons: Dr. Joseph Berry (Golden, Colorado), Dr. Nicolae Bârsan (Tübingen, Germany), Dr. Daniel Beysens (Paris, France), Dr. Arno Böhm (Ludwigshaven, Germany), Mr. Michael Bonello (Sydney), Dr. Gerrit Boschloo (Uppsala), Dr. Tobias Boström (Narvik, Norway), Professor Michael Brett (Edmonton, Canada), Professor Michael Cortie (Sydney), Professor Jan-Olof Dalenbäck (Gothenburg, Sweden), Mr. Jim Franklin (Sydney), Dr. Martha Garrett (Uppsala), Dr. Angus Gentle (Sydney), Mr. Bengt Götesson (Uppsala), Mr. Eddy Joseph (Mudgeeraba, Australia), Professor Björn Karlsson (Lund, Sweden), Professor George Kiriakidis (Heraklion, Greece), Mr. Rex Lehmann (Sydney), Professor Sten-Eric Lindquist (Uppsala), Mr. Steve Lynch (Sydney), Mr. Geoff McCredie (Sydney), Dr. Abbas Maaroof (Aarhus, Denmark), Dr. Norbert Mronga (Ludwigshaven, Germany), Professor Gunnar Niklasson (Uppsala), Ms. Annica Nilsson (Uppsala), Dr. Torbjörn Nilsson (Mölndal, Sweden), Professor Boris Orel (Ljubljana, Slovenia), Professor Lars Österlund (Uppsala), Dr. John Ridealgh (Lathom, United Kingdom), Professor Arne Roos (Uppsala), Dr. Neil Sbar (Faribault, Minnesota), Mr. Steve Selkowitz (Berkeley, California), Professor Girja Sharan (Ahmedabad, India), Dr. Nariida Smith (Sydney), Dr. Paul Swift (Dublin, Ireland), Mr. Torben Thyregod Jensen (Hørsholm, Denmark), Professor Ewa Wäckelgård (Uppsala), Professor Volker Wittwer (Freiburg, Germany), Dr. Ric Wuhrer (Sydney), and Dr. Shuxi Zhao (Uppsala). We are sorry for any omissions here; please be assured that your help was much appreciated even if you are not on this list.

We would also like to thank our families for their endless patience and support.

Geoff B. Smith
Claes G. Granqvist
Sydney and Uppsala

REFERENCES

1. Climate Change 2007—The Physical Science Basis. Working Group I Contribution to the Fourth Assessment Report of the IPCC (Intergovernmental Panel on Climate Change), Cambridge University Press, Cambridge, UK, 2007; http://ipcc-wg1.ucar.edu/wg1/.
2. B. N. Goswami, V. Venugopal, D. Sengupta, M. S. Madhusoodanan, P. K. Xavier, Increasing trend of extreme rain events over India in a warming environment, *Science* 314 (2006) 1442–1445.
3. http://www.urbanheatislands.com.
4. A. L. Perry, P. J. Low, J. R. Ellis, J. D. Reynolds, Climate change and distribution shifts in marine fishes, *Science* 308 (2005) 1912–1915.
5. http://earthobservatory.nasa.gov/wildfires San Diego.
6. J. Zillman, G. Pearman, Climate change: Global and local implications, Presentation to the Australian Academy of Technological Sciences and Engineering, March 1, 2007; http://www.atse.org.au/uploads/ccsb010307.pdf.
7. PIERS, High Performance Commercial Building Systems, Lawrence Berkeley Laboratory, Berkeley, CA, U.S.A., 2003; http://buildings.lbl.gov/hpcbs/pubs/HPCBS-FinalProgPPT10-28-03.pdf.
8. R. Lehmann, private communication (2009); http://www.SkyCool.com.au.
9. K. Vallentin-Hansen, Velux A/S, Hørsholm, Denmark; private communication.
10. Sage Electrochromics, Inc. (2009); http://www.sage-ec.com/pages/projgallery_comm.html.

CHAPTER 1

Green Nanotechnology
Introduction and Invitation

1.1 WHAT IS NANOTECHNOLOGY?

Nanotechnology makes use of materials whose structures have characteristic features on the nanoscale (i.e., on the scale of 10^{-9} meter [a nanometer, nm]). Obviously, this size is a very small one compared to objects we have around ourselves, but it is not particularly small on an atomic scale [1]. Indeed, characteristic distances between atoms in a solid are of the order of 10^{-10} meter (a tenth of a nanometer, also called an Ångström), so a piece of material whose side is one nm may contain hundreds or up to a thousand atoms. This means that, normally, a nanomaterial shows some resemblance to a normal solid comprising the same atoms, but it is typically modified to achieve some superior property such as higher strength, different electromagnetic properties, permeability to a fluid, or some other quality. A higher strength may mean that less material is needed to accomplish a given task; different electromagnetic properties may mean that we can harness the sun's rays more efficiently in a solar energy conversion device or that we can build better electrical generators, and achieving a desired permeability may lead to improved filtering technology to remove undesired substances from water or air. Of course, these examples can be multiplied almost ad infinitum. Figure 1.1 illustrates the size of various things and sets "nano" in the proper perspective.

Thus, nanotechnology enables us to achieve material properties that were previously impossible, impractical, or too expensive for use on a scale large enough to have a global impact on energy use and supply. Almost certainly nanomaterials will continue to surprise us for years to come with unexpected characteristics and new applications. This also makes them attractive for scientists looking for stimulating research avenues.

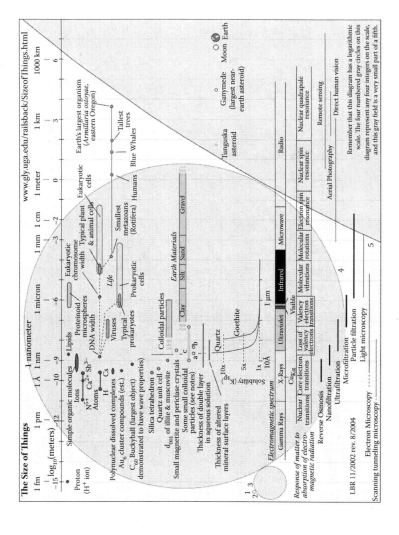

FIGURE 1.1   The size of nanoobjects in the context of the size of things in general. (From http://www.gly.uga.edu/railsback/SizeofThings.html. Used with permission of Professor Bruce Railsback, Department of Geology, University of Georgia.)

In the past, more complex materials—apart from new chemical compounds, including new polymers—were prepared by alloying, impurity doping, and making composites involving microscale or larger size additives. These processes were the main sources of the new materials that then led to emerging technologies and major advances in human capabilities and lifestyles. While a few examples of nanomaterials for decorative applications can be found in antiquity, as we will come back to briefly in the opening of Chapter 3, a scientific approach for their development started only some 30 to 40 years ago. The science of nanomaterials, what we call *nanoscience*, has accelerated rapidly over the last two decades, thereby opening up new vistas and many opportunities for materials design. This research field is so vast and diverse that at present it is largely unexplored. Such diversity can be counter-productive, though, and means that good scientific understanding and a strategic approach to materials development are essential, because empirical possibilities at the nanoscale are almost endless. To draw an analogy, it is as though we had suddenly developed the practical ability to travel the universe and needed to decide where to head first, knowing little of what would be out there. The aim with this book then is not only to inform, but also to supply a map for the first stages of the journey, to point out the likely best directions and modes of travel, and to list the tools to take along. It is a guide book for those wishing to join in the urgent process of opening up the important and exciting new territory where nanoscience and energy get together.

This book thus provides strategies for the use of nanoscience to give unprecedented abilities for tuning material properties to human needs, to keep economies humming along without intolerable degradation of the environment. The dominant energy technologies that will emerge from those described in this book will ultimately have to be manufactured and implemented on huge areas, such as 20 or 100 million square meters per annum, to achieve the dual goals of largely improved energy efficiency at affordable prices and significant energy from renewables. These are commodity scales, such as those applying in current flat glass and metal sheet output. Commodity-scale volumes usually only arise if the product is cheap enough. It is fascinating that we are thus exploiting our ability to engineer at the tiniest scale of approximately 10^{-16} square meters to achieve advanced materials and technologies that can be affordably produced on scales of 10^8 square meters each year. In fact, commodity volumes of systems containing nanoscale features are already in existence, including in the latest computer chips and in multilayer coatings on windows for saving energy.

Why this emphasis on large area? It will be needed for two reasons apart from economics: first, because various renewable energy sources are highly dispersed, and second, because in order to save energy it will be necessary to use novel materials and coatings for windows, walls, and roofs of most of our buildings and appliances, and for our various means of transport.

1.2 WHAT IS GREEN NANOTECHNOLOGY?

We use "green" to denote that the technology is environmentally benign and sustainable; the technology is intended to contribute to the solution of some environmental problem, or, at a minimum, it should perform better than alternative "nongreen" technologies.

Hence, a green technology, as the term is used in this book, is developed in response to a challenge that humankind faces. What are these challenges? Any list is, of course, subjective. One list, by Nobel Laureate Richard E Smalley, gives the top ten problems as follows [3]: (1) energy, (2) water, (3) food, (4) environment, (5) poverty, (6) terrorism and war, (7) disease, (8) education, (9) democracy, and (10) population. One can question and challenge the contents and ordering in this list. For example, with a smaller population, the need for energy, clean water, and food would not be so huge. And with better education one may expect persons to make more rational choices and make better use of the energy that is available. Nevertheless, we can use the list as a point of departure for "green nanotechnology" and make the provisional definition as "nanotechnologies for providing energy, clean water and a good environment in a sustainable way." Our strongest focus will be on energy issues, on what Smalley called "The terawatt challenge" [3]. The critical demand on energy has many reasons, one obvious one being that the population grows steadily year by year [4,5] and another being the understandable desire of people in the less-developed regions of the world to obtain a higher standard of living.

Energy is needed in all sections of society, and it is customary to make a division into buildings, industry, and transport. The energy used in buildings is the largest. As found from a study in 2007 by the United Nations Environmental Programme, worldwide about 30 to 40% of all primary energy is used in buildings [6]. The corresponding number for the European Union and the United States is 40%. Some 70% of the electrical energy in the United States is spent in buildings [7]. For hotter regions, such as Singapore and Hong Kong, about half of the primary energy is used in buildings, and in an extreme climate such as the one in Kuwait more that 75% of the primary energy goes into buildings [8]. The energy expenditure in the built environment is for heating, cooling, lighting, and ventilation; in particular, the energy for space cooling has skyrocketed during the past few decades, and air conditioning is what drives the demand on peak electricity production in several parts of the world.

Our goal in this book is to show how one can use nanotechnology to harmonize or tune the properties of materials so that they take account of the physical attributes of the environment, and the responses of the human body to light and heat, in a holistic way. Examples of these attributes are the solar spectrum, the role of gases in the atmosphere on its transmission of radiation, and the response of our eyes to diffuse and direct light. Understanding these properties is a prerequisite to designing the materials needed for saving energy or extracting energy from the environment, so this topic is treated early on, in Chapter 2. We may note that the initial impacts of nanomaterials in the solar energy field are not entirely new but date back to the 1970s [9,10].

Water—including supply of clean water and adequate water for crops—is another looming environmental challenge for the world; it is number two on Smalley's list [3]. The relevance of nanotechnology to water is a major topic in its own right. Some of the materials for solar energy applications and for energy efficiency discussed in later chapters have applications in water collection and supply, purification, desalination, and in greenhouses operating with salty water. We will restrict further discussion of water-related green nanotechnologies to these, though it is worth noting that nanomaterials have emerging direct roles in water cleaning and in waste water reuse since they can filter out not only inorganic salts but also bacteria [11]. Another application is in regard to nanostructured membranes that are being explored for desalination via reverse osmosis in order to reduce the power requirements [12].

The connection between water and food on one side and health on the other is obvious. However, the relationship between energy and health may be equally strong. It has been stated that the warming and precipitation trends due to man-made climatic changes during the last three decades are already claiming more than 150,000 human lives annually [13]. Furthermore, these changes in the climate are expected to be accompanied by more common and/or extreme events such as heat waves, heavy rainfall, and storms and coastal flooding. There is also the danger that nonlinear climate responses will lead to even more climate changes, and that rapid ones—such as breakdown of the ocean "conveyor belt" circulation, collapse of major ice sheets, and/or release of large quantities of methane at high latitudes—will lead to intensified warming [14].

1.3 SOME BASIC ISSUES IN NANOSCIENCE

What is the difference between the science of nanomaterials and micromaterials and atomic and molecular science? Some researchers demarcate nanoscience at 100 nm, and this book will mainly be concerned with structures smaller than that. But we will push toward the microscale if it is relevant for the technologies of interest or for our understanding of nanoscience and its benefits.

We shall see that nanotechnology may involve using molecules in new ways, but molecular science itself is not nanoscience with the exception, possibly, of research on some large biomolecules. With energy technology and materials science, there is a continuum of object and structure scales that one can use, starting at around one nanometer and reaching up to several micrometers. Many "green" entities from nature, such as the leaf illustrated in Figure 1.2, have features spanning this range [15].

The principles of atomic and molecular science are essential ingredients in nanoscience, and a good grasp of microscience and solid state physics is important, too. To appreciate what the nanoworld has to offer, it is thus useful to have an understanding of how properties evolve from the atomic scale and up to the bulk. A practical issue might be how the optical properties of polymers differ when they are doped with particles of another material having diameters of 50 nm, 150 nm, and 2 µm. We will see later that such size changes lead to dramatic shifts in the way light interacts with the doped polymer.

It is worth noting that one can develop new microscale physics by nanoengineering within or on microstructures, and that microphysics often has nanoaspects. One example from energy science is polymer microparticles with internal nanostructures that make them behave chemically and thermally unlike normal polymer solids containing the same molecules. Another example is the pointed tip of mass-produced micropyramids on a polymer; using injection molding with sufficient care, these tips can display sharp enough features to yield interesting nanoeffects that influence their lighting response [16]. Yet other examples of nanoscale effects in microscale devices will follow.

According to quantum physics and chemistry, electrons on atoms and molecules occupy discrete and separated energy levels, and light can cause electrons to "jump" between these levels. Condensed matter physics and solid state chemistry deal with electron energy bands which cover a range of energies and where the quantum energy levels are so closely spaced that they can be treated as continuous. Different bands may be separated by energy gaps or may overlap. A nanostructured solid spans these two situations so that its energy level scheme depends on the actual number of atoms in the object and the type of material (for example, if it is a semiconductor or a metal). If a nanoparticle is very small, all of its energy levels may still be separated by finite gaps, and it is then electronically like a special "artificial molecule." But as the particle size gets larger, the energy levels eventually form bands just as in a normal solid. The size at which this happens is important to know for nanoengineering. In metals and semiconductors it is usually between 2

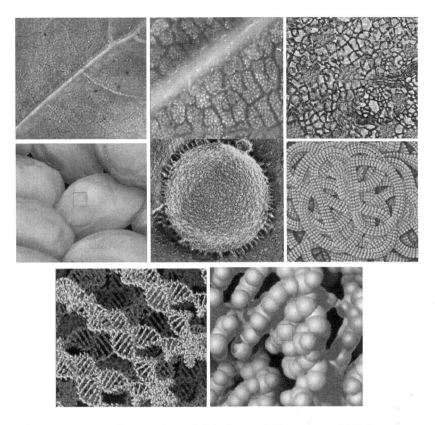

FIGURE 1.2 *A color version of this figure follows page 200.* Images at successive magnifications of a green leaf. The linear scale in neighboring images falls by a factor of ten, starting at 1 cm and ×10 magnification and ending at 1 nm and ×100 million magnification. The last three images are in the domain of nanotechnology, with the final one showing molecular orbitals. (From http://micro.magnet.fsu.edu/primer/java/scienceopticsu/powersof10/index.html. Reprinted with permission from Dr. Mike Davidson, Florida State University National Magnet Laboratory.)

and 6 nm. In gold and silver, for example, it is surprisingly small, only 2 to 3 nm. Nanoparticles of semiconductors such as CdS can have energy-level schemes which vary continuously with particle size up to around 6 nm, and this can be seen in their color differences as sizes change.

One should note that metals and other good electrical conductors are special and are still poorly understood in nanoparticles with sizes below about 2 nm. This lack of understanding is very significant, since metals play an exceptionally important role in nanotechnology. Their outermost occupied energy levels are those that carry the charges which provide conduction, and the associated quantum wave functions are far removed in character from their atomic origins. Electrons in these orbitals are largely shielded from atomic cores and hence relatively free to move around inside the solid. In contrast, electrons in outer occupied orbitals of insulators tend to be tightly bound to their atom cores and hence jump between atoms far less frequently.

Materials with internal nanostructures can have inner surfaces that sum up to very large areas. The individual boundaries separating one medium from another are so small (10^{-13} to 10^{-17} m^2) that a nanostructured surface on, say, a metal sheet or glass pane—or in the interior of a nanostructured solid such as opal or a butterfly wing, as in Figure 1.3 [17,18]—can have a huge actual surface area, one that is orders of

FIGURE 1.3 *A color version of this figure follows page 200.* Examples of brilliantly colored nanoporous systems whose internal pattern of holes and solid matter leads to unique optical effects. The photos show the external color on the left and the associated nanostructure on the right; they refer to "peacock eye" butterfly wings (upper) and a famous opal gem, "the flame queen" (lower). (Images used with permission from the following sources: Upper left: The Peacock Eye butterfly © Copyright Lynne Kirton [image licensed under the Creative Commons Licence. http://creativecommons.org/licenses/by-sa/2.0/]; upper right: H. Ghiradella, *Appl. Opt.* 30, 3492–3500 (1991); lower panel: (c) en.wikipedia.org/wiki/Flame_Queen_Opal [image licensed under Creative Commons ShareAlike 3.0 en.wikipedia.org/wiki/Creative_Commons].)

magnitude larger than the area that contains it. This feature is very useful in a number of important energy and environmental technologies, such as in electrochromic devices, batteries, supercapacitors, and nanofiltration; it can also lead to large energy savings in manufacturing and chemical processing. The butterfly wing is strong and yet so delicate. And, just as for this wing, artificial porous or composite nanostructures make it possible to have materials with very high strength-to-weight ratio. If such materials were to become commonplace in construction, one would need much less iron, steel, and concrete, which are very energy- and resource-intensive to produce, and one would also use much less energy in transport and for erecting structures. The change could be dramatic since the material throughput devoted to the built environment accounts for some 70% by weight in developed countries, and construction and demolition waste from buildings is about 60% of the total nonindustrial waste stream [19]. Nature has "figured out" that porous structures can be good for strength, and such structures are everywhere around us— and even in us, including in the cores of our own bones.

Internal surface properties can affect apparent bulk (i.e., "whole-of-solid") properties in various ways. For example, when a static or changing electric field is applied to cause electrical flow, electrons will collide with internal surfaces and lose energy gained from the applied field at a faster rate than they normally do inside a bulk-like, homogeneous metal. These electrons can also appear to be heavier than normal if they have to take a tortuous path between electrodes, as would be necessary in the nanoporous gold in Figure 1.4 [20]. (The electrons are not really "heavier" but act as though they are because of the extra time it takes to traverse the circuit.) These factors mean that nanostructures can have a high electrical resistance, and hence that resistive heating increases.

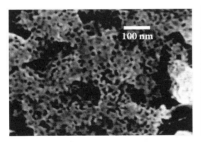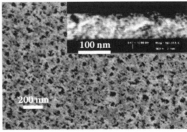

FIGURE 1.4 Nanoporous thin gold layer seen from the top at two magnifications (largest magnification on the left). The inset on the right is a cross-section view of the film's nanostructure. This is a highly tortuous but very strong porous network. (From A. I. Maaroof et al., *J. Opt. A: Pure Appl. Opt.* 7 [2005] 303–309. With permission.)

Thus, nanostructures within metals almost always lead to an increase in electrical resistance. This may seem disadvantageous, but it implies that one can create "new conductors" from old by using a nanostructured network. Indeed, the nanoporous gold shown in Figure 1.4 behaves optically and electrically so differently from what one would expect from normal gold that, not knowing what it was and just seeing the data, one would think it was a unique, or at least a different, material.

Another approach to "new conductors" is an area of ballooning interest. It involves bonding between nano-objects and molecules as a way to create new structures, as well as new means of charge transport and new photo-electric effects. In some cases the organic "glue" can be biologically based, for example, DNA [21]. Bridging two metal or semiconducting nanoparticles with an organic molecule is of relevance for molecular electronics, and attaching fluorescent dye molecules to nanoparticles is of much interest in photovoltaics and lighting devices. What is sought is, in effect, new types of electronic solids for which the nanoparticles play the role of "atoms" in normal solids while charge transfer between these "nanoatoms" is controlled. In principle, this possibility opens up a "new electronic world" not just through the above processes but because the number of bonds or interconnects linking the nanoatoms also can be engineered. In some nanostructures—both man-made and naturally occurring—one can find junctions with up to 15 separate interconnecting branches. Such structures are said to be "hyper-dimensional" with regard to their charge transport properties. This is yet another example of how the geometries and topologies available at the nanoscale can lead to new exploitable properties.

Ultimately, nanotechnology and nanoscience are about new and better functionalities, within the scope of this book, for energy supply and savings. But it should not be forgotten that for this purpose one must also be able to integrate nanostructures into larger material components and systems while not losing control of component dimensions at the nanometer scale. These micro- and macro-aspects of nanoscience have received rather scant attention in the voluminous literature on nanomaterials, but it is a core issue for new products. Processing conditions to form large area sheets or coatings can feed back into nanostructures, sometimes beneficially, sometimes adversely. One common problem is the ease and strength by which nanoparticles stick together due to van der Waals' forces [22]. Fortunately, this problem is often manageable by treating particle surfaces. Process heat can also cause changes, especially for low melting point metals such as gold and silver. In fact, particles of Ag and Au can change shape on thermal treatment and thin films of these metals can develop new surface nanostructures, or even radically transform into separate nanoparticles when hot. The resulting property changes can then be dramatic.

1.4 NANOSCIENCE, DIMENSIONALITY, AND THIN FILMS

So far we have used *nanoscale* as a general term, but it is important to realize that in practice it must be considered separately in each of the three dimensions. A nanoparticle or quantum dot has nanoproperties in all three dimensions, whereas a nanowire has ordinary properties along the wire and nanoproperties normal to its axis. Single thin layers of a nanomaterial as well as stacks of such layers, sometimes called *quantum wells*, have nanoproperties in only one dimension.

This last case takes us into the domains of thin-film science and thin film deposition, which are exceptionally important for energy technology and will feature often in this book. A number of deposition technologies are discussed in Appendix 1. But when does thin film science qualify as nanoscience? Thin film engineers have produced layers with thicknesses in the nanorange, and even precise atomic monolayers, for decades [23]. Again we do not want to be rigid about what qualifies as "nano." However, based on our personal experience, we propose the following categories as areas in which nanoscience has an impact on the properties of thin films:

- Films with internal nanostructure
- Films with properties arising from surface excitations or surface nanostructure, and
- Stacks of many layers each having a thickness on the nanometer range

The first group is largest and includes nanocomposite structures, fully dense films in which crystalline grain sizes are of nanoscale, and layers of nanoparticles. The third group is growing rapidly in importance, and embraces mirrors made entirely of plastic, lenses, and other unusual filters and mirrors.

Modern thin-film process control falls into a broader definition of nanoscience, which includes growth, development, and manipulation of structures to desired nanodimensions. This, of course, implies precision metrology and imaging at the nanoscale, and such measurement methodology is both a core aspect of nanoscience and an enabling feature of nanotechnology. Improvements in these capabilities from the mid-1980s are essentially why nanoscience "took off" in the late 1990s. As an important technological example, we note that optimum functionality for various solar energy-related applications may require thickness control of thin films at the nanometer or subnanometer level. Such control is commonly available today, even in mass production, given due care.

Film thickness has to be measured as part of the quality control, and the observed thicknesses can then be used to model the optical properties in order to see if they match what is observed. Thus, optical techniques play a vital role for quantifying layer thicknesses with nanometer or subnanometer precision.

1.5 OUTDOING NATURE IN EXPLOITING COMPLEXITY

Nanoscience is about understanding complexity, and its associated technologies are based on the ability to engineer and use complexity. Complex systems are the nanoscientist's "bread and butter," and nano-structured materials are often referred to as "complex mediums" [24].

The diversity of form and function in nature exemplifies the complexity that nanoscale engineering has to offer, and also its value. Nature has experimented with and explored complexity at the nanoscale for hundreds of millions of years, and has achieved high functionality through the evolutionary processes of mutation and natural selection. Examples of this functionality include color variations that provide camouflage and/or promote procreation, multilayered seashells with superior strength, nanostructured insect eyes that efficiently harvest light at oblique incidence, and the photoelectrochemical nano and microfactory commonly known as photosynthesis that converts sunlight, water, and carbon dioxide into plant matter.

One can obtain many useful ideas from these remarkable natural structures; the scientific field for exploiting the analogies is referred to as *biomimetics*. It is even possible to progress at a much faster pace than in nature by understanding the connection between nanostructures and properties and by establishing new growth protocols. These are core matters which underpin the contents of this book. Scientists and engineers are not restrained by the random forces and local supplies of chemicals to which nature has had to adapt. There are, however, some important constraints. Thus, availability of specific mineral resources may limit materials options, and it is essential that ethical, health-related, and safety concerns are taken very seriously. Nanomaterials are often new, so their health impacts are not fully understood [25]. One of the most widely studied nanoparticles, the carbon nanotube, displays features that resemble those of asbestos—a material with well-known dangers to human health [26]. And the emergence of "nanomachines" will pose ethical dilemmas if their function becomes too lifelike [27]. It may seem ironic that it is mankind's need to deal with and prevent undesirable environmental changes in the natural world that forces us to develop nanoengineering skills which are superior to those deployed in nature itself.

The key value in using complexity is that it opens many doors for modifying and optimizing physical and chemical properties, and it exploits both material and spatial diversity. Complexity can be confronting, like going into a maze for the first time. However, one does not have to be a mathematical genius to deal with it. Some of the techniques for understanding complex system behavior are relatively simple and give useful, often accurate, answers as we shall see in Chapter 3. Just as unit cells are the building blocks of crystalline solids, most complex nanostructures have underlying building blocks and design rules for how they come together. One such building block is the fractal, or self-similar, pattern which is common in nature [28,29]. Knowing these basic building blocks, one can usually progress toward some useful modeling of system properties. Another powerful tool for gaining insights into complex systems is computer simulation—in particular, finite element analysis. This has only recently made an impact on nanotechnology and is usually limited in scope by the computational power at hand. Some of the important technologies we will address later rely on such simulation.

The shape of boundaries is an important source of diversity and, additionally, nanospheres and nanoellipsoids have different optical responses. Solid nanostructures of interest include simple planar multilayers, nanowires, rods, spheres and ellipsoids, nanopyramids, hexagons, hemispheres, nanoshells, and many others. To this "zoo" we can add forms created by growing, etching, or indenting to leave nanovoids such as in inverse opals, nanogrooves of various shapes, and complex nanowire "weaves." Self-assembly, or spontaneous formation, is of much interest, and some desired networks can be made from particle components simply by mixing.

The final aspect of complexity to be addressed here is that one almost always deals with more than one component in defining how nanomaterials respond to external stimuli. The very simplest nanosystem—a single nanoparticle sitting on a substrate or maybe floating in a liquid—is never a "one material system." Its properties depend on the shape, the constituent material, and the materials that it contacts. As a simple example, 50-nm-diameter nanoparticles may scatter light when in water and render it "milky," but these particles can be invisible to the eye in a liquid or solid of higher refractive index than water.

Sometimes one component in a mixture may be there only for production reasons and have no effect on the desired structural, electrical or optical outcome, but generally this is not the case. Instead the effective, or measured, performance of the combined system depends on the properties of all the constituents, and only rarely can this new property be found by any simple averaging. A lot of physics theory in this field (and simple examples appear later) is directed at solving these "effective" combined properties, as they are what is ultimately used and measured.

For example, we will see that the average optical properties of nanoparticle arrays, of interest for solar absorption, are sensitive to the properties of the embedding medium or, if the nanoparticles sit on a surface, to the properties of the two mediums above and below the interface. Recent solar cell developments based on thin layers of polycrystalline silicon can exploit nanostructured interfaces or nanoparticle coatings in various ways to reduce costs as discussed in Chapter 6.

1.6 ENERGY SUPPLY AND DEMAND

To gain insights into the scope for "green" nanotechnologies, it is useful to first consider some broad energy-related issues. Energy technologies are commonly grouped into two main classes, both of which are covered extensively in this book:

- Technologies for production, supply, and storage of energy, and
- Technologies for reducing the demand for energy

Most people are more or less familiar with the various ways of supplying energy, but the many ways of reducing the need for traditional energy sources are not so well understood. Of course, it is important to use energy-efficient lamps, have good thermal insulation in ceilings and walls, drive fuel-efficient cars, and turn off computers when they are not in use. But what about the importance of the types of windows we install, and the impact of novel paints and new thermal insulators, perhaps including phase-change materials? What can we accomplish with new passive cooling technologies using the sky as a heat sink, new structural materials and better manufacturing methods, superior energy storage techniques, and improved motors?

Supply and demand functions can be blurred or integrated in some emerging nanotechnologies. A simple example is a glass roof with integrated solar cells, which provides energy but also reduces the energy demand via the provision of daylight. A more sophisticated and futuristic example is multilayered "switchable" glazings in which some layers use the sun for providing power to optically switch other layers so that heat or glare in the building are kept down [30].

Energy supply for powering buildings, transport, and industry currently relies mainly on nonrenewable power sources including coal-fired plants and nuclear power stations coupled to electricity grids, natural gas via pipelines, and oil via refineries and tankers. Supply also includes a growing contribution from renewable sources including wind, water, biofuels and solar cells as shown in Figure 1.5. Energy-demand technologies influence how much power is needed to run our homes, offices, schools,

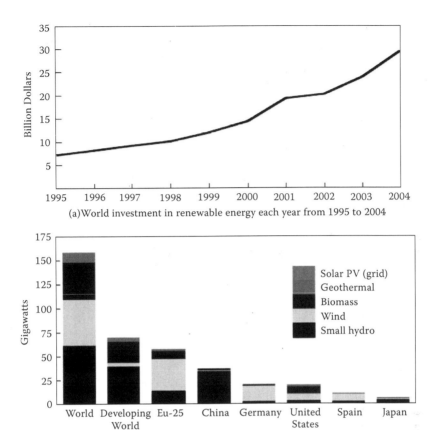

(a) World investment in renewable energy each year from 1995 to 2004

(b) Composition and capacity of renewable power in the world, developing world, Europe and countries which have installed the most

FIGURE 1.5 *A color version of this figure follows page 200.* Panel (a) reports annual world investment (in billions of U.S. dollars) in renewable energy since 1995. Panel (b) shows renewable power capacities in GW for the world and for select groups of nations and individual countries in 2004. (Adapted from REN21 [Renewable Energy Policy Network for the 21st Century], Renewables 2005: Global Status Report, The Worldwatch Institute, Washington, DC, 2005; http://www.ren21.net.)

and shops; to operate factories, mines, waste treatment plants, and farms; and to transport people and goods, locally or across the world.

Solar radiation, wind, waves, and other sources of renewable energy are dispersed and intermittent regarding watts per square meter at any time, as will be detailed in Chapter 2. In contrast, the power density exiting the core of a nuclear reactor, or in a coal-fired high-pressure steam vessel used to drive a steam turbine, is orders of magnitude larger and relatively steady in time. Wind, waves, and solar radiation are thus not good for base-load power unless cost-effective energy storage is available or power networks cover very large distances, which is why these renewables are sometimes seen as top-ups with a cap on the percentage they can supply.

Alternative energy can be captured, converted, and stored in useful amounts. For this to work, large areas are needed for wind turbines, wave-powered generators, solar thermal collectors, or solar cells, and efficient storage is essential if these sources are going to be used for supplying base-load power. Storage efficiency is measured by weight or volume per unit of stored energy, and storage must enable us to deal with both the long-term (season) and short-term (hour or minute) dynamics of renewable energy sources. Fortunately, the areas and volumes of collectors that are needed are not impossibly large. For example, even with existing solar technology, two to three square meters are enough to supply a typical home with hot water, and its south-facing roof area (north-facing, if in the southern hemisphere) is usually more than sufficient to cover its electrical power needs, which generally scale with floor area and, hence, roof area. The challenge is to reduce costs and increase availability, so the required areas must come down and materials must be less expensive, more durable, and of lighter weight. Reduced area will come about if conversion efficiencies from solar energy to heat or electrical power improve, or if efficiency measures result in less energy being required. In Chapters 3 and 6 we will show where new nanoscience has led to major collector efficiency gains in the past.

In fact, it was the raised interest in solar energy during the oil crises of the 1970s that gave the first real boom in green nanotechnology and led to research on optical and thermal properties of nanoscale metal embedded in insulating oxide layers and of nanostructured metal surfaces for use in flat plate solar collectors [10,32–35]. And today (2010), again, concerns about the oil supplies and energy security are key drivers for new solar energy research. But this time we have, in addition, the even more significant threat posed by global warming [36]. Fortunately, nanoscience has come a very long way since the 1970s, and has put a dazzling array of new experimental tools at our disposal, especially for imaging at tiny scales. There have also been many advances in physics and chemistry for understanding the growth of nanostructures and their properties. The emergence of nanoscience as a popular mainstream discipline during the late 1990s and 2000s is very fortunate, given the coincident realization that we must deal urgently with the energy challenge.

In addition to the well-known and intensely discussed threat from global warming, there is another less widely appreciated warming effect emerging from heat evolution in cities. This effect is of growing importance since not only is the world's population increasing but the portion living in big cities is going up and already surpasses 50%. In particular, the growth of megacities leads to strong "urban heat islands" which contribute to the need for energy spent on cooling. For the specific case of Greater London, the urban cooling load was estimated to be 25% above that of surrounding rural areas [37] and will be relatively higher still as the climate warming continues. We come back to the urban heat islands in Chapter 7 where cooling techniques are discussed.

This book is devoted equally to the ways nanoscience can change the demand and supply sides of the energy equation. It is in on the demand side where most progress in carbon dioxide abatement is feasible in the short term [38]. The potential and relative ease of demand-side improvement have been both badly underestimated and widely ignored due to entrenched practices. New materials and new science, if adopted, could impact energy savings and lifestyles to such an extent that, in 20 or 30 years, people will look back on current practices in bemusement. If we do not move quickly, however, the looks back may be looks of anger. Encouraging signs of changes have emerged recently, but education, more stringent regulation, and ready availability of new products are needed if these changes are going to influence how buildings are constructed and operated. Easy retrofits must also be available. Better choices of materials for constructing, insulating, and coating our buildings and cars indeed can save vast quantities of energy. As we shall outline with proven examples below, simple changes in the materials used on certain types of buildings have reduced their need for power by more than 50% without any loss in amenity and comfort. Coupled with environmentally aware building design and sensors and controls, nanomaterials may enable buildings that are nicer to occupy, healthier and safer, and require only a fraction of their present power.

Some topics in this book have both supply and demand aspects—for example, lighting and cooling. The topic of lighting, which will be discussed in detail in Chapter 5, involves the natural daylight resource plus lamps and light fittings. Daylight, in contrast to solar energy, is so abundant that one normally has to limit the amount that is let in. Unfortunately, practices went too far in its exclusion during the last 50 years due to an infatuation with new lamps. While lighting and cooling are energy-demand technologies and daylight is a natural resource, it is conventional to consider daylighting also as an energy-demand issue. This is so because it is used passively and enables a reduction in power for lighting while also improving the attractiveness of building spaces. It is important to first understand daylight as a natural resource, and this is one of the topics of Chapter 2. We shall see that it is so abundant and strong that in many climates its management is a challenge, but one which various nanostructures are well up to in windows and skylights. Savings on cooling might simply involve reducing the solar heat gain into a building by passive means to cut down on energy for air conditioning, or savings may entail active devices to tap into the environment and use radiative cooling toward the clear night sky to store up "coolness." We will consider both lighting and cooling as demand-side technologies, even though sky cooling uses a largely untapped natural resource: the deep cold of outer space. The physics behind the utilization of this resource will emerge in Chapters 2 and 3.

For convenience we will thus limit the term "supply" in the context of renewables to solar electrical power, solar heat, wind, waves, and renewable fuel sources, along with associated storage techniques. These are the topics of Chapters 6 and 7. However, it is important to keep in mind that the environment also supplies abundant useful energy in other forms, especially natural light and the ability to pump a lot of heat away at night using little or no power. Tides, waves, and geothermal sources arc also renewable resources, while local breezes for cooling at night are useful passive aids for saving energy in many populated locations, especially those near the seaside.

Studies, such as the one illustrated in Figure 1.6 [39], on the various future contributions to reduction and stabilization of CO_2 emissions indicate that the largest single impact is expected to come from demand-side reductions, though overall there will be about equal contributions from all renewable energy sources combined and all energy savings technologies combined. The importance of energy efficiency is emphasized also in the most recent (2009) studies on energy in the world [38]. Figure 1.6 indicates that efficient coal burning, gas, nuclear power, and carbon sequestration also will be needed to stabilize the CO_2 content in the atmosphere.

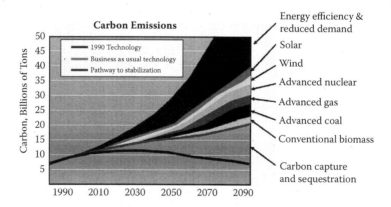

FIGURE 1.6   A "wedges plot" of the potential impact of various technologies on reduction and stabilization of carbon emissions from 1990 to 2090. (Adapted from R. Socolow et al., *Science* 305 [2004] 968–972.)

Technological developments must take into account the costs and the benefits of investing in a product or system, but cost and energy savings are not the only factors that influence the decision to buy [40]. The purchasers' choice set—that is, the competition—is the starting point, and they will consider a variety of attributes of each product on offer. Novelty, durability, functionality, aesthetic appeal and design, impact on lifestyle, ease of installation, environmental impact, and government incentives are additional factors affecting the choice of energy technologies. The situation is changing somewhat, though, with the market for "green" products moving from niche to mainstream. The competition is thus shifting to the differences among "green" products themselves, and this leads to improved performance, lower cost, and elimination of products with poor quality. For example "green" buildings fulfilling certain standards in the United States command ~6% higher effective rents (adjusted for building occupancy) than comparable "nongreen" buildings, and the selling prices of the green buildings are higher by ~16% [41].

If one cannot afford both solar power and better energy efficiency, then a choice has to be made. The energy savings per dollar invested is one starting point for comparison. As an example, the energy savings per dollar for making a roof solar reflective and insulating in warm to hot climates by use of high-performance nano or standard insulation, relative to investing in energy production by solar cells, probably differs by a factor between 20 and 40 to the benefit of the former option for many building types. Unfortunately, comparisons of this kind are rarely made. Devices such as solar cells are more "glamorous" and are seen by many scientists as a larger challenge, so they currently attract much more R&D funding. This is a pity since nanomaterials for the demand-side effort also involves challenging new science, and, more importantly, offer vastly more energy savings per dollar invested in the near term. Both energy production and energy saving are needed, and it is the authors' hope that one influence of this book will be to create a more balanced investment of scientific and technical efforts between these two components.

Chapter 5 gives ample examples of how attributes other than energy savings dictate the choice of an energy technology, here related to daylight through windows, skylights, and roof glazings. The primary attraction is not related to energy savings but to the impact daylight has on the functionality and appeal of interior spaces [42]. The energy saving challenge is not to conserve on lamp use (though electricity indeed is saved) but to achieve better use of daylight without excessive heat gain or loss. A "multifunctional mindset" is clearly needed here, and we will see that nanotechnology is great for achieving multifunctionality. Examples of windows possessing four to five attributes that add appeal

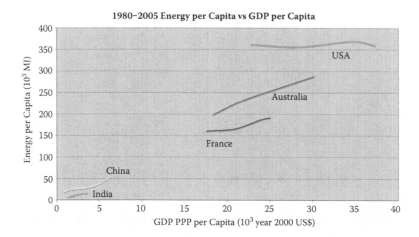

FIGURE 1.7   Time series plot of energy per capita versus GDP PPP (gross domestic product adjusted to give local purchasing power parity) from 1980 to 2005 for select developed and developing countries. Plotted by the authors from data supplied by the U.S. Energy Information Administration. (Adapted from EIA, International Energy Outlook 2007, Energy Information Administration, U.S. Department of Energy, Washington, DC, 2007; http://www.eia.doe.gov/emeu/international/ energyconsumption.html.)

in addition to the benefits of daylight and a view will be discussed. Pointing these out will influence the customer's choice and should be part of marketing efforts.

1.7 ENERGY AND DEVELOPMENT

Perhaps the greatest challenge comes from the historic connection between rising prosperity and increasing energy use. This is exemplified in Figure 1.7, which shows their linked evolution for select countries over the 25 years from 1980 to 2005 [43]. The implication of this plot is that bridging the wealth gap—obviously an important aspiration for many billions of people in the world—demands technologies that will allow this to happen with a minimal future rise in energy demand, which is shown along the y-axis. Because global net cuts are essential, the plots for developed countries must have their vertical axis projections turn down sharply in the future. Dips in energy use have happened occasionally in the past for short periods but need to be sustained and deep in coming years. The big question is then, Can we do this without crippling economies? And can we do it not only for today's (2010) population of 6.8 billion but for a population that is expected to grow and stabilize at 10 billion [4,5]?

Various strategies have been proposed for bridging the wealth gap. The best are those ensuring that many low-cost technologies arise, which will enable us to deal with the problems while growing economically. This is unfamiliar territory for humanity. Each point in Figure 1.7 represents energy used per unit of gross domestic product, which is termed *energy intensity*. It usually falls slowly in developed countries, but unfortunately at a smaller pace than the normal economic growth rate, meaning that total energy needs and, hence, CO_2 emissions keep rising. In simple terms, if economies are to grow by 3% per annum, then energy use must be reduced by more than 5% per annum per unit of GDP to meet targets for emission cuts. Historically, the developed world has generally struggled to do better than 2.0 to 2.5%.

Nanotechnology will contribute significantly to the difficult but essential goal of creating large cuts in the world's energy intensity. The economic activity associated with the required changes and the many new products that will emerge in this field should, with the right policies and planning, provide much of the economic growth needed to sustain economies in the awkward transition period we are now entering.

REFERENCES

1. P. Morrison and the Office of Charles and Ray Eames, *Powers of Ten: About the Relative Size of Things in the Universe*, Scientific American Books, New York, 1982.
2. http://www.gly.uga/railsback/SizeofThings.html.
3. R. E. Smalley, Future global energy prosperity: The terawatt challenge, *MRS Bull.* 30 (2005) 412–417.
4. U.S. Census Bureau., Historical estimates of world population, 2009; http://www.census.gov/ipc/www/worldhis.html.
5. World population 2009; http://www.un.org/esa/population/publications/sixbillion/sixbilpart1.pdf.
6. UNEP, Buildings and Climate Change: Status, Challenges and Opportunities, United Nations Environment Programme, Paris, France, 2007.
7. B. Richter, D. Goldston, G. Crabtree, L. Glicksman, D. Goldstein, D. Greene, D. Kammen, M. Levine, M. Lubell, M. Sawitz, D. Sperling, F. Schlachter, J. Scofield, D. Dawson, How America can look within to achieve energy security and reduce global warming, *Rev. Mod. Phys.* 80 (2008) S1–S107; http://www.aps.org/energyefficiencyreport/.

8. M. A. Darwich, Energy efficient air conditioning: Case study for Kuwait, *Kuwait J. Sci. Engr.* 32 (2005) 209–222.

9. C. G. Granqvist, Radiative heating and cooling with spectrally selective surfaces, *Appl. Opt.* 20 (1981) 2606–2515.

10. G. A. Niklasson, C. G. Granqvist, Surfaces for selective absorption of solar energy: An annotated bibliography 1955–1981, *J. Mater. Sci.* 18 (1983) 3475–3534.

11. R. Zhang, S. Vigneswaran, H. Ngo, H. Nguyen, A submerged membrane hybrid system coupled with magnetic ion exchange (MIEX®) and flocculation in wastewater treatment, *Desalination* 216 (2007) 325–333.

12. F. Fornasiero, H. G. Park, J. K. Holt, M. Stadermann, S. Kim, J. B. In, C. P. Grigoropoulos, A. Noy, O. Bakajin, Nanofiltration of electrolyte solutions by sub-2 nm carbon nanotube membranes, *Proceedings of the 11th Annual NSTI Nanotechnology Conference*, Boston, MA; Lawrence Livermore National Laboratory LLNL-PROC-402246, 2008.

13. J. A. Palz, D. Campbell-Lendrum, T. Holloway, J. A. Foley, Impact of regional climate change on human health, *Nature* 438 (2005) 310–317.

14. J. F. B. Mitchell, J. Lowe, R. A. Wood, M. Vellinga, Extreme events due to human-induced climate change, *Philos. Trans. Roy. Soc. A* 364 (2006) 2117–2133.

15. http://micro.magnet.fsu.edu/primer/java/scienceopticsu/powersof10/index.html.

16. A. Gombert, B. Bläsi, C. Bühler, P. Nitz, J. Mick, W. Hossfeld, M. Niggemann, Some application cases and related manufacturing techniques for optically functional microstructures on large areas, *Opt. Engr.* 43 (2004) 2525–2533.

17. (a) The Peacock Eye butterfly. http://creativecommons.org/licenses/by-sa/2.0/. (b) H. Ghiradella, Light and color on the wing: Structural colors in butterflies and moths, *Appl. Opt.* 30, 3492–3500 (1991). (c) en.wikipedia.org/wiki/Flame_Queen_Opal.

18. X. Hue, M. Thomann, R. J. Leyrer, J. Rieger, Iridescent colors from films made of polymeric core-shell particles, *Polymer Bull.* 57 (2006) 785–796.

19. J. E. Fernández, Materials for aesthetic, energy-efficient, and self-diagnostic buildings, *Science* 315 (2007) 1807–1810.

20. A. I. Maaroof, M. B. Cortie, G. B. Smith, Optical properties of mesoporous gold films, *J. Opt. A: Pure Appl. Opt.* 7 (2005) 303–309.

21. A. Rakitin, P. Aich, C. Papadopoulos, Yu. Kobzar, A. S. Vedeneev, J. S. Lee, J. M. Xu, Metallic conduction through engineered DNA: DNA nanoelectronic building blocks, *Phys. Rev. Lett.* 86 (2001) 3670–3673.

22. W. H. Marlow, van der Waals energies in the formation and interaction of nanoparticle aggregates, in *Gas Phase Nanoparticle Synthesis*, edited by C. G. Granqvist, L. B. Kish, W. H. Marlow, Kluwer, Dordrecht, the Netherlands, 2004, pp. 1–27.

23. D. M. Mattox, V. H. Mattox, Eds, *50 Years of Vacuum Coating Technology and the Growth of the Society of Vacuum Coaters*, Society of Vacuum Coaters, Albuquerque, NM, 2007.

24. W. S. Weiglhofer, A. Lakhtakia, *Introduction to Complex Mediums for Optics and Electromagnetics*, SPIE Press Monograph, Vol. PM123, SPIE Press, Bellingham, WA, 2003.

25. K. Sellers, C. Mackay, L. L. Bergeson, S. R. Clough, M. Hoyt, J. Chen, K. Henry, J. Hamblen, *Nanotechnology and the Environment*, CRC Press, Boca Raton, FL, 2009.

26. C. A. Poland, R. Duffin, I. Kinloch, A. Maynard, W. A. H. Wallace, A. Seaton, V. Stone, S. Brown, W. MacNee, K. Donaldson, Carbon nanotubes introduced into the abdominal cavity of mice show asbestos-like pathogenicity in a pilot study, *Nature Nanotech.* 3 (2008) 423–428.

27. K. E. Drexler, *Nanosystems: Molecular Machinery, Manufacturing, and Computation*, John Wiley & Sons, New York, 1992.

28. B. Mandelbrot, *The Fractal Geometry of Nature*, Freeman, San Francisco, CA, 1982.

29. M. F. Barnsley, *Fractals Everywhere*, Academic Press, New York, 1988.

30. S. K. Deb, S.-H. Lee, C. E. Tracy, J. R. Pitts, B. Gregg, H. M. Branz, Stand-alone photovoltaic-powered smart window, *Electrochim. Acta* 46 (2001) 2125–2130.

31. REN21 (Renewable Energy Policy Network for the 21st Century), Renewables 2005: Global Status Report, The Worldwatch Institute, Washington, DC, 2005; http://www.ren21.net.

32. C. G. Granqvist, G. A. Niklasson, Ultrafine chromium particles for phototermal conversion of solar energy, *J. Appl. Phys.* 49 (1978) 3512–3520.

33. C. G. Granqvist, O. Hunderi, Selective absorption of solar energy in ultrafine metal particles: Model calculations, *J. Appl. Phys.* 50 (1979) 1058–1065.

34. G. A. Niklasson, C. G. Granqvist, Ultrafine nickel particles for photothermal conversion of solar energy, *J. Appl. Phys.* 50 (1979) 5500–5505.

35. G. B. Smith, A. Ignatiev, G. Zajac, Solar selective black cobalt: Preparation, structure and thermal stability, *J. Appl. Phys.* 51 (1980) 4186–4196.

36. IPCC, Climate Change 2007: Mitigation. Contribution of Working Group III to the Fourth Assessment Report of the Intergovernmental Panel on Climate Change, edited by B. Metz, O. R. Davidson, P. Bosch, R. Dave, L. A. Meyer, Cambridge University Press, Cambridge, U.K. and New York, 2007.

37. M. Kolokotroni, Y. Zhang, R. Watkins, The London heat island and building cooling design, *Solar Energy* 81 (2007) 102–110.

38. World Energy Outlook 2009, International Energy Agency, Paris, France, 2009; http://www.worldenergyoutlook.org.

39. G. Pearman, personal communication. Unpublished plot based on the approach in R. Socolow, S. Pacala, J. Greenblatt, Wedges: Early mitigation with familiar technology, in Proceedings of GHGT-7, the Seventh International Conference on Greenhouse Gas Control Technology, Vancouver, Canada, September 5–9, 2004; see also http://www.princeton.edu/~cmi/.

40. J. J. Louviere, D. J. Street, L. Burgess, A 20+ years' retrospective on choice experiments, in *Marketing Research and Modeling: Progress and Prospects*, edited by Y. Wind, P. E. Green, Kluwer Academic, New York, 2003, chap. 8, pp. 201–214.

41. P. Eichholtz, N. Kok, J. M. Quigley, Doing well by going good? Green office buildings, Center for the Study of Energy Markets, Berkeley, CA, Working Paper CSEM WP-192 (2009); http://www.ucei.berkeley.edu/PDF/csemwp192.pdf.

42. L. Heshong, R. L. Wright, S. Okura, Daylighting impact on retail sales performancs, *J. Illum. Engr. Soc.* 31 (2002) 21–25; Daylighting impacts on human performance in school, *J. Illum. Engr. Soc.* 31 (2002) 101–114.

43. EIA, International Energy Outlook 2007, Energy Information Administration, US Department of Energy, Washington, DC, 2007; http://www.eia.doe.gov/emeu/international/energyconsumption.html.

CHAPTER 2

In Harmony with the Environment

Nature's Energy Flows and Desired Materials Properties

Figure 2.1 can serve as a pleasant point of departure and a stimulant for important questions. In a painting by Prince Eugen of Sweden called *The Cloud,* solar light is shown reflected from a cloud on an otherwise blue sky over a pastoral landscape outside Stockholm, Sweden, in the late 1800s [1]. As scientists we may ask how much of the incident solar light goes through the atmosphere if there are clouds and if there are no clouds? And what part of this solar radiation can our eyes sense? And why do clouds temper the climate and make it less hot at midday and less cold at midnight? These are very important questions! Answering them will in fact enable us to engineer material properties and make the most of the natural flows of energy that occur in the environment. This is what the present chapter is about.

Various elements need to be brought into harmony to make, for example, buildings both great to live in and energy efficient and to make solar cells and solar thermal collectors work well. A list of important elements would read as

- The response of humans, animals, and plants to heat and light
- The spectral properties of solar and thermal radiation
- The spectral properties of the atmosphere
- The spectral response of materials to solar and thermal radiation

We use the word "spectral" to describe how radiation is characterized by its intensity or energy density at each wavelength λ. The focus is on radiation, because it has a dominant influence on the earth's energy balance and internal energy flows as considered next.

FIGURE 2.1 *A color version of this figure follows page 200.* This oil painting on canvas by Prince Eugen of Sweden (1865–1947) dates from 1895 and is known as *The Cloud*. It is in the collection of the Gothenburg Museum of Art in Sweden. (From http://www.waldemarsudde.se/xsaml_molnet_g.html. Photo by Lars Engelhardt. Reprinted with permission.)

2.1 GLOBAL ENERGY FLOWS

Figure 2.2 shows that the sun's energy is distributed in an intricate manner involving many energy forms and locations where energy transfer takes place [2]. Radiation is the dominant environmental energy resource and also the major means to remove heat from the earth. It is inferred that 29% (37,800/130,000) of the total solar energy incident on the earth's atmosphere is reflected back into space by clouds, oceans, lakes and rivers, and snow and ice. The remainder goes into the various systems shown in Figure 2.2. Apart from the 0.08% (100/130,000) involved in photosynthesis and producing energy stored as organic matter, it enters initially as heat, which in turn leads to wind, water evaporation, melting ice, and general warming. Wind energy corresponds to 0.69% (900/130,000) of the incident solar energy.

For the earth's environment to be stable, energy inflow must equal energy outflow plus energy stored. The outflow and storage, as

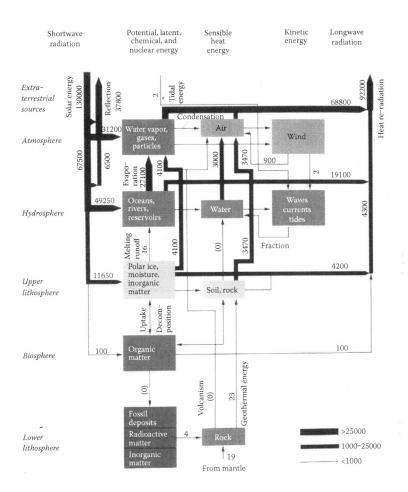

FIGURE 2.2 Renewable energy flows on earth in Gtoe (Gigatons oil equivalent, where 1 toe = 4.19 x 1010 Ws). (From B. Sørensen, *Renewable Energy: Its Physics, Engineering, Environmental Impacts, Economics and Planning*, Third Edition, Elsevier, Amsterdam, the Netherlands, 2004. With permission.)

percentages of solar inflow, are made up of thermal radiation (70.9%), reflected solar energy (29.0%), and organic storage (0.08%). Changing any of these three to any significant extent will upset the earth's long-term energy balance unless, by some stroke of luck, two or more change and their effects balance out.

Figure 2.2 is a detailed description of a very complicated energy system, and a simplified and more pictorial view of the energy flows on earth appears in Figure 2.3 [3]. Here, the average radiation balance is separated into the incoming and reflected solar radiation and the

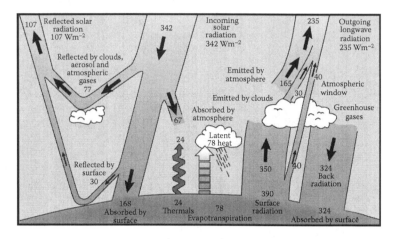

FIGURE 2.3   Short-wave and long-wave radiant flows to and from the earth's surface and the atmosphere. (From J. T. Kiehl, K. E. Trenberth, *Bull. Am. Meteorol. Soc.* 78 [1997] 197–208. © American Meteorological Society. Reprinted with permission.)

incoming and outgoing thermal radiation. The incoming and reflected solar radiation is short wave, while the incoming and outgoing thermal radiation is long wave. This spectral separation underpins most of the technologies we deal with in this book. An interesting feature in Figure 2.3, which is not often realized, is that the average energy flux in the incoming and absorbed long-wave thermal radiation from the atmosphere, about 324 Wm⁻², actually is larger than that in the incident and absorbed solar energy, which is 168 Wm⁻². This is due to greenhouse gases, without which the earth's surface would be inhospitably cold. The total energy flux emitted by these important minor atmospheric gases averages to 489 Wm⁻²; about 50% of this radiates outward and hence helps cool the earth. The well-known "greenhouse problem" results from an increased concentration of greenhouse gases, which leads to the incoming average thermal radiation flux rising above 324 Wm⁻². These gases do this by reducing the average flux of 40 Wm⁻² going through the "atmospheric window" as noted in Figure 2.3 so, in effect, the atmosphere gets "blacker" to thermal radiation. To restore energy balance, the only option for the earth is to increase its surface thermal flux, but then the surface and atmosphere must get hotter.

2.2 RADIATION IN OUR AMBIENCE: AN OVERVIEW

We are surrounded by electromagnetic radiation with different wavelengths all the time, day and night, as pointed out above. "Green" technologies make good use of this radiation, and it is important to understand the nature of this ambient radiation, including its spectral properties. Figure 2.4 introduces a number of important aspects of the electromagnetic radiation around us in a unified manner [4].

The most fundamental property of ambient radiation ensues from the fact that all matter sends out electromagnetic radiation, which is conveniently introduced by starting with the ideal blackbody whose emitted spectrum—known as the Planck spectrum—is uniquely defined once the temperature τ is known. Planck's law is a consequence of the quantum nature of radiation. The outgoing flux of photons at wavelength λ per unit area and per unit wavelength increment is called the *Planck exitance*. The associated emitted power spectral density in Wm^{-2} per wavelength increment is easily obtained by multiplying the exitance by the energy per photon, which is hc/λ or hf where h is Planck's constant, c is the speed of light, and f is the frequency. The actual formula constituting Planck's law is introduced later in this chapter when we link the cooling of real materials placed outdoors to the spectral properties of the atmosphere. Planck's equation is very useful and easy to apply in spreadsheets or computer programs when modeling energy flows in solar energy and building technology. At the moment we simply focus on the qualitative features of the ideal radiator.

Part (a) in Figure 2.4 depicts Planck spectra for four temperatures between –50°C and +100°C. The vertical scale denotes power per unit area per wavelength increment or $(GWm^{-2})m^{-1}$ (hence the unit GWm^{-3}). The spectra look bell-shaped and are confined to the $2 < \lambda < 100 \ \mu m$ spectral range. The peak in the spectrum is displaced toward shorter wavelength as the temperature goes up, which is referred to as Wien's displacement law; the peak lies at about 10 μm for room temperature. Not surprisingly, this peak position for room temperature thermal emission is no accident but corresponds precisely to where the atmosphere is most transparent to thermal radiation. In other words, human thermal comfort is directly related to the spectral properties of the atmosphere.

Blackbodies do not exist in reality, but the Planck formalism nevertheless is very useful since *thermal radiation* from a material is obtained by multiplying the Planck spectrum by a numerical factor—the emittance—which is less than unity. In general, the emittance is a function of λ. Most materials in nature have an emittance of about 0.9, and hence Planck's law gives a good approximation to thermal radiation. The most important exceptions are the metals, whose emittance can be as low as 0.01. But irrespective of the value of the emittance, the thermal radiation for energy efficiency in buildings and for human comfort is confined to the wavelength range given above.

Next we consider *solar radiation*. Part (b) of Figure 2.4 reproduces a solar spectrum for radiation immediately outside the earth's

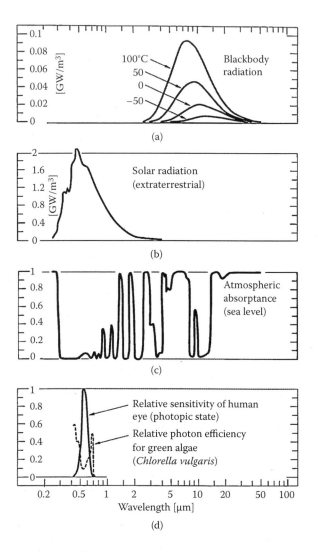

FIGURE 2.4 Spectra for radiation in our natural surroundings, showing (a) blackbody radiation from surfaces at "normal" temperatures, (b) solar spectrum outside the atmosphere, (c) absorption in the atmosphere during clear weather, and (d) spectral response of the human eye and of green algae. (From C. G. Granqvist, *Appl. Opt.* 20 [1981] 2606–2615. With permission.)

atmosphere. Once again the curve has a bell shape. If we return for a moment to Planck's law and let it represent a temperature of ~6000°C, we would get approximately the same curve shape as for the solar radiation. Therefore, the sun's surface temperature is taken to be ~6000°C, although the temperature inside the sun is orders of magnitude higher in order to keep the energy generation by fission going. It is important to observe that the solar spectrum is limited to $0.25 < \lambda < 3$ μm, so that there is almost no overlap with the spectra for thermal radiation. Hence, it is possible to have materials with properties that are entirely different for thermal and solar radiation. The integrated area under the curve gives the sun's power density at the top of the atmosphere, known as the "solar constant," and specifically being 1353 ± 21 Wm^{-2}. This is the largest possible power density on a surface oriented perpendicular to the sun (in the absence of atmospheric scattering).

Most technical systems, taken with the widest meaning, are located at or near to ground level, and it is of obvious interest to consider to what extent *atmospheric damping* influences solar irradiation and net thermal emission. Part (c) of Figure 2.4 illustrates a typical absorption spectrum vertically across the full atmospheric envelope during clear weather conditions. The spectrum is complicated and comprises bands of high absorption—caused mainly by water vapor, carbon dioxide, and ozone—as well as bands of high transparency. Thus, it is the minor components of the atmosphere that govern the damping, whereas the main constituents—oxygen and nitrogen—only play a minor role. Thus, if one alters the mix of trace gases in the atmosphere there is a risk that the average temperature at ground level is shifted.

It is evident that most of the solar energy can be transmitted down to ground level, and only parts of the ultraviolet (UV) radiation at $\lambda < 0.4$ μm and infrared (IR) radiation at $\lambda > 0.7$ μm are strongly damped. The maximum power density perpendicular to the sun's rays is about 1000 Wm^{-2}. Human skin exposed to the sun is protected in two main ways: from UV-induced chemical damage by absorption in atmospheric ozone, and from excess heat gain by absorption in the water vapor in the atmosphere. The latter is important due to the high water content in our bodies, and care is thus needed in outer space to block the solar IR radiation, which is normally removed by the atmosphere. On high mountains solar IR radiation may help counter the cold, but care is needed.

Thermal radiation from a surface exposed to the clear sky is strongly absorbed by the atmosphere except in the $8 < \lambda < 13$ μm range, called the "atmospheric window" or "sky window," where the transmittance can be large as long as the humidity is not too high. This "window" is the main channel by which the earth ultimately loses the energy it gains from the sun and is thus crucial to all life on earth. The absorption at the short-wavelength side is dominated by water vapor and the absorption at the long-wavelength side is dominated by carbon dioxide. Further details on the way this spectrum influences the heat balance on earth and our ability to provide coolness follow when we discuss how to tailor material properties to radiate preferably through this window. We will then also consider how to engineer systems to make the most of the directional infrared properties of the atmosphere. This is feasible since the atmospheric window gradually closes up as the atmosphere gets thicker, and rays passing close to the horizon "see" a much thicker atmosphere than those traveling vertically up. Thus, very little radiation gets away into space at near-horizontal angles. The conspicuous absorption peak within the atmospheric window ensues from ozone, but CO_2 and water vapor also contribute to attenuation in the sky window. The average transmittance in the window is about 87% for vertical radiation under clear conditions. If the atmospheric window is diminished, as a result of increasing carbon dioxide or increasing water vapor (cloudiness), the result will be a rise of the earth's temperature, that is, global warming [5].

Figure 2.4d illustrates two biological conditions of relevance for technical and other applications. The solid curve shows the relative sensitivity of the human eye in its light-adapted (photopic) state; the bell-shaped graph extends across the $0.4 < \lambda < 0.7$ μm interval and has its peak at 0.555 μm. We use this curve explicitly when considering lighting from both the sun and from lamps. The physical unit to determine lighting level is called lux, and the amount of light produced by a lamp is measured in lumen. Both units are based on this physiological curve. Clearly, a large part of the solar energy comes as invisible IR radiation. It is typically just over 50% of the total solar energy, and its main effect is to heat.

The dashed curve in Figure 2.4d indicates that photosynthesis in plants makes use of light with wavelengths in approximately the same range as those for the human eye, which is relevant for greenhouse applications. Plants also transmit most nonvisible solar energy; otherwise, they would risk getting too hot and would have to evaporate excessive amounts of water.

Figure 2.4 emphasizes that ambient radiation is *spectrally selective*, that is, confined to specific and usually well-defined wavelength ranges, which is of great importance for a number of green nanotechnologies as we will see many times later in this book. However, these are not the only characteristic features of this radiation, and another one is the *angular properties* of the radiation. For example, thermal emission typically takes place in all available directions, whereas solar incidence in the absence of clouds is from a well-defined direction. Both spectral and angular properties can be taken advantage of. Still another characteristic feature of the ambient radiation is its *variation over the day and season*. Solar irradiation comes during the day while thermal emission takes place all the time, both from the earth's surface and from the atmosphere. Thus, incoming radiation does not cease at night, and thermal radiation from the atmosphere still warms the earth enough to stop most exposed surfaces from getting unbearably cold and prevent plants from dying.

Finally, we repeat some of the lessons learned above:

- The quantities of the minor gases, water, CO_2, and ozone in the atmosphere are crucial to plant and animal life on earth due to the delicate control they exert on the intensity and spectral distribution of incoming solar and thermal radiation, and on outgoing thermal radiation; clearly, there are high risks connected with any significant perturbation of this mix, and
- The solar spectrum and the Planck thermal spectrum near room temperature—both modulated by the minor atmospheric gases—plus the spectral sensitivity of the human eye and of plants provide the basis for green nanotechnologies benefiting from spectral selectivity

2.3 INTERACTION BETWEEN RADIATION AND MATERIALS

2.3.1 Fundamentals Based on Energy Conservation

When electromagnetic radiation impinges on a material, one part can be transmitted, a second part is reflected, and a third part is absorbed. Energy conservation yields, at each wavelength, that

$$T(\lambda) + R(\lambda) + A(\lambda) = 1 \qquad (2.1)$$

where T, R, and A denote the fractions that are transmitted, reflected, and absorbed, respectively, that is, the transmittance, reflectance, and absorptance. Another fundamental relationship, also ensuing from energy conservation and referred to as Kirchhoff's law, is

$$A(\lambda) = E(\lambda) \qquad (2.2)$$

with E being emittance, that is, the fraction of the blackbody radiation that is given off at a particular wavelength. Equation 2.2 is of practical relevance mainly for $\lambda > 2\ \mu m$.

These simple formulas make it possible to define the properties of a number of materials for green nanotechnologies, as we will come back to repeatedly in subsequent chapters. Since we often deal with quanta of radiation, called *photons*, an alternative interpretation of the energy conservation equation, involving probabilities, is also worth noting: each photon striking a material has a probability of being reflected, transmitted, or annihilated (i.e., absorbed), and since all probabilities must sum to one we recover Equation 2.1.

A photon has a momentum defined by the ray's direction, its wavelength in air, and the property of the solid or liquid material in which it is traveling (known as the *refractive index*). When the photons go from one medium to another, or reflect off an interface, the component of each photon's momentum parallel to the interface is not changed. This last fact leads to the famous rule governing refraction of light known as Snell's law, which is a core issue in all solar, lighting, and vision technologies.

It is important to know the magnitude of the absorptance, as it is critical for most technologies in the transformation from electromagnetic energy to some other useful form of energy such as heat, electrical power, chemical energy, or fluorescence. The absorptance is often obtained indirectly from data on $R(\lambda)$ and $T(\lambda)$ by rearranging Equation 2.1 to read

$$A(\lambda) = 1 - R(\lambda) - T(\lambda) \tag{2.3}$$

Thermal energy is the most common outcome of light absorption, but other forms of energy are possible and important for the discussion of the energy systems that will follow. These include

- Transport of electrons or holes in an external circuit, for example in a solar cell
- Chemical energy, as in plant photosynthesis, photocatalysis for self-cleaning and disinfection treatment, or conversion of water to hydrogen and oxygen, and
- Fluorescence, for which the output is photons with lower energy, and hence longer wavelength, plus some heat

If these transformations are the goal, then loss as heat should be minimized.

2.3.2 Directionality and Polarization Dependence

The spectral quantities $A(\lambda)$, $R(\lambda)$, and $T(\lambda)$ depend on the incident direction (ϕ, λ) of the light, as defined in Figure 2.5, and on its polarization as

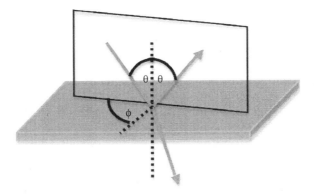

FIGURE 2.5   Definition of a light ray's polar (θ) and azimuthal (ϕ) directions in three dimensions using the plane of incidence.

indicated in Figure 2.6. Figure 2.5 shows that in order to define the polar and azimuthal angles—θ and ϕ, respectively—one first has to establish a plane of incidence, which is given by the value of ϕ. Polarization is determined by the direction in which the electric field component of the radiation points; it is accomplished by a polarizer, which is a device or material able to block light with electric fields in a particular direction.

If rotation of the incidence plane in Figure 2.5 to any angle ϕ about the normal makes no difference optically, one can drop ϕ and the quantities of interest become $T(\theta,\lambda)$, $R(\theta,\lambda)$, and $A(\theta,\lambda)$, with θ being the angle of incidence. We caution, though, that some useful nanomaterials for "green" technologies require that ϕ is kept; such materials are said to be anisotropic. If the surface is very smooth, and the material is optically clear, the situation reduces to that in Figure 2.6. Here, the two key

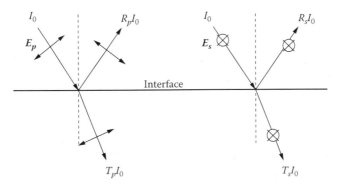

FIGURE 2.6   Directions of the electric field E for s and p polarized radiation onto an interface. I_0 is the intensity of the incoming light, and T and R denote transmittance and reflectance, respectively.

light polarizations, or electric field directions, are defined with respect to the plane of incidence. If the electric field component is polarized in the plane of incidence one speaks of p-polarized light, and if the electric field component is perpendicular to this plane—and, hence, parallel to the surface—one has s-polarization. These labels are used as subscripts to designate T, R, and A for each polarization component. At oblique incidence R, T, and A are polarization dependent, so outgoing radiation will be polarized even when incident light is unpolarized. If oblique light is unpolarized, then R and T simply are arithmetic averages of (R_p, R_s) and (T_p, T_s), respectively.

When bright light, such as sunlight, strikes a smooth surface obliquely one may experience uncomfortable brightness, and this is so even for materials with weak reflection at normal incidence, such as glass and polymethyl methacrylate (PMMA). The reason is that the reflectance generally goes up for increasing angles with regard to the material's normal.

2.4 BEAM AND DIFFUSE RADIATION

2.4.1 General Considerations

Sunlight and solar energy are usually a combination of one part that arrives undeviated and another part that has been scattered by clouds and other atmospheric constituents. Our most commonly experienced diffuse light is probably the blue light from a clear-sky hemisphere. We will refer to the direct light as "beam radiation" and the scattered component as "diffuse radiation." Lighting from lamps and light-emitting diodes also requires that one considers this combination. All rays are parallel in beam radiation, whereas the rays span a range of directions in diffuse radiation. If a material is smooth and specular, each transmitted or reflected ray has one possible direction as given by Snell's law. But if there is some scattering from either the surface or the inside of the material, then each reflected or transmitted photon has a probability of going off in many directions. Most of the naturally occurring materials—for example, human skin, soil and rocks, and plant material—reflect and transmit diffusely, but the surface of a lake in still air is almost specular and thus capable of creating images. Diffuse transmitters and reflectors can appear to emit just like light sources and are often treated as such in ray-tracing computer programs. For example, leaves or grass may appear to glow like weak green lamps in the late afternoon. Measurement of diffuse radiation is a very important topic. If the incident radiation is diffuse, properties such as absorption must be averaged over a range of incident directions. And it has already been noted that thermal radiation from a surface is diffuse.

We normally expect surface roughness or interior doping of clear solids or liquids with microparticles to scatter light. For example, add a few drops of milk to water (particles that are microspheres of fat) and see how the liquid turns "milky"! Or feel and look at matt nonglossy paints (with microparticles typically of TiO_2)! But nanoscale surface roughness and doping with nanoparticles are different, and scattering can be negligible if the scale is small enough. Thus nanostructured surfaces can appear optically smooth, and materials doped with nanoparticles can appear clear. But then the inhomogeneities must be less than 5 to 50 nm in size, depending on material, as we will see later.

These are very useful features of nanomaterials and render them of interest for a variety of energy-saving products. This is so because the products must not only perform well technically—they also have to look good. If the nanomaterials are used in a normal window, for example, they should not distort vision. There is a transition from beam to diffuse optical behavior at some critical maximum nanoscale, and most features over 100 nm will scatter light unless embedded in a medium whose optical properties are very close to those of the nanostructured component.

2.4.2 Energy Flows in Diffuse and Nonparallel Radiation Beams

It is obviously important to understand how to describe energy flows when rays are not parallel and how to measure energy flux and visual brightness in diffuse beams. There are special physical units to describe and measure spreading and converging light, and these units are needed also to represent the brightness of light emitters or diffuse surfaces. The importance of these topics is indeed obvious since most interior lighting is diffuse, the exception being daylight on clear days entering windows without blinds or curtains. The intensity of the incident light as a function of direction determines lighting levels on all surfaces within a building and can also create lighting patterns on walls and floors and determine whether the lighting is visually comfortable or not.

We first consider an everyday example: how to measure the intensity of the blue light from a clear sky separately from the direct solar energy component. The blue color is caused by solar rays scattered by atmospheric molecules. To understand diffuse light measurements one needs to consider two things:

- The geometric character of the source or scatterer, and
- The placement or orientation of the detector relative to the source

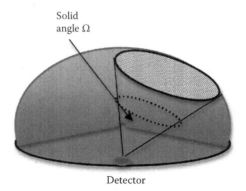

Solid
angle Ω

Detector

FIGURE 2.7　Solid angle Ω defined by a section of the sky and as seen by a ground-based detector.

A clear blue sky is almost uniform across the hemisphere, so rays are converging more or less uniformly from the sky vault onto any horizontal detector. But our eyes, which are also detectors, take in only some forward component (so to collect most of the blue sky light we would have to lie on our backs). Thus, detectors, including our eyes, will get different answers that are in direct proportion to the area of the sky hemisphere that is sampled and whether the light is coming from ahead or from the side.

We first disregard detector orientation and consider the sampling area aspect, as it may be easiest to grasp. Radiation impinges from a cone onto a detector, as illustrated in Figure 2.7. Such cones define a special geometric unit needed for lighting and diffuse radiation; it is called the solid angle Ω and is measured in steradians. A hemisphere contains 2π steradians. The general definition is $\Omega = A(r)/r^2$, where $A(r)$ is the surface area projected onto a sphere of radius r by the cone or solid angle. The rays inside this cone have a certain total energy flux (in watts) and light flux (in lumens).

The natural unit for quantifying energy flow in diffuse radiation is energy flux per unit solid angle, which is called *radiant intensity* Γ_R and is measured in watts per steradian [W Sr^{-1}]. The equivalent for light is called *luminous intensity* Γ_L and is expressed in lumens per steradian (lm Sr^{-1}), which is also called *candela* (cd). Detectors of optical, solar, and thermal radiation usually record intensity or energy flux incident onto them per unit area, that is, in Wm^{-2} falling onto the detector. In lighting, the equivalent is luminous intensity measured in lm per m^2, or lux. Most light meters give outputs in lux.

The levels of Wm^{-2} and lux are end result measures ensuing from all of the incoming radiation—beamlike as well as diffuse—and they are important parameters for surfaces which are irradiated or lit up, such as solar collectors or desk tops. But there are many cases for which the diffuse intensity has to be quantified for each incoming direction; these cases include finding out to where the light goes after reflection from a surface (such as in glare analysis), or assessing how much of the radiation is absorbed by a surface (such as for solar heat collectors and solar cells). By summing up the contribution from each direction one then finds the intensity or lux level. Diffuse intensity relates to how many watts per m^{-2} of radiation are incident from a given solid angle, with the incidence direction defined by two angles, denoted θ and ϕ as in Figure 2.5. In science and engineering vocabulary, the physical quantity diffuse intensity is called *radiance* for energy and *luminance* for light. The unit for radiance is thus Wm^{-2}Sr^{-1}, and the unit for luminance is (lm)m^{-2}Sr^{-1}, or more briefly (cd)m^{-2}. The latter unit in essence refers to brightness, and in lighting design one sometimes speaks of "nits" instead of (cd)m^{-2}.

The uniformly bright sky of Figure 2.7, with about the same radiance or luminance in whichever direction we look, is a special case. A source of radiation for which the light comes in with constant luminance from all directions in a hemisphere is said to be *Lambertian*. However, most real situations of interest have radiation sources that are far from Lambertian, so one needs a framework in which to define and measure radiance or luminance for any direction. To that end we shrink the solid angle to a very small size around the radius pointing in the direction (θ,ϕ). For practical measurements, this may mean having optics which only collects light within a cone with an aperture that is 1–2° wide and aiming the instrument in the direction of interest.

Figure 2.8a shows a schematic for measuring radiance or luminance. Here, the focusing system collects incoming radiation over a small aperture or acceptance angle of light along the axis of the instrument, and the aiming point is on the surface or the light source or light diffuser. A measurement of intensity or lux level for diffuse radiation proceeds differently, as shown in Figure 2.8b; in contrast to the case of measuring radiance or luminance, the detected light is usually accepted from the full hemisphere.

For mathematical and computer modeling, one can use the differential limit of solid angle $d\Omega(\theta,\phi)$ defined as

$$d\Omega(\theta,\phi) = \sin\theta\, d\theta\, d\phi \quad [\text{Sr}] \tag{2.4}$$

The increase in $d\Omega(\theta,\phi)$ as the angle θ to the vertical increases is important in practice. The radiant or luminous flux, being differentially small within $d\Omega(\theta,\phi)$, is $d\Phi$, so that the diffuse energy or light flux in each direction becomes

$$\Gamma(\theta,\phi) = \frac{d\Phi}{d\Omega} \ [\mathrm{W\ Sr^{-1}}] \ \text{or} \ [\mathrm{lm\ Sr^{-1}}] \tag{2.5}$$

It is important to be able to relate total intensity I in $\mathrm{Wm^2}$ as in Figure 2.8b to radiance $L_R(\theta,\phi)$ in each direction, and for light to relate illuminance in lux to luminance $L_L(\theta,\phi))$ as in Figure 2.8a. One cannot just divide Γ by detector area and sum over all angles. Such a procedure would give the wrong answer (though the correct physical unit). Why is that? What is involved here is a very important matter relating to energy accounting! To get the right answer one needs to return to the second issue, put aside earlier, namely detector or surface orientation. It can be assumed for simplicity that the detector is pointing vertically up or simply define θ with respect to the detector normal as in Figure 2.9. When a beam is incident in this manner the energy flux per unit area of surface—which is what our detector measures—is reduced because the illuminated area is larger than if the same beam was incident vertically.

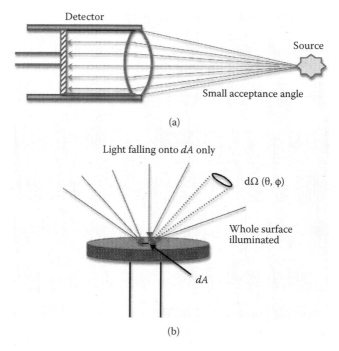

(a)

(b)

FIGURE 2.8 (a) Schematic of a radiance or luminance measurement involving a narrow acceptance cone around one direction. (b) Schematic of an energy intensity (in $\mathrm{Wm^{-2}}$) or lux measurement involving all incoming directions from a hemisphere. If the lux level is uniform over the detector area A, then the final intensity is the same as that onto dA.

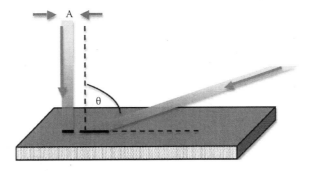

FIGURE 2.9   Beams of radiation from two directions incident on a surface. The irradiated areas are indicated in bold.

Intensity is power per unit area, so if a beam with cross-section A_0 and intensity $I_0 = \Phi/A_0$ is incident obliquely, it illuminates an area A_d, as seen in Figure 2.9, so that $A_d = A_0/\cos\theta$. Thus, the energy per unit area hitting the detector surface is reduced to $I_d = I_0\cos\theta$.

We now return to the case of a diffuse beam falling onto the detector. To link the total intensity and lux level to $L_R(\theta,\phi)$ and $L_L(\theta,\phi)$, respectively, in each direction one should recall that radiance and luminance involve three things for each direction: incident power, solid angle, and area. Along the direction (θ,ϕ) one has $L(\theta,\phi) = dI(\theta,\phi)/d\Omega = d^2\Phi(\theta,\phi)/dAd\Omega$, and hence

$$L(\theta,\phi)\cos\theta = [dI(\theta,\phi)/d\Omega]\,\cos\theta = dI_d(\theta,\phi)/d\Omega \qquad (2.6)$$

Equation 2.6 can be used to define $L(\theta,\phi)$ and also gives $dI_d(\theta,\phi) = L(\theta,\phi)\cos\theta\,d\Omega$. Thus, one can now sum over all solid angles to find total incident intensity I_d (assuming that I_d is constant for each element dA). Using Equation 2.4 for $d\Omega$, the result for intensity (or lux level) becomes

$$I_d = \int_0^{2\pi} d\phi \int_0^{\pi/2} d\theta \sin\theta \cos\theta L(\theta,\phi) \qquad (2.7)$$

What does $L(\theta,\phi)$ look like after reflection or transmission of a parallel incident beam of light off or through a nonsmooth surface such as a matt paint or a pigmented polymer sheet? Most such surfaces are partially specular, that is, the scattered intensity tends to be largest in the direction of the specular or mirror rays as illustrated in Figure 2.10 for reflection [6]. The difference with regard to a Lambertian, or uniformly scattering, material is also shown. Surfaces which scatter nonuniformly

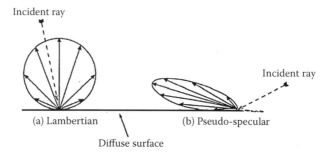

FIGURE 2.10   Reflected radiation from a diffuse surface being (a) a
Lambertian or perfect diffusing reflector and (b) a "pseudo-specular"
reflector. (From A. Earp, Luminescent Solar Concentrators for Fibre
Optic Lighting, Ph.D. thesis, University of Technology, Sydney, Australia,
2005. With permission.)

are sometimes referred to as "pseudo-specular." Incidentally, Equation
2.7 gives that a Lambertian surface, with $L(\theta,\phi) = L$ independent of
direction, must have $I_d = L/\pi$.

Summing or integrating reflected or transmitted light over the full
exit hemisphere at each wavelength yields total diffuse reflectance and
transmittance. This is measured with devices called, unsurprisingly,
integrating spheres. Information on such instruments can be found in
the literature [7–9].

2.5 HEMISPHERICAL ABSORPTANCE

Spectral hemispherical absorptance $A_H(\lambda)$ of a material is a useful quantity for much energy-related work with a bearing on the environment.
Consider first an equal flux of photons from any direction with total
incident power P_0 falling onto a small area dA. This is the inverse of
the Lambertian emitter, now with all rays pointing inward. As just discovered, the total incident intensity at any point is then $I_d = [P_0/dA] =
L/\pi$. Next, we need the monochromatic power $P(\theta,\phi,\lambda)$ incident from
the direction (θ,ϕ) and falling onto a sample with an area dA. With no
explicit dependence on axial rotation, one obtains for each wavelength
from Equations 2.4 and 2.6 that $dI_d = P(\theta,\phi)/dA = L(\theta,\phi) \cos\theta\, d\Omega$. Using
$L = \text{constant} = P_0/[\pi(dA)]$ now gives

$$P(\theta,\phi,\lambda) = (P_0 \cos\theta\, d\Omega)/\pi \qquad (2.8)$$

where $(\cos\theta\, d\Omega/\pi)$ is the fraction of the total incident energy converging
in the (θ,ϕ) direction.

It is easy to check the integral

$$\int_{hemisphere} d\Omega \frac{\cos\theta}{\pi} = 1 \qquad (2.9)$$

so integrating over the full $P(\theta,\phi)$ gives P_0, as required for energy conservation. This check also shows the importance of the $\cos\theta$ factor. The power absorbed at θ at each wavelength is $2A(\theta,\lambda)P(\theta)$, as obtained after integrating over all ϕ in Equation 2.8. The hemispherical absorptance now becomes

$$A_H(\lambda) = \frac{P_{absorbed}}{P_0} = 2\int_0^{\pi/2} d\theta \cos\theta \sin\theta A(\theta,\lambda) \qquad (2.10)$$

for uniformly incident radiation, where Equation 2.9 has been used. An opaque surface has $A(\theta,\lambda) = 1 - R(\theta,\lambda)$, so Equation 2.10 leads to

$$A_H(\lambda) = 1 - R_H(\lambda) \qquad (2.11)$$

with $R_H(\lambda)$ being the hemispherical reflectance. Simple modifications of these equations to any angle range can also be used when dealing with focused beams of light, such as those occurring with solar concentrator mirrors and lenses, to find the relevant integrated absorptance based on measured or modeled $P(\theta,\phi,\lambda)$.

Under a clear sky one can consider solar energy as coming from only one direction at any instant. But in many cases there is also a significant diffuse component of the solar radiation, and then one must integrate over all incident directions, using the actual incoming intensity, to model the instantaneous solar absorptance A_{sol}. This is often cumbersome, though, and the usual practice for comparing the optical quality of materials in applications such as windows, skylights, solar cells, and solar thermal collectors is to use data only for normal incidence. The practice is fraught with clear limitations, though, and the directional properties become very important for system performance over time, so indeed the angular features are of considerable interest for many nanostructures. Think, for example, of a standard vertical window for which the angle of incidence of solar energy and daylight often is very high, except during early mornings and late afternoons for east- or west-facing orientations, respectively. Thus, one needs to know, or at least be able to confidently model, the directional dependencies.

2.6 SOLAR AND DAYLIGHTING
PERFORMANCE PARAMETERS

Figure 2.4 introduced the spectral properties of thermal, solar, and luminous radiation. The term "luminous" tells us that the radiation is visible to the human eye in its light-adapted (photopic) state. These properties can be utilized to determine a number of key parameters for energy-related performance, as discussed next. They include the total solar energy or daylight that is reflected, transmitted, and absorbed at a particular angle of incidence, as well as the thermal emittance. Clearly, these parameters are important and can be used in many connections, such as in building codes and for comparing products. They involve averages weighted by the solar spectrum intensity, or by the Planck spectrum, which can be done on spreadsheets as sums over small uniform wavelength steps [10] or, equivalently, as integrals.

Solar transmittance $T_{sol}(\theta)$ and reflectance $R_{sol}(\theta)$ are given by

$$T_{sol}(\theta) = \int d\lambda \, T(\lambda,\theta) \, S(\lambda)/\int d\lambda \, S(\lambda) \qquad (2.12)$$

$$R_{sol}(\theta) = \int d\lambda \, R(\lambda,\theta) \, S(\lambda)/\int d\lambda \, S(\lambda) \qquad (2.13)$$

where $T(\lambda,\theta)$ and $R(\lambda,\theta)$ are the measured spectral reflectance and transmittance, respectively, for an angle of incidence θ, and $S(\lambda)$ is the incident solar spectrum, which depends on the thickness of the atmosphere through which sunlight has passed. The common practice for comparisons is 1.5 times the atmosphere's thicknesses, which corresponds to the sun standing 37° above the horizon (known as air mass 1.5 or AM 1.5). Solar absorptance is then obtained from $A_{sol}(\theta) = [1 - R_{sol}(\theta) - T_{sol}(\theta)]$.

The luminous, or visible, transmittance $T_{lum}(\theta)$ and reflectance $R_{lum}(\theta)$ of a surface illuminated by solar radiation are calculated from the integrals

$$T_{lum}(\theta) = \int d\lambda \, T(\lambda,\theta) \, S(\lambda) \, Y(\lambda)/\int d\lambda \, S(\lambda) \, Y(\lambda) \qquad (2.14)$$

$$R_{lum}(\theta) = \int d\lambda \, R(\lambda,\theta) \, S(\lambda) \, Y(\lambda)/\int d\lambda \, S(\lambda) \, Y(\lambda) \qquad (2.15)$$

with $Y(\lambda)$ being the spectral response of the light-adapted human eye, that is, the photopic sensitivity [11]. Luminous absorptance is given by $A_{lum}(\theta) = [1 - R_{lum}(\theta) - T_{lum}(\theta)]$. If the light is not from solar radiation but from a lamp, one must use a source spectrum that is different from $S(\lambda)$, and such information is usually available from the lamp manufacturer. However, Equations 2.14 and 2.15 are appropriate for daylighting.

In many practical cases it is of interest to limit solar heat gains but still allow adequate daylight to get through, and then one can make use of the fact that $Y(\lambda)$ is weak for $\lambda > 700$ nm and that it vanishes beyond 780 nm, as seen in Figure 2.4. Blocking as much as possible of the solar spectrum beyond ~700 nm all the way to where it ends at ~2,500 nm will thus lead to a large reduction in heat gain. This type of spectral selectivity is widely used for glazings and is well suited to control with various nanostructures. It also links us, for the first time, to an important general issue for energy efficient lighting called the "luminous efficacy," denoted K. It is defined as the amount of light (in lumens) divided by the quantity of accompanying heat (in watts).

Energy efficient lighting has high luminous efficacy, but it is important to know what is practical to aim for. First we caution that a very high efficacy may in fact be unsatisfactory, and this is so because one also needs to consider the color of the light and that the color can change when light passes through windows, especially those with coatings based on nanomaterials. Thus, we redefine the main goal to mean high efficacy achievable with near-neutral color. Such color-neutral surfaces look white if highly reflective and gray if weakly reflective. Our eyes are most sensitive to green light, and the peak in $Y(\lambda)$ corresponds to $\lambda = 555$ nm, as noted above. Illumination at this wavelength only gives $K = 683$ lm W^{-1}. But white light has a much lower value of K, because our eyes are less sensitive to red and blue. At night, our eyes have their maximum sensitivity shifted to 507 nm, and then K can be as large as 1700 lm W^{-1} because of the enhanced sensitivity of the eyes.

We will mainly consider light-adapted eyes (photopic vision). The best neutral color is obtained with a light spectrum which is close to $Y(\lambda)$, and K is then about 200 to 220 lm W^{-1}. Since daylight makes up only around half of the solar spectrum, K is ~100 lm W^{-1} under a clear sky. This efficacy is higher than for most ordinary lamps in use today. The better lamps of this kind approach 100 lm W^{-1}, and they may well surpass this value in the future, as will be discussed later in Chapter 5. Very powerful lamps, such as those used in sports stadiums, do surpass the luminous efficacy of daylight already at present. For lamps, the K value is usually defined as the light output in lumens divided by the electrical power input, and, ideally, one would like nearly all of the latter converted to light. In order to match the K value of daylight, one needs ~50% of the electricity to be converted to light. This is far off for most lamps, but we will see later how nanoscience can help improve the light output.

2.7 THERMAL RADIATION AND SPECTRAL PROPERTIES OF THE ATMOSPHERE

Earlier we divided the solar spectrum into two parts in order to discuss daylighting and vision. For some energy-related applications it is, analogously, useful to divide the Planck thermal radiation spectrum into different wavelength ranges, as discussed at length below. But the key parameter for describing thermal radiation from most surfaces is the blackbody emittance.

2.7.1 Blackbody Emittance

We start by considering heat transfer, which adds contributions from thermal radiation, conduction, and convection. The total heat flow P_{tot} from an object at the temperature τ is given by

$$P_{tot} = U\,(\tau - \tau_a) \tag{2.16}$$

where τ_a is the ambient temperature, taken to be lower than τ, and the factor is called the U value. U depends on temperature and other parameters such as wind speed and humidity. In well-insulated and designed systems—such as some double glazings and evacuated glazings—U may be dominated by emittance. The radiative component to U should be minimized in some cases, such as in a solar thermal collector, while it should be maximized in other situations such as for cooling of a solar-illuminated roof or for cooling by exposure to the sky at night.

The specific spectrum of the emitted radiation depends on the material at the surface, in particular its infrared optical properties. The net radiant loss is what counts, and this is determined by the emittance as well as by the temperature of the surroundings which send thermal radiation toward the surface. In an outdoors situation, the atmosphere is the main contributor to incoming radiation, but surrounding buildings and vegetation may also be important so surface orientation has to be considered. Roofs generally give off more radiation than walls and hence they cool much faster at night—but also heat up more in the day. The "surrounds" are anything in the forward hemisphere as viewed from the emitting surface.

The Planck spectral photon exitance $M(\lambda,\tau)$ is given by

$$M(\lambda,\tau) = \frac{2\pi c}{\lambda^4 \left[e^{\frac{hc}{\lambda k_B \tau}} - 1 \right]} \quad [(\text{photons/s})/\text{m}^2]/\text{m} \tag{2.17}$$

where τ is in Kelvin, and k_B is Boltzmann's constant. The emitted power per unit area per unit wavelength, or Planck spectral radiant exitance denoted $P(\lambda,\tau)$, is Equation 2.17 multiplied by the energy per photon, that is,

$$P(\lambda,\tau) = \frac{hc}{\lambda}M(\lambda,T) = \frac{c_1}{\lambda^5\left[e^{\frac{c_2}{\lambda\tau}} - 1\right]} \ (\text{Wm}^{-2})/\text{m} \qquad (2.18)$$

with c_1 = 3.71415 x 10^{-16} Wm^2 and c_2 = 1.4388 x 10^{-2} Km. $P(\lambda,\tau)$ is the blackbody spectrum introduced earlier in Figure 2.4. Integrating $P(\lambda,\tau)$ over all wavelengths gives the total power P emitted by a black body at temperature τ, which gives the well-known Stefan–Boltzmann equation

$$P = \sigma_{SB}\tau^4 \qquad (2.19)$$

with σ_{SB} = 5.67×10^{-8} Wm^{-2} K^{-4}.

A surface only has a net heat loss by radiation if the surrounds, which are also radiating toward it, are at a lower temperature. In the cases of present interest, the surrounds will normally be the ambient atmosphere at τ_a. The simple equation commonly used to handle this assumes that a homogeneous hemispherical black body environment at ambient temperature is sending in radiation so that

$$P = \sigma_{SB}(\tau^4 - \tau_a{}^4) \qquad (2.20)$$

However, this simple relation needs to be modified for real materials and also to account for the infrared properties of the earth's atmosphere. And this is where the fun bit comes in, since outdoor surfaces can lose heat when they are at temperatures lower than their surrounds, that is, for $\tau < \tau_a$. No basic physics is violated, of course, but the underlying reason for this nonintuitive result is at the core of how the earth stays cool, and it is also a cornerstone for understanding global warming.

2.7.2 The Sky Window

Look again at Figure 2.4. In the wavelength range between 8 and 13 μm, what we called the "atmospheric window" or "sky window," there is almost no incoming radiation from the atmosphere. In other words a surface can "see" outer space, which is at a very low temperature, for these wavelengths. Figure 2.11 describes the same thing in more detail and demonstrates with regard to the absorption by trace gases—that is, water vapor, CO_2 and some ozone (not shown)—that this window occurs at wavelengths where their combined absorption does not completely block the outgoing radiation, which is from 8 to 13 μm [12]. If an object would radiate only at these wavelengths and be perfectly

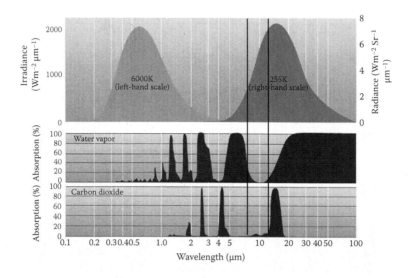

FIGURE 2.11   Solar spectrum just outside the atmosphere compared with the radiation spectrum from a "black" cool object at 255 K. Also shown are absorption bands of atmospheric water vapor and CO_2. The "atmospheric window" for wavelengths between 8 and 13 μm is outlined in bold. (From Australian Bureau of Meteorology, The Greenhouse Effect and Climate Change, Melbourne, Australia [2009]; http://www.bom.gov.au/info/TheGreenhouseEffectAndClimateChange.pdf. With permission.)

insulated from the surrounds, it could possibly cool to exceptionally low temperatures. We shall see later in Chapter 7 just how far below the ambient temperature one can get in practice by using nanomaterials.

Figure 2.11 also gives a good explanation of global warming due to increased CO_2: When its concentration goes up there will be more absorption in the atmospheric window, meaning that the amount of energy that can be radiated into outer space is cut down. And so the earth's temperature must rise.

The generalization of Kirchoff's law in Equation 2.2 for any emission direction is $A(\theta,\lambda) = E(\theta,\lambda)$. Using Equation 2.10, the total radiant power emitted into a hemisphere per unit area of sample surface becomes

$$P_{rad}(\tau) = \int_0^\infty d\lambda A_H(\lambda)P(\lambda,\tau) \qquad (2.21)$$

and dividing P_{rad} by total output P_0 gives the hemispherical emittance E_H at the temperature τ. Then, one needs to subtract the incoming thermal radiation from P_{rad} to obtain the net loss by radiation. Directional issues are important for determining incoming radiation because the direction-dependent spectral emittance of the atmosphere $E_a(\theta_z, \lambda)$ depends strongly on the angle θ_z to zenith (i.e., the vertical). The dependence is particularly important for the sky window range, which follows because as rays approach the horizontal the atmosphere gets thicker and hence absorbs and radiates more. This also means that very little radiation can escape the earth for $\theta_z > 70°$. At wavelengths outside the sky window, the atmosphere is already very "black" so it cannot absorb or emit much more. These angular properties will be used to advantage in Chapter 7.

The directional dependence of the emittance within the sky window, denoted $E_{a,sw}$, is a complicated issue, but previous work [13,14] has made good use of the expression

$$E_{a,sw}(\theta_z) = 1 - \left[1 - [E_a(0,\lambda)]_{sw}\right]^{1/\cos\theta_z} \approx 1 - 0.87^{1/\cos\theta_z} \qquad (2.22)$$

where the inner bracket indicates an average over the sky window, whereas the emittance was set to unity outside the sky window. This is for a very dry atmosphere. Inside the sky window range, the atmospheric emittance is approximated to be spectrally flat with an average value of 0.13 for the vertical direction. Care is needed, however, if materials used for cooling have sharp spectral features within the sky window or if variable conditions, involving more water vapor, are present. Directional properties are sensitive to water vapor content, for which explicit equations follow, and accurate spectral and directional models of the atmosphere are then needed as given by

$$E_a(\lambda, \theta_z) = 1 - [T_a(\lambda, 0)]^{1/\cos\theta_z} \qquad (2.23)$$

where $T_a(\lambda, 0)$ is the spectral transmittance in the zenith direction for a particular humidity.

The directional dependence of E_a, which largely arises within the sky window, can be very advantageous for cooling applications, but then one must go beyond simple relationships of the type $E_H = A_H$. For example, what if we block all incoming radiation with $\theta_z > \theta_{z,max}$? The integral in Equation 2.10 can be generalized to define a new limited-range absorptance or emittance simply by reducing the upper limit on θ from $\pi/2$. This gives the average absorptance $A_{\theta_z,max}(\lambda)$ at a certain wavelength for radiation into a cone ending at the surface of interest and having an aperture $\theta_{z,max}$, that is,

$$A_{\theta_{z,max}}(\lambda) = \frac{P_{absorbed}}{P_0} = 2 \int\limits_{0}^{\theta_{z,max}} d\theta \cos\theta \sin\theta A(\theta,\lambda) \qquad (2.24)$$

The average of the atmospheric absorptance in the sky window, as given by Equation 2.23, starts at around 0.13 for $\theta = 0$ and rises significantly as this angle increases. The other direction-dependent terms in the integral yielding the cooling power also add different weights, as the angle changes due to the combined effect of increasing solid angles between successively wider cones and the decreasing projection factor $\cos\theta$.

Overall, this truncation of some incident radiation can lead to a very large difference between the atmospheric absorption for an optimized cone angle and the value for hemispherical exposure. It means that the total incoming radiation from the atmosphere is dramatically reduced if one blocks the high-angle radiation, so net radiant output and total cooling can be much higher. To achieve this in practice is straightforward but not simplistic, as we shall see later in Chapter 7, because one has to ensure that any blocking system does not itself send in much radiation or absorb much of that coming from the emitter.

The amount of water in the atmosphere is of much importance for its absorptance and emittance. An empirical relation to describe the emittance averaged over the whole hemisphere and over the Planck spectrum in terms of the dew point temperature τ_{dp} can be written as [15]

$$E_a = 0.711 + 0.56\left[\frac{\tau_{dp}}{100}\right] + 0.73\left[\frac{\tau_{dp}}{100}\right]^2 \qquad (2.25)$$

The moisture content, thus, is an important parameter to consider and impacts at all thermal wavelengths. Well-known computer codes exist for modeling $E_a(\theta_z,\lambda)$, and details for doing this can be found in the literature [15].

2.8 DYNAMICAL ENVIRONMENTAL PROPERTIES

2.8.1 General Considerations

A key challenge for renewable energy sources and energy-saving technologies that harmonize with the environment is to cope with large and naturally occurring variations in local energy flows during the day and between seasons. Natural dynamics must not be perceived as problems, though, because they have many psychological benefits. Thus, sunset and sunrise in clear weather are generally regarded as uplifting, while moving cloud patterns add much to the visual appeal of the outdoors (but present problems to solar power stations). Flowers, crops, fruit, and foliage come and go with the season.

Systems, as well as materials, are important to account for and benefit from nature's variability. Here are a few examples:

- The focusing of solar energy systems, which usually need to track the sun
- Glazings on energy-efficient buildings in warm climates, with smaller windows or more shading on their west façades
- Glazings with different solar transmittance according to the angle of incidence or, if "smart," with a transmittance which varies as the incident solar flux changes, and
- Climatization systems utilizing cold air during the night

In any practical situation one needs to strike a balance between the vitalizing effects of outdoors dynamics and the comfort and functional needs of interiors.

2.8.2 Solar Energy and Daylight Dynamics: The Sun's Path

The sun moves in an arc across the sky, as shown in Figure 2.12 [16]. Its position is described in terms of either the angle θ_z to the zenith or the altitude angle $\alpha = (90° - \theta_z)$. The rotation angle of the sun's vertical projection about the earth's normal is given by the angle A_z, or azimuth angle ϕ.

The sun's arc in Figure 2.12 moves with the season because the earth's axis tilts relative to its direction to the sun, as shown in Figure 2.13, by a declination angle which varies between +23.45° and –23.45°. The daily variation of this angle is given by

$$\delta = -23.45 \sin\left[(J + 284)(360/345)\right] \qquad (2.26)$$

where J is the (Julian) day in a year, with $J = 0$ at January 1 and $J = 365$ on December 31.

A complete path diagram for the sun's position in the sky in terms of altitude angle α and azimuth angle ϕ can be generated for any location, and an example is given in Figure 2.14 [17]. Such diagrams are very useful in daylighting work.

The angle of incidence θ onto the surface of interest is one of the most important parameters for solar energy utilization, in window and skylight

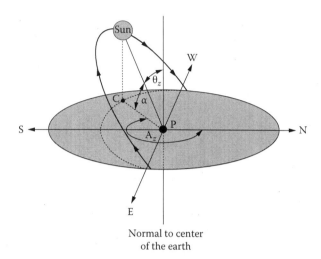

Normal to center
of the earth

FIGURE 2.12   Sun path during a day. In this schematic the sun rises and
sets just north of the east–west axis, so it is moving into spring or out
of late summer. (From http://www.itacanet.org/eng/elec/solar/sun1.pdf.
With permission.)

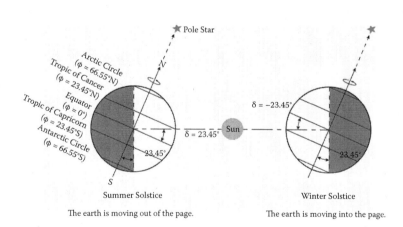

FIGURE 2.13   Earth's declination δ at summer and winter solstice. The
sun is then directly over the tropics of Capricorn or Cancer. (From http://
www.itacanet.org/eng/elec/solar/sun1.pdf. With permission.)

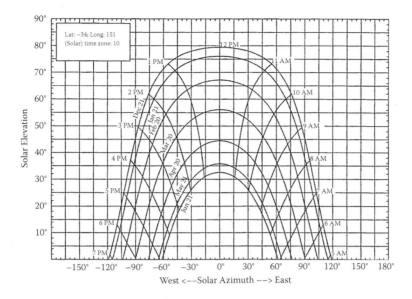

FIGURE 2.14 Example of a sun path diagram for Sydney, Australia. Curves denote time of day. (From http://solardat.uoregon.edu/SunChartProgram.html.)

technology, and for daylighting; it determines the total solar flux or incident lumens. For a horizontal surface, the angle of incidence is the altitude angle α and the incident solar flux I_{sol} has three components according to

$$I_{sol} = I_{pb} \cos (90° - \alpha) + I_D + I_{ref} \qquad (2.27)$$

where I_{pb} is the direct parallel beam intensity, I_D is the diffuse intensity, and I_{ref} is any solar energy reflected from the ground or surrounds. For arbitrarily oriented and tilted surfaces, α has to be replaced by angle of incidence θ in Equation 2.27. Box 2.1 shows how one can compute the sun's position, and hence the important angles α, ϕ, and θ. The equations shown allow modeling of device performance over a day, month, or year, and they also indicate how to control the orientation of a sun-tracking device. The equations are easily implemented in a spreadsheet.

2.9 MATERIALS FOR OPTIMIZED USE OF THE SPECTRAL, DIRECTIONAL, AND DYNAMICAL PROPERTIES OF SOLAR ENERGY AND SKY RADIATION

So far this chapter has laid a foundation on which one can build a basic understanding of the delicately balanced radiative and thermal

---

BOX 2.1   EQUATION FOR SOLAR POSITION AT ANY TIME IN ANY LOCATION, AND FOR ANGLE OF INCIDENCE ONTO A SURFACE ORIENTED IN ANY DIRECTION

The equations require the following input angles: the longitude L, the earth's declination angle δ from Equation 2.26, and the "hour angle" θ_H so the time of the day can be used. θ_H in degrees is given by

$$\theta_H = 0.25 \text{ x (time from noon in minutes} + \Delta t) \quad \text{(B2.1.1)}$$

A time correction Δt in minutes must be added for the actual position at L_{loc} relative to the longitude L where official noon and solar noon coincide. Δt is the difference in minutes to noon solar time, when the sun is closest to overhead at the actual position and is somewhere along its longitude. This is given by

$$L - L_{loc} = 0.25 \text{ x } \Delta t \quad \text{(B2.1.2)}$$

Equations B2.1.1 and B2.1.2 show that there is a shift of 4° per minute in θ_H from solar noon.

The two sun angles follow from

$$\sin(\alpha) = \cos(L)\cos(\delta)\cos(\theta_H) + \sin(L)\sin(\delta) \quad \text{(B2.1.3)}$$

$$\cos(\phi) = \frac{\sin(\alpha)\sin(L) - \sin(\delta)}{\cos(\alpha)\cos(L)} \quad \text{(B2.1.4)}$$

One can then calculate, for any time of day, the angle of incidence θ on any surface. The direction of the normal to the surface specifies its orientation. It can be defined by an angle of rotation Ψ of a vertical surface about the north–south axis, followed by a tilt to the horizontal by an angle θ_T. The end result is

$$\cos(\theta) = \cos(\alpha)\cos(\phi - \Psi)\sin(\theta_T) + \sin(\alpha)\cos(\theta_T) \quad \text{(B2.1.5)}$$

---

environment in which we exist. Now we consider, in general terms, how this knowledge can be used to engineer materials and structures to make solar and other renewable technology work most efficiently and create buildings that are both energy efficient and pleasant to live and work in. Subsequent chapters will then examine how specific materials can meet the generic requirements specified next.

We will look at the total inflow of electromagnetic energy, which consists of solar energy incident from the sun and from long-wavelength radiant energy incident from the atmosphere. The balanced energy out-flow consists of reflected solar energy and thermal radiation given off by surfaces depending on their temperature.

2.9.1 Opaque Materials

Spectral selectivity is an important feature of the radiation around us, as already emphasized repeatedly. Now we will make use of this selectivity, specifically of the lack of overlap between the solar spectrum and the blackbody or Planck radiation spectrum shown in Figures 2.4 and 2.11. This will enable us to

- Maximize the capture of solar energy for heat production, and
- Maximize the rejection of solar energy to keep surfaces cool

Spectrally selective solar absorbers are designed for the first of these tasks, while materials having exactly the opposite spectral properties are the simplest approach to the second task.

A good solar thermal collector must have a surface which absorbs strongly at all solar wavelengths but reflects strongly, and hence emits weakly, at thermal wavelengths. A sharp transition at a wavelength around 2.5 μm is thus needed, as seen for the ideal absorber in Figure 2.15a. The second task, in contrast, requires strong reflection of all incident solar energy and maximum loss by thermal radiation. The ideal performance is also shown in Figure 2.15a; it has a sharp, but opposite, transition at the same wavelength. Spectral selectivity is demanding because, ideally, it requires a switch from one optical extreme to another between neighboring wavelength regions.

The next applications require that we divide the solar and thermal spectra into parts. Solar radiation consists of ultraviolet (UV) light, the visible range with its component colors, and the near-infrared (NIR) which is invisible. The NIR range contains somewhat over 50% of the incoming solar energy and hence is very important for energy-related applications. The desired response to each spectral range varies with the application, and aesthetics or visual appeal is also a key aspect. Further useful subdivisions of the full spectrum can include breaking the NIR into its main energy contribution between 0.75 and 1.6 μm, and the rest.

We now consider color, with or without maximum solar reflection. Ideal spectra for surfaces with the same color differ for warm and cold climates, and the main issue is whether the NIR solar radiation is reflected (in warm climates) or absorbed (in cold climates). Ideally, these surfaces have NIR properties that are opposite to each other, as shown in Figure 2.15c. The thermal radiation properties should be opposite,

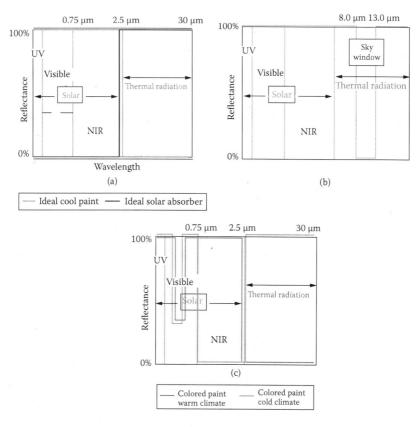

FIGURE 2.15 *A color version of this figure follows page 200.* Ideal spectral reflectance for opaque surfaces. Panel (a) refers to a solar absorber and a surface (paint) for cooling applications; the sharp transition is at ~2.5 μm in wavelength. Panel (b) shows ideal spectral reflectance for maximizing radiation loss to the clear sky; the reflectance is 100% except where the atmosphere is transparent (between 8 and 13 μm). Panel (c) applies to colored surfaces (paints) that look the same but have widely different thermal performance; the properties are ideal for warm and cold climates. Ranges for ultraviolet (UV), visible, near-infrared (NIR), and thermal radiation are shown.

too, as also shown, with "cool paints" for warm climates needing high emittance and "warm paints" for cold climates needing low emittance. It is important to realize that these two coatings will look exactly the same, but their thermal performance will be radically different.

The final opaque system of interest is coatings, usually backed by metals, with ideal spectral selectivity for cooling under the clear night sky. The spectral range for thermal radiation is of obvious interest and the outgoing radiation—and hence the emittance—should be as large as possible. For daytime application it is important that the coatings also reflect the sun. At the moment we only consider surfaces which radiate uniformly in all directions, but later it will be seen that angle-dependent effects can also be very important for maximizing radiation loss and temperature drop. Figure 2.15b illustrates a reflectance spectrum for maximizing heat pumping to the sky at subambient temperatures. There is high emission in the wavelength range where the sky is transparent and hence sending in little radiation. The reflectance is high for other wavelengths in the thermal spectrum so that incoming radiation from the atmosphere is reflected off. This ideal surface can cool to very low temperatures, often even below the freezing point of water, if nonradiative heat gains and dew formation are minimized.

2.9.2 Transparent Materials

The ideal spectral selectivity for windows and skylights depends on the climate. Windows in all climates should provide daylight with near-neutral color and a clear view, as discussed in Chapter 4. Skylights, in general, should be hazy, as discussed in Chapter 5, but will have spectral properties similar to those of windows with regard to their hemispherical properties. For warm climates, solar heat gain has to be minimized so all incoming solar NIR should be blocked; for cold climates, in contrast, the solar NIR should be admitted. Low thermal emittance is preferable in both cases to give thermal insulation. Low emittance leads to low U value, which helps keep in heat during winter and keep out heat during summer.

The ideal window transmittance spectrum for cold climates and warm climates are shown in Figures 2.16a and 2.16b, respectively. Again there is a change between two spectral extremes at a wavelength in between solar and thermal radiation. These ideals show two options for visible wavelengths: high transmittance and moderate transmittance. The latter will reduce glare as well as solar heat gain. In a cold climate it may be desirable to let in solar energy, but glare can be a serious problem because the sun spends much time at low elevation angles. The blocking in the NIR can occur in two ways, by absorption or reflection, but the blocking at thermal wavelengths has to be via reflectance in order to

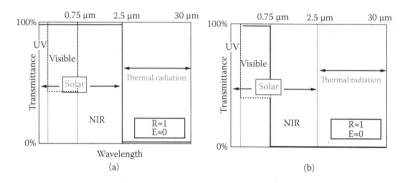

FIGURE 2.16 *A color version of this figure follows page 200.* Ideal spectral transmittance for windows in (a) cold and (b) hot climates; the dotted option at visible wavelengths is for glare reduction. Ranges for ultraviolet (UV), visible, near-infrared (NIR), and thermal radiation are shown.

have a low emittance. We return to these matters in Chapter 4, where specific window coatings are discussed.

2.9.3 Other Generic Classes of Optical Properties for Radiation Control

Apart from the various types of spectral selectivity introduced above, there are some other generic material properties of importance for controlling visible, solar, and thermal radiation. They will be discussed in detail in later chapters, but a brief overview of angular selective control and of "switchable" optics is useful here.

Angular selective control can provide considerable energy benefits in windows in addition to pleasing daylight and an outside view. It can give different optical or thermal responses according to the direction of the incoming radiation, whether it is solar or thermal. The sun is at high elevations for most of the day, except at very high or low latitudes and during early morning and late afternoon (of course there is a seasonal dependence too, except near the equator). Useful visible radiation comes in across the whole sky vault, especially when there are clouds, while the view to the outside mainly is in a near-horizontal or downward-looking direction. One aim of angular selective materials is thus to let in daylight and provide a clear view while blocking the solar heat coming from higher angles.

Macroscopic approaches to angular selectivity use, for example, eaves, awnings, slatted blinds and deciduous plants. Unfortunately, many of these approaches distort or cancel the outside view. Nanostructures, in contrast, can provide angular selectivity without disturbing the visual contact to the outside and thus are superior for normal windows; particular examples follow in Chapter 4. Angular selectivity is also very useful in skylights, mirror light pipes, and roof glazing because of the high solar fluxes onto these devices. They face upward and therefore need structures that are different from those for vertical windows in order to account for the different average angles of incidence.

Angular selectivity also is important for thermal radiation management and night cooling because, as we will see in Chapter 7, the atmosphere's radiative properties change as ray directions change relative to the zenith. Both macroscopic and intrinsic material options are of interest for thermal radiation.

Switchable optics, also referred to as "smart optics," is considered next. A variety of materials can have their optical properties changed according to weather and daylight conditions and thereby optimize visual and thermal comfort as well as energy use. The aim is often to switch as far as practicable between the spectral extremes in Figures 2.15 and 2.16, but it is also of interest to have intermediate optical properties. The overall class of materials, and their associated technologies, are labeled "chromogenic" [18]. The particular types of materials are designated, according to the driving mechanism by which their radiative properties are altered, as

- Photochromic (incident UV radiation)
- Thermochromic and thermotropic (temperature change)
- Electrochromic (electrical charge or potential), and
- Gasochromic (hydrogen plus catalyst)

Spectral changes are most important in chromogenics, but in some cases there can be a switch between specular (visually clear) states and hazy states in which radiation is scattered. This is the case for thermotropic materials.

Specific materials, both bulk and nanostructured, suited to the optical goals spelled out in Section 2.9, are the subject of the next chapter, where the physical processes that lead to the desired spectral, directional, and dynamic properties are outlined, too. Understanding the relevant materials will enable us to computer model optical responses in various micro- and nanostructures from simple basic principles and hence to explore how close one might get practically to the ideal limits. Thus, we will use a combination of basic optical and materials physics together with computer modeling and integrate this with established—and sometimes new—synthesis techniques for coatings in order to engineer the desired materials. Nanomaterials have removed a number of constraints in what one can achieve optically, which is why they are so important in this field.

2.10 THERMAL AND DENSITY GRADIENTS IN THE ATMOSPHERE AND OCEANS

The oceans and the atmosphere store huge amounts of energy that originally, of course, mostly came from the sun. Thermal gradients in the ocean and the atmosphere can be used to generate energy, as considered next. The connection to the built environment—the theme of this book—might seem tenuous, but advances in high-rise architecture and demands on new sites for new buildings may change the situation in the future, so a brief outline is justified.

The oceans' surfaces heat up by solar absorption and retain about 15% of the incident energy. These surface layers then become less dense, and hence warmer water stays on top and can have a temperature as high as 20°C to 25°C in tropical and temperate climate zones. One kilometer below the surface, the water typically is 5°C to 10°C colder. If this lower and colder water is pumped to the surface, it can be used in special types of heat engines which also employ the hot water from the top to form steam at reduced pressure. The colder water is used to condense this steam, which results in power production at low efficiency. The surface water is thus a large heat source and the bottom water a heat sink. This principle for energy generation is called ocean thermal energy conversion or OTEC. The conversion efficiency to electric power is low, but the amount of available energy is vast. This raises an important principle with regard to renewables, discussed in some detail in Box 2.2 at the end of this chapter, namely, that a low efficiency can be tolerated provided that costs are low, which also usually means the resource has to be both large and easily accessible. Lower efficiency with fossil fuel, however, cannot be tolerated.

The atmosphere, just as the ocean, gets colder when the distance from ground is increased, and the reason again is that the sun generates heat at ground level. The lapse rate with height is −6.5°C/km for dry air, so 10 km up, at the top of the troposphere (where most international air traffic takes place), the air temperature is about −65°C. This fall in temperature can be used for energy generation if one is able to access the upper air for heat rejection while taking in hot air at ground level, preferably heated to a high temperature by the sun which can be done by enclosing the air under glass or polymer.

Heating of air decreases its density, and air circulation then can be accomplished in systems operating under principles similar to those in OTEC, although the higher temperature difference with air heating can yield superior efficiency. Buoyancy forces drive the solar heated air upward and the air pressure p_{air} falls off exponentially with height H according to the relation

$$p_{air}(H) = p_{air}(0)\left[1 + \frac{d\tau}{dH}\frac{H}{\tau(0)}\right]^{-\frac{g(MW)_{air}}{R_g(d\tau/dH)}} \quad (2.28)$$

where R_g is the universal gas constant, g the acceleration due to gravity, and $(MW)_{air}$ the molecular weight of air.

The main method considered to date to entrap hot air and bring it to a sufficient height is with a specially built "chimney," but the only practical construction until now had a height of only 195 m, meaning that the thermal difference was a meager 1.3°C. As seen from the thermodynamic principles in the following box, much higher chimneys would be needed for attractive efficiencies.

2.11 PERFORMANCE OF ENERGY SYSTEMS: THERMODYNAMICS AND VALUE

Materials properties, ideal or not, is not the only thing to aim for, but it is also necessary to know how to assess the value and performance of the various energy technologies these materials enable. Indeed, getting very close to optical ideals with real materials does not guarantee that the resultant systems will be commercially viable or have widespread appeal, though it does enhance the prospects. There are three fundamental and strongly interconnected issues to address, that is,

- Performance limits
- Renewable resource availability, and
- Economic and social return on the investment

The first and third items are discussed briefly from the viewpoint of thermodynamics in Box 2.2. Resource availability is treated authoritatively elsewhere [2]. However, an interesting and seldom emphasized perspective on resource availability is the utilization factor, that is, the ratio of yearly produced energy in the installation divided by the installed energy (installed power per hour times the number of hours per year). This factor, of course, depends not only on the type of installation but also on its geographical location. Figure 2.17 gives a Swedish perspective on electric energy production from a number of conventional and renewable energy sources [19,20]. Clearly, the intermittent nature of wind and solar energy gives much lower utilization factors than those for hydro and nuclear energy.

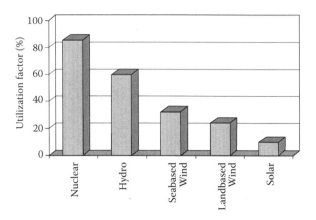

FIGURE 2.17 Utilization factors for some conventional and renewable sources of electricity. (From M. Leijon et al., *Renewable Energy* 28 [2003], 1201–1209. With permission.)

BOX 2.2 POWER SUPPLY, ENERGY EFFICIENCY, AND COST: A "WHOLENESS" PERSPECTIVE ON RENEWABLE AND CONVENTIONAL ENERGY

Efficient production of electrical power and the efficient use of electrical energy are fundamentally connected. Later chapters focus on either energy supply or energy demand, so we take the opportunity here to present a unifying overview of energy generation, energy use, resources, and costs. Efficient generation of energy and efficient use of energy are subject to the same fundamental scientific constraints, summarized by the laws of thermodynamics. The links also arise because thermodynamics is about changes that occur when energy is transformed from one form to another. Such changes can usually go both ways: forward and backward. For example, pressure can be raised or lowered, light can produce electricity, and electricity can produce light. The latter will usually be an energy supply issue, but scientifically we can treat them in the same way.

Solar cells are semiconductor junction devices which turn light into electrical power, but junction devices called LEDs or OLEDs turn electrical power into light. LEDs and OLEDs also waste some power input as heat. Solar cells and LEDs have in common that they perform better if one can keep their junctions cooler (but not too cold). Another example is thermoelectric devices. They can take in heat to produce electrical power or use power to pump heat, and it is important to reduce waste heat output in both of these functions.

For power production, basic science tells us that inefficiencies or waste of some input energy is unavoidable. It is important to see "both sides of the coin" when using natural resources to supply or save energy. For example, using solar heat to produce electricity one tends to focus on the heat input, but just as important is the fact that engines for generating electrical power need a supply of coldness to complete their cycle. This may be, for example, the air or a local lake. Emerging solar thermal systems, especially those using low hot-side temperatures, work by exploiting colder natural resources including deep ocean water (OTEC) or cold at elevations above ground level. Other even better natural sources of coldness are available, but as yet not exploited, which could be used to enhance the performance in renewable and nonrenewable energy systems, as we will see later. It is a general principle of thermodynamics that the larger the temperature difference between hot and cold sides, the larger is the efficiency of conversion of heat to power, and there is an upper limit to conversion efficiency. Conversion efficiency η is simply output power divided by input power, that is, $\eta = P_{out}/P_{in}$. If heat energy Q (in joules) is the source of power, then $P_{in} = dQ/dt$. The limit to efficiency using heat is called the Carnot efficiency and is given by

$$\eta_{Carnot} = 1 - \frac{\tau_c}{\tau_h} \qquad (B2.2.1)$$

with input temperature τ_h and exhaust or waste heat at τ_c. Thus, if one wishes to operate at lower input temperatures than in conventional power stations, which is the case for a number of emerging low-cost renewable systems, then lowering τ_c will add value.

Practical systems operate at lower efficiencies, with one useful rough guide being

$$\eta_{CA} = 1 - \left[\frac{\tau_c}{\tau_h}\right]^{1/2} \qquad (B2.2.2)$$

Here η_{CA} is called the Curzon–Ahlborn efficiency [21]. It is also to be noted that conventional stations waste around 55 to 65% of the input heat (from burning coal, gas, or nuclear fuels). Thus, a power station supplying 1 GW of electrical power pumps 2 to 3 GW of heat into the local environment. This waste heat represents many tons of CO_2 emissions and has much residual value. It can be used for heating, or even be used to generate more power, albeit at lower efficiency. It requires much cooling as currently handled, typically lots of water, which is getting scarce or needed for other purposes in some locations.

All of the main renewable energy systems that follow operate in both directions. One direction involves input of natural energy (say solar energy as heat or light) for power production and the reverse use electric power, or other natural resources, for cooling, lighting, and heating. Efficiency should be maximized in both directions, but it is not the only consideration when using natural resources, which are free and abundant. The reduced efficiency may be acceptable if there are large cost advantages and satisfactory availability. Unlike in power stations using fossil fuel, waste heat from renewables conversion does not add to what would have been in the environment anyway. In summary, conventional energy supply and demand systems must become more efficient, while renewable systems should focus on lowest cost and highest availability.

Better efficiency of conversion may reduce the cost of renewables, but not necessarily. For example efficiency gains in solar cells usually come at a very significant cost premium, so the most widely used cells still have relatively low efficiencies. This does not mean that the lowest cost cells are the best either. Thus, power from them may actually cost more due to their shorter lifetime and their need for very large areas. Another example has to do with lighting, and the lowest up-front-cost light bulbs are the most expensive in total, while more costly but more efficient and longer-life LEDs or compact fluorescent lamps cost far less overall. Thus, care is needed in evaluating the value of any renewable or energy-saving technology, and clearly a large number of factors are involved.

REFERENCES

1. http://www.waldemarsudde.se/xsaml_molnet_g.html.
2. B. Sørensen, *Renewable Energy: Its Physics, Engineering, Environmental Impacts, Economics and Planning*, 3rd edition, Elsevier, Amsterdam, the Netherlands, 2004.
3. J. T. Kiehl, K. E. Trenberth, Earth's annual global mean energy budget, *Bull. Am. Meteorol. Soc.* 78 (1997) 197–208.
4. C. G. Granqvist, Radiative heating and cooling with spectrally selective surfaces, *Appl. Opt.* 20 (1981) 2606–2615.

5. IPCC, *Climate Change 2007: Mitigation*. Contribution of Working Group III to the Fourth Assessment Report of the Intergovernmental Panel on Climate Change, edited by B. Metz, O. R. Davidson, P. Bosch, R. Dave, L. A. Meyer, Cambridge University Press, Cambridge, U.K. and New York, 2007.

6. A. Earp, Luminescent Solar Concentrators for Fibre Optic Lighting, Ph.D. thesis, University of Technology, Sydney, Australia, 2005.

7. P. Nostell, A. Roos, D. Rönnow, Single-beam integrating sphere spectrophotometer for reflectance and transmittance measurements versus angle of incidence in the solar wavelength range on diffuse and specular samples, *Rev. Sci. Instrum.* 70 (1999) 2481–2494.

8. J. C. Jonsson, A. Roos, G. B. Smith, Light trapping in translucent samples and its effect on the hemispherical transmittance obtained by an integrating sphere, *Proc. Soc. Photo-Opt. Instrum. Engr.* 5192 (2003) 91–100.

9. A. Roos, Materials performance and system performance, in *Performance and Durability Assessment: Optical Materials for Solar Energy Systems*, edited by M. Köhl, B. Carlsson, G. Jorgensen, A. W. Czanderna, Elsevier, Amsterdam, the Netherlands, 2004, chap. 2, pp. 17–55.

10. ASTM, ASTM G173-03 standard tables of reference solar spectral irradiances: Direct normal and hemispherical on 37° tilted surface, *Annual Book of ASTM Standards*, Vol. 14.04, American Society for Testing and Materials, Philadelphia, PA, 2003; http://rredc.nrel.gov/solar/spectra/am1.5.

11. G. Wysecki, W, S. Stiles, *Color Science: Concepts and Methods, Quantitative Data and Formulae*, 2nd edition, Wiley-VCH, New York, 2000.

12. Australian Bureau of Meteorology, The Greenhouse Effect and Climate Change, Melbourne, Australia (2009); http://www.bom.gov.au/info/TheGreenhouseEffectAndClimateChange.pdf.

13. C. G. Granqvist, A. Hjortsberg, Radiative cooling to low temperatures: General considerations and application to selectively emitting SiO films, *J. Appl. Phys.* 52 (1981) 4205–4220.

14. E. M. Lushiku, A. Hjortsberg, C. G. Granqvist, Radiative cooling with selectively infrared-emitting ammonia gas, *J. Appl. Phys.* 53 (1982) 5526–5530.

15. M. Martin, P. Berdahl, Summary of results from the spectral and angular sky radiation measurement program, *Solar Energy* 33 (1984) 241–252.

16. http://www.itacanet.org/eng/elec/solar/sun1.pdf.

17. http://solardat.uoregon.edu/SunChartProgram.html.

18. C. M. Lampert, C. G. Granqvist (eds.), *Large-Area Chromogenics: Materials and Devices for Transmittance Control*, SPIE–The International Society for Optical Engineering, Bellingham, WA, 1990.
19. M. Leijon, H. Bernhoff, M. Berg, O. Ågren, Economical considerations of renewable electric energy production: Especially development of wave energy, *Renewable Energy* 28 (2003) 1201–1209.
20. M. Leijon, A. Skoglund, R. Waters, A. Rehn, M. Lindahl, On the physics of power, energy and economics of renewable energy sources: Part I, *Renewable Energy* 35 (2010) 1729–1734.
21. F. L. Curzon, B. Ahlborn, Efficiency of a Carnot engine at maximum power output, *Am. J. Phys.* 43 (1975) 22–24.

CHAPTER 3

Optical Materials Science for Green Nanotechnology
The Basics

Let us start by looking at the famous Lycurgus cup from the Roman era [1]. It is shown in Figure 3.1a. The appearance is green, but if a small lamp is put inside the cup the transmitted light looks red. The more modern vase from Victorian times in Figure 3.1b looks red both in transmission and in reflection. Both objects consist of glass to which gold has been added so that nanoparticles have been formed as inclusions. Objects of equal beauty were made at other times, too, and, for example, copper nanoparticles were used to create stunning luster on some Renaissance pottery [2]. This chapter will dwell on the theoretical models for understanding the beautiful colors formed in the glass and why they may look so different, depending on the illumination.

More important, however, the present chapter provides insights into the materials and optical science that form the basis of much green nanotechnology. This science explains how the technology works and how it can be improved, and perhaps it even points to where the next breakthroughs in technology might arise. A fundamental question in all technology is about the limits of its performance, and hence the goals to strive for. The underlying science answers these questions; to mention one example, it tells that basic silicon solar cells cannot achieve efficiencies better than 29%. But new science and new materials can change the goals to aim for—sometimes dramatically— and that is what nanoscience is now doing. Practical, economic, and health-related considerations set limits to the technology, and nanoscience can also explain how to best deal with those. Occasionally new science can also lead to completely new technologies and innovations, sometimes called "disruptive technologies" and "disruptive innovations" because they fundamentally change the notions of what is feasible [3,4].

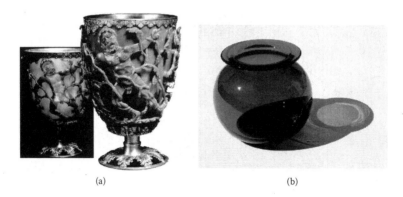

(a)   (b)

FIGURE 3.1 *A color version of this figure follows page 200.* Two pieces of glassware, both doped with gold nanoparticles. Part (a) is the Roman Lycurgus cup from the British Museum and part (b) is an old Victorian vase owned by one of the authors (GBS). [Part (a) is from http://www. britishmuseum.org/explore/highlights/highlight_objects/pe_mla/t/the_ lycurgus_cup.aspx. Reproduced with permission.]

Nanoscience, solid state physics, and materials design lie at the core of green nanotechnology, and this chapter outlines the science that enables us to create materials for optimum use of the radiative energy flows in the environment, which were introduced in Chapter 2. This exercise takes into account the specific application, human comfort and visual needs, local climate, and location on earth, and it often also considers aesthetic appeal. The discussion in Chapter 2 was in general terms and considered how materials interact with and emit radiation. We now turn to the way *specific* materials respond to radiation. An understanding of the material parameters that govern the radiative properties of solids will then allow us to design nanostructures to achieve responses tailored to the environment. Furthermore, new energy-related functionalities, not possible with standard materials, will emerge.

The contents of this chapter are mainly conceptual, although we do include key equations and models, as well as some important data, which can easily be adapted for use in spreadsheets, MATLAB®, Mathematica, or simple computer programs. More advanced topics are generally given in boxes, which the reader may skip over without losing the most essential contents. Models are explained in general terms, with derivations and details available in the references or in some proprietary software for those who want to further explore the origins and applications of these models. We thus hope there is much of value to the reader who is interested just in qualitative aspects as well as to those who like to utilize quantitative models. But whatever the reason for studying this book, it is essential that the reader gets to know some key numbers, and understands their origins, in order to come to grips with what the green nanotechnologies have to offer. Such key numbers will, of course, be given and discussed.

Let us reiterate the significance of understanding the fundamentals of green technologies: It is important for the home owner who perhaps wants to install solar cells or a solar water heater, just as it is important for the researcher, or for a start-up enterprise or an existing company that wants to expand into a new field. The researcher, the industrialist, and the layperson all need to make the right decisions. Otherwise they will have to rely on sales pitches—and could be led astray.

3.1 LIGHT AND NANOSTRUCTURES

3.1.1 Local Fields and Far Fields

We first consider the refractive index n, which is an essential parameter for optical materials. It is dimensionless and commonly introduced using Snell's law applied to clear optical materials such as glass, transparent polymers such as PMMA, and polycarbonate and clear liquids. This index describes refraction, or light bending, at interfaces between different materials, and air or vacuum is characterized by $n = 1$. Snell's law is a consequence of the wave nature of light and of momentum conservation. The combination of wave character and geometry is vitally important when dealing with nanostructures, whose scale typically is much smaller than the wavelength. For visible and infrared radiation, the wavelengths of interest range from 400 nm to 50 μm, so nanoscales of, say, 40 nm are smaller than the wavelength by a factor of at least 10. This is very useful as it often takes us into a so-called "quasi-static" regime where the complexities of a full wave treatment can be avoided. In essence, this means that the electric and magnetic fields of light are, at any instant, almost constant on average across the main feature sizes, just as they are in electrostatics.

Nanooptical science predominantly deals with physical or wave optics, rather than geometric or ray optics. This means that the field patterns, and hence waves near and inside the nanostructure, do not resemble the waves observed outside. However, in practice one can often use simple ray models to describe what is seen well away from a nanostructure, but we first have to find a recipe for describing the "effective homogeneity"—that is, effective refractive index—which is determined by what is going on inside the nanostructure. This is, indeed, a crucial issue in nanooptics and its many spin-off technologies.

The wave and electric field behavior within a few nanometers or inside the nanostructure is different from the one observed after bouncing light off it with a standard optical instrument such as a microscope and

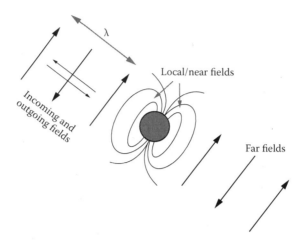

FIGURE 3.2   Local fields induced around a nanoparticle by the electric field, of wavelength λ, in an incident light beam. The external waves traveling away from the particle form the far field.

a spectrophotometer. This difference should not come as a surprise, since the structure itself imposes field variations on a scale that is much smaller than the scale outside the nanostructure, which is ~λ. Optical science connects what happens inside the nanostructure to what results outside, and armed with this knowledge one can design the desired materials.

Optical measurements normally take place in the so-called "far field" where wave patterns tend to be relatively simple. However, the electric field pattern is much more complex very close to or inside the nanostructure; this region is called the "near field." The change in field patterns comes about as emerging "local" fields combine to form the external waves we see. Some modern instruments can probe the near field, but in general optical measurements are confined to the far field.

Imaging at nanoscales with light was thought impossible, but such imaging can now be done provided that the object under study is close enough to the detector. Figure 3.2 illustrates how this is done, and shows the local fields produced near an illuminated nanoparticle. These local fields are often referred to as "evanescent" as they do not go anywhere. We will come back to them often.

3.1.2 Refractive Index and Absorption Coefficient

The optical properties of a material describe how light waves travel inside it, and they determine what happens to light or solar energy at the interface between two different materials. Knowing the wavelength inside the material and the rate by which a material absorbs radiation allows most properties of interest to be determined. Key attributes in free space—in addition to wavelength λ_0, frequency f, and speed of light c—are wave vector $k_0 = 2\pi/\lambda_0$, angular frequency $\omega = 2\pi f$, and field amplitudes. The fields oscillate back and forth at the frequency ω. For optically isotropic materials, the refractive index determines the internal wavelength to be $\lambda = \lambda_0/n$, and the internal light speed to be c/n. The refractive index depends on frequency, so one can write $n(\omega)$. The relationship between n and ω is called a dispersion relation.

When radiant energy is absorbed, the electric field amplitude falls off according to

$$|E(z)| = E(0)\left|\exp(i2\pi n(\omega)z / \lambda_0)\exp(-2\pi k(\omega)z / \lambda_0)\right| =$$
$$E(0)\exp(-2\pi k(\omega)z / \lambda_0) \tag{3.1}$$

where z is the distance the light has traveled inside the material. This equation is written in a form that allows introduction of a generalized refractive index representing both propagation and loss and given by $N(\omega) = n(\omega) + ik(\omega)$, where $i = \sqrt{-1}$. The complex wave vector is Nk_0 so that the wave propagates as $E(z) = E(0)\exp(iNk_0z)$. k is called the extinction coefficient, and n and k are referred to as "optical constants."

Energy flux and intensity I depend on $|E^2|$, and hence the absorption coefficient α satisfies

$$I(z) = I(0)\,e^{-\alpha z} \tag{3.2}$$

It also follows that

$$\alpha(\omega) = \frac{4\pi k(\omega)}{\lambda_0} \tag{3.3}$$

In many practical applications it is important to control the absorption coefficient, and methods of enhancing or reducing α can lead to significant cost savings. This is of considerable interest in technologies for solar heating, solar electricity, windows, and roof and wall paints. One simple example is for solar cells, where an increase in α permits a lower thickness and hence cheaper and less energy-intensive manufacturing. Nanostructures provide a number of ways of tuning the absorption coefficient.

Wavelength-dependent variations of n and k play a crucial role for the possibilities to achieve spectral selectivity. Figure 3.3 gives some examples and illustrates $n(\lambda)$ and $k(\lambda)$ in the wavelength range for solar radiation for one metal (silver) and one insulator (titanium dioxide) [5].

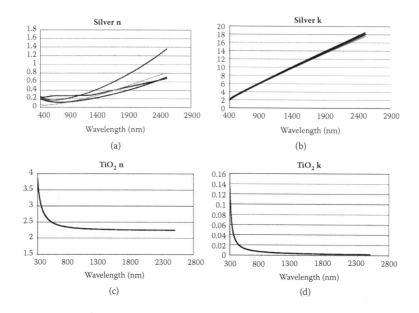

FIGURE 3.3 Spectrally dependent optical constants $n(\lambda)$ and $k(\lambda)$ for variously prepared films of Ag (panels a and b) and TiO$_2$ (panels c and d) across the solar spectrum. (From G. B. Smith, unpublished work.)

Both of these materials have many applications in green nanotechnology, as shown repeatedly below. Silver is seen from Figures 3.3a and 3.3b to have $n < 1$ for short wavelengths, while n goes up progressively for long wavelengths; k, on the other hand, increases almost linearly in proportion with the wavelength. The sets of data refer to variously prepared films (different deposition techniques, substrate materials and substrate temperatures). Silver grows differently on glass than on tin oxide, for example, while the energy and momentum of impinging atoms, which vary according to the deposition method, also influence film structure. (Relevant thin film deposition methods are described in Appendix 1). It is immediately clear that the optical properties can depend strongly on the technique for making the film. Titanium dioxide, as evident from Figures 3.2c and 3.2d, has a refractive index that falls off gradually with increasing wavelength but n is rather constant in most of the shown range; k drops rapidly with increasing wavelength and it is apparent that TiO$_2$ is almost nonabsorbing for wavelengths in the luminous and solar ranges. Again, there is some dependency on manufacturing technique [6].

The actual values of n and k depend on how charges inside the solid respond to optical fields. When light shines on a material, electrons or ions redistribute so that local separations of positive and negative charges occur. The illuminated material is then said to be polarized, and it behaves as though it has many internal dipoles which oscillate at optical frequencies with $P_d(\omega)$ being the resulting dipole moment per unit volume. Basic electromagnetic texts [7,8] describe how $P_d(\omega)$ is related to the complex relative dielectric constant $\varepsilon(\omega) = \varepsilon_1(\omega) + i\varepsilon_2(\omega)$ and hence to the complex refractive index N. At normal incidence, N is electromagnetic admittance, that is, $Y(0)$ is equivalent to N. We label $Y(0) = Y$ in the future and $N = \sqrt{\varepsilon}$. The actual permittivity is given by $\varepsilon_0\varepsilon(\omega)$, with $\varepsilon_0 = 8.85 \times 10^{-12}\ \mathrm{C^2 N^{-1} m^{-2}}$ being the permittivity of free space. A useful relationship between the real and imaginary components of N and ε is thus

$$\varepsilon_1 = n^2 - k^2 \qquad\qquad (3.4a)$$

$$\varepsilon_2 = 2nk \qquad\qquad (3.4b)$$

The magnetic field in light usually has no influence in optics since optical magnetic dipoles do not form in most solids (except at very long wavelengths that are not of much interest in green nanotechnology [9]). However, recent important work [10,11] has shown that some nanomaterials can have a magnetic response under illumination, even though their constituent bulk materials do not. The formation of tiny current loops is then responsible; a complex relative magnetic permeability $\mu(\omega)$ different from unity is needed to account for these optical magnetic dipoles. The complex refractive index is linked since $N = \sqrt{\varepsilon/\mu}$, with $N_0 = \sqrt{\varepsilon_0/\mu_0}$ being the refractive index of free space which is real. One has $\mu_0 = 4\pi \times 10^{-7}\ \mathrm{C^{-2}\,N\,s^2}$, but in most materials one can put $\mu = 1$.

If $\mu < 0$ is achievable along with $\varepsilon < 0$ (i.e., both are negative), then a negative refractive index becomes possible. This has been demonstrated for optical wavelengths and in certain nanostructures, though the effect is weak [12]. However, evidence for $n < 1$ is now convincingly shown at microwave frequencies. A strongly negative refractive index—if ever achievable at low cost and over a wide enough optical wavelength band—would be very exciting for solar energy applications because of the way it could simplify and enhance focusing of light [13].

3.2 SPECTRAL PROPERTIES OF UNIFORM MATERIALS

It is very useful in green nanotechnology to have simple mathematical equations to describe the spectral variations of optical properties of materials such as those in Figure 3.3. These data can then be used in computer models or spreadsheets. Typical optical properties are introduced next for a number of classes of materials. Handbooks list refractive index data for most materials of interest [14–16], and such information is also available in manuals for some optical instruments. Caution is needed when these data are used, though, since nanomaterials can behave differently from their bulk counterparts in various ways, as we will see later. Sometimes it may even be necessary to make separate investigations to determine the properties of the components in a nanomaterial.

3.2.1 Insulators and Liquids

A number of useful transparent insulators, including glass, have a very low extinction coefficient for visible wavelengths and sometimes also in the near infrared, and thus the refractive index is the parameter of main interest. When the refractive index of a material is quoted without reference to a specific wavelength, it usually regards $\lambda \sim 500$ nm. This choice is natural since the eye has its peak sensitivity to light at the green wavelength of 555 nm for normal lighting (cf. Figure 2.4) and at 507 nm when it is dull. However, n varies with wavelength in all materials, as evident for the two materials in Figure 3.3, and this variation must be considered in solar, heat management, lighting, display, and agricultural applications, which cover the very wide range from the UV (350 nm) throughout the thermal IR (50 µm).

The refractive indices of a number of liquids—including water and alcohol—are low for $\lambda \approx 500$ nm, and these materials are of much interest, too. One reason for this is that fluids increasingly find uses in some nano-based energy applications, such as for harvesting energy directly using photochemistry or in roofs of greenhouses which distill salty water into fresh. A second reason for the interest in liquids is that the chemical literature commonly reports optical properties of nanoparticles immersed in solution (in which they are easily measured). Unfortunately, the results from such immersion measurements can be different from those when the particles are embedded in solids, which generally have higher indices than liquids. For example, dilute nanoparticles dispersed in a solution may lead to a cloudy appearance due to optical scattering, while a similar concentration of the same nanoparticles in an oxide or a polymer can yield a clear material capable of transmitting light without scattering. This difference is easily explained and relates to the difference in refractive indices between the particle and the embedding matrix, and in Sections 3.8 and 3.9 we shall look in more detail at how the index of a medium surrounding a nanoparticle influences its optical response.

Some applications of green nanotechnology require high-index host media ($n > 2.0$) and some require low-index media ($n < 1.4$). Nature gives many options as seen from Table 3.1 which contains a selection of data for useful insulating materials—including polymers and liquids—with

TABLE 3.1   Mid-Visible Refractive Indices for Some Insulating Materials
of Interest for Green Nanotechnologies

Inorganic Compound	Refractive Index	Polymer	Refractive Index	Liquid	Refractive Index
MgF_2	1.40	Teflon	1.31	Water	1.33
SiO_2	1.46	PMMA	1.49	Ethanol	1.36
Al_2O_3	1.8	Polyethylene	1.51		
ZnO	2.0	Nylon	1.54		
ZrO_2	2.2	Polycarbonate	1.58		
ZnS	2.3	Polystyrene	1.59		
TiO_2	2.5				

low, medium, and high index values at $\sim 500 < \lambda < 550$ nm. The values are for bulk materials, and it is possible to have low-index nonscattering materials by making them nanoporous. An extreme example of a solid material with a refractive index approaching unity is found in silica aerogel, which is discussed in Section 8.3.

Insulators can be transparent in a wide wavelength range, but eventually they show a significant value of k and hence attenuation of energy at both short (UV) and long (IR) wavelengths. Sharp rises in $k(\lambda)$ usually occur in the UV once photons have enough energy to excite electrons to cross the insulator's band gap. This onset of absorption is why nanoparticles of TiO_2 and ZnO can act as excellent sunscreens; as nanoparticles they are too small to scatter at visible wavelengths and they are thus invisible, but they still absorb the harmful solar rays at the shortest wavelengths.

In the infrared, at $\lambda > 3$ μm, molecular and crystal lattice vibrations can lead to significant absorption and hence a large extinction coefficient near wavelengths characteristic of various molecular or lattice vibrations. For polymers there is a complication that can limit their uses in solar-energy-related applications, namely that absorption due to light molecule oscillations—especially of the C-H bond—occurs at solar wavelengths in the NIR range, with a series of absorption peaks starting just beyond the visible. Polymers in which thermal IR absorption at $\lambda > 3$ μm is weak are of special interest for sky cooling systems, as discussed in Chapter 7.

3.2.2 Conductors, Semiconductors, and Superconductors

Conductors, semiconductors, and, possibly, superconductors have important roles in green nanotechnology related to energy. Their applications include current transport, electronic and optoelectronic devices such as solar cells and light-emitting diodes (LEDs), management of solar radiation, thermal control, power generation, and energy and fuel storage. Multiple functionalities can be taken advantage of in some cases, and, for example, both electronic and optical properties are used in devices such as solar cells and LEDs.

Uniform conductors of interest include most industrial metals and many compound transparent conductors. Semiconductors include Si and GaN. Conductors generally have dramatic changes in their refractive indices in the visible or IR ranges as the wavelength is varied. There are two physical reasons behind this:

- Band gaps lie at visible or NIR energies, typically between 0.5 and 4 eV.
- Carrier densities, especially in metals, are sufficiently high that the materials start to absorb very strongly at long enough wavelengths, and the carriers can also lead to high reflectance.

This role of the carriers is crucial to much emerging green nanotechnology and brings us to the "hot" topic of plasmonics, which is given a special treatment in the following text.

Some key materials for energy applications—such as indium oxide, zinc oxide, and tin oxide—are insulators or semiconductors in their pure state, but they can become good conductors when suitably doped; then they behave electrically like metals but optically like glass. In particular, there is a lot of research on $In_2O_3{:}Sn$, known as indium tin oxide or ITO, and on $ZnO{:}Al$. This class of materials will be discussed in detail in Section 4.4.

A number of important polymers, such as doped polypyrrole, polyacetylene, and polyaniline, qualify as good conductors and are of much interest for organic solar cells and organic light-emitting diodes (OLEDs). Nanostructures of such polymers will be important for achieving improved performance—for example, optimum electrode spacing—because excited charge carriers need to be collected quickly or they will be lost.

Superconductors have zero electrical resistance but only at very low temperatures and yet remain to have a major impact in the field of energy. Their ability to transport and store electrical power with very low loss, especially for dc applications, may be of great importance in the future, and generators and electric motors with superconducting windings remain of interest.

3.2.3 Chromogenic Materials

An important class of materials for energy systems is the chromogenic one which, as noted in Chapter 2, can be switched by some external means. Some of these are uniform and others are composites, but they are considered as one group in this introductory overview. A material capable of changing from being an insulator to a conductor clearly is very useful for optical and energy-related applications. This means that n and k in the visible and IR, and hence the optical response, can be altered by a large amount. Chromogenic materials, also known as "switchable materials" or "smart materials," have uses in energy-efficient and comfort-enhancing windows, in color-changing paints, and for lighting purposes; they can also be used in sensors to manage energy flows and monitor the environment. If the external influence is a temperature change, the material is said to be thermochromic (such as VO_2); if it is electrical, the material is electrochromic (such as WO_3); if by adsorbed gases, it is gasochromic (such as some Mg alloys); and if energetic light itself drives the change, it is photochromic (as in some spectacle lenses). There are still other emerging "chromics," such as the one discussed in Box 3.1.

3.3 PLASMONIC MATERIALS IN GENERAL

Plasmonics attracts much interest today, and the number of publications in this field has skyrocketed. There is at least one scientific journal dedicated specifically to plasmonics. This area of nanoscience now opens up a whole new world of optical engineering. Some of what plasmonics has to offer would have been considered science fiction only a few years back, as would many of the recent developments in smart materials, some of which can be switched between insulating and plasmonic states.

Plasmonics concerns charge density waves inside or on surfaces of materials. We use a broad definition of a plasmonic material as one with

$$\varepsilon_1(\omega) < 0 \qquad\qquad (3.5)$$

which requires that $k > n$. This is not the only possible definition, though, and some authors reserve the word "plasmonics" for electronic conductors.

There are two distinct classes of plasmonic materials, differing as to the origin of the charge waves underlying the optical response. The most widely studied class, which is already applied in green nanotechnologies, includes electronic conductors such as metals. However, there is growing interest in insulators, which are plasmonic by a different physical mechanism: ionic motion or lattice vibrations known as phonons. The latter class has a more limited role in emerging energy technologies, but a special subclass involving nanoparticles has the potential to help save vast amounts of energy now used for cooling and to help with water management. Furthermore, these systems do not just deposit the heat

BOX 3.1   MAGNETOCHROMICS

There are numerous possibilities to create innovative "smart materials." For example, recent developments in nanoscience used magnetically coated polymer nanobeads embedded in lattice patterns on microparticles to achieve optical switching with a magnetic field [17]. This technology may be labeled "magnetochromics" because it uses magnetism to change the color, as seen in Figure B.3.1.1. The science behind this color change relies on photonic crystals and their rotation, and is discussed later in this chapter. This is an example of optical switching using nanoparticles, which is in its infancy but holds great promise if the relevant particles can be easily dispersed in large-scale materials at low cost. "Smart materials" will feature strongly in later chapters.

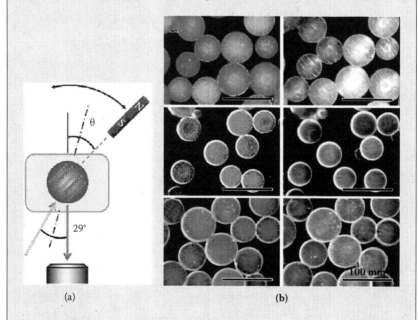

(a)   (b)

FIGURE B3.1.1   *A color version of this figure follows page 200.* Magnetochromics based on nanoparticle arrays and magnetically rotatable photonic microspheres. Part (a) shows the experimental set-up, and part (b) illustrates the color changes that occur as the angle θ has different values. (From J. Ge et al., *J. Am. Chem. Soc.* 131 [2009] 15687–15694. With permission.)

FIGURE 3.4   Surface plasmon charge waves induced on the surface of a metal. Resulting electric field lines are shown. E and H denote electric and magnetic fields, respectively. Evanescent fields will be discussed in detail in Section 3.11 below. (After http://nanohub.org/resources/1748.)

elsewhere in the earth's ecosystem, like vapor compression cooling, but can get rid of the heat altogether by sending it into outer space. Given that the globe is warming, it is obvious that a low-cost, low-materials usage process to pump heat into space is of high value. These applications are discussed in Chapter 7.

Thus, both electronic and ionic plasmonics are of interest. When the two types of charge waves occur at surfaces, they are sometimes referred to as surface plasmons and surface phonons. The latter terminology is not necessarily associated with $\varepsilon_1(\omega) < 0$ and hence our choice of terminology.

What does a negative value of ε_1 mean? It seems to say that when one applies an electric field, then electrons move the wrong way—to where the field is negative. To make sense of this, consider the nature of a plasmon: It is an oscillating wave of charge polarization, as depicted at a metal surface in Figure 3.4, and arises in the electron case as conduction electrons slosh back and forward when the optical field switches sign, while ions stay fixed as they cannot follow the field at frequencies corresponding to solar radiation [18]. If this oscillation is around 180° out-of-phase with the applied field cycle, then the net affect is that $\varepsilon_1(\omega) < 0$. Plasmonic behavior only occurs below a critical frequency which, as we will see, depends on carrier density and other material properties.

In the world of nanoplasmonics one also needs to allow for the possibility of a magnetic response, including optical diamagnetism in which the induced magnetic moments have $\mu(\omega) < 0$. In this case each nano-magnet reverses direction at ~180° out-of-phase with the cycle of the applied field, and the effect is due to induced current loops.

3.4 MATERIALS FOR ELECTRON-BASED PLASMONICS: MIRRORS FOR VISIBLE AND INFRARED LIGHT

Good conductors are scientifically peculiar mainly because they become plasmonic. A number of questions on their uses in energy applications, and in things we see in everyday life, are relevant to these oddities. Thus,

- Why do metals reflect light so strongly?
- Why are the metals gold and TiN "golden" in color, whereas silver and aluminum reflect all colors nearly equally?
- How can a good conductor such as ITO or ZnO:Al be transparent to visible light?
- Why do metals have low thermal emittance and silver/dielectric thin film multilayers have exceptionally low emittance?
- What are the best mirror materials for solar thermal power generation and for lighting?
- And why do metals conduct heat so well?

The answers lie in an understanding of the link between the electronic conductivity and the optical and thermal response, which in turn leads us into the fascinating world of electron-based plasmonics. Next, we focus on how a plasmonic material's complex refractive index and dielectric constant are related to the density of charge carriers and hence to electrical conductivity. Later we will look into the details of how plasmonic materials can also form special charge density waves at surfaces of illuminated conductors. These waves are called surface plasmons and, though not occurring on everyday smooth metal surfaces, they have a unique role in energy technology and nanoscience.

Optical transitions can lead to local movement of charge or absorption of energy. They occur between occupied and unoccupied quantum levels. The unique situation for current-carrying electrons in metals comes about because the two distinct quantum states they jump between when optically excited have negligible energy separation, while normal, nonplasmonic optical transitions in solids involve a finite-energy jump between two separated quantum levels. The latter are referred to as interband transitions. This consideration enables us to put all optical transitions in a material into a single, simple, and useful mathematical model, which can describe the overall spectral response in plasmonic and nonplasmonic materials as a function of frequency. This is the Lorentz–Drude (LD) model [19] for the dielectric constant, which can be written

$$\varepsilon(\omega) = 1 + \sum_q \frac{A_q}{\omega_q^2 - \omega^2 - i\omega\tau_{,q}\omega} \qquad (3.6)$$

Each characteristic mean jump is labeled with subscript $q = 0, 1, 2, \ldots$ and tends to dominate in a particular frequency range. States in two different energy bands define q and are separated on average by energy ω_q, except for $q = 0$ where excitation of the current carrying electrons occurs within the same band near the Fermi level (i.e., the highest energy of the occupied states). Thus, $\omega_0 = 0$ for the latter transitions, which gives the so-called Drude term. The electrons involved in the Drude term are those that give a metal its conductivity. The factors A_q give the strength of each transition and $A_0 = \omega_P^2$, with the plasma frequency ω_P, depending only on conduction electron density N_e and effective electron mass m^* as

$$\omega_P^2 = \frac{N_e e^2}{m^* \varepsilon_0} \tag{3.7}$$

with e being the electron charge. The higher the charge density, the higher the frequency at which a conductor starts reflecting.

We note in passing that the earth is surrounded by a "material" that becomes noticeably plasmonic and hence reflective at night for very low radio frequency waves; it clearly must have a very low value of N_e. This is the ionosphere, which consists of rarefied plasma of mobile charge in which N_e builds up during the day due to the solar energy flux, mainly its most energetic parts.

The parameter ω_P is of central importance for many aspects of green nanotechnology, and $\omega_{\tau,q}$ is relevant, too; it is a measure of the broadening of the interband terms. The latter parameter can be thought of as a quantum uncertainty in the transition energy and depends on the widths of the two bands contributing at each transition. The Drude relaxation term is different in character, though, and ω_τ ($= \omega_{\tau,0}$) depends on the lifetime τ_e of the excited conduction electrons before they are scattered and lose the energy gained from the applied optical field. The carrier lifetime is

$$\tau_e = \frac{2\pi}{\omega_\tau} \tag{3.8}$$

and ω_τ is called the relaxation frequency. It is the chief underlying cause of conversion of radiant energy into heat and hence a very important property for energy-related applications. It is also the source of electrical resistance. In Section 3.10 we shall see that ω_τ is larger in nanostructures than in bulk metals. We shall also see how to keep it low, if desired, in some nanostructures.

The link between the Drude parameters and the dc conductivity σ is given by

$$\sigma = \frac{N_e e^2 \tau_e}{m^*} \tag{3.9}$$

Experimental values of τ_e, obtained from dc electrical and optical data, tend to be somewhat different but still close [20].

Applying Equation 3.6 to real materials makes it clear that it is important to consider a wide wavelength range even if the prime interest is for frequencies well away from interband transitions. In fact, materials usually do not become plasmonic at $\omega = \omega_P$ but at a much lower frequency as a result of the residual impact of interband transitions. Here, we caution the reader that some textbooks neglect interband transitions in discussions of the Drude model, and the frequency for the onset of plasmonics is then incorrectly stated to be at ω_P for real materials.

For plasmonic thin films and nanomaterials of interest in the solar range, one may need to explicitly include only one or two interband terms with the lowest energies, say $q = 1$ and perhaps $q = 2$, for accurate results. In gold, for instance, it is usually only needed to use the ω_1 term centered at 3.4 eV [21], but more terms would be demanded to treat optical properties in the UV.

The neglected high-energy and more "remote" transitions still contribute to ε_1 via residual background polarization. The net effect is a modified version of the LD model in Equation 3.6 for practical use in the visible, NIR, and thermal ranges according to

$$\varepsilon(\omega) = \varepsilon_\infty^* + \sum_{q=1,(2)} \frac{A_q}{\omega_q^2 - \omega^2 - i\omega_{\tau,q}\omega} - \frac{\omega_P^2}{\omega^2 + i\omega_\tau\omega} \qquad (3.10)$$

This simplified LD model can give remarkably good fits to optical data on useful plasmonic materials if data are gathered over solar or solar plus thermal wavelengths, that is, from ~0.35 to ~30 μm. Its universal character at long wavelengths can be appreciated by the qualitative similarities in the modeled thin-film reflectance and transmittance spectra in Figure 3.5, which are shown against wavelength scaled to the plasma frequency given by

$$\lambda_P = 2\pi c/\omega_P \qquad (3.11)$$

Layer thicknesses are identical and small for the data shown, which tends to visually amplify the impact of the strong plasmonic response in gold. High reflectance can be obtained also in films of ITO and the metal phase of VO_2, but only with much larger thicknesses than for gold [22]. The quantitative differences at long wavelengths are attributable to differences in the proximity of ω_P and ω_τ; at short wavelengths the differences are due to the location of the lowest energy interband term.

The value of the "constant" ε_∞^* depends on how many interband terms are explicitly included in Equation 3.10. With one or two terms, ε_∞^* may run up to values of 3 or 4 [23,24] but with none it is higher in

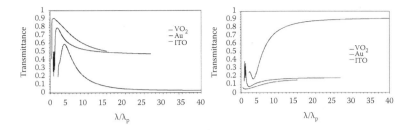

FIGURE 3.5 Spectral transmittance and reflectance of 12-nm-thick films of three important plasmonic materials as a function of wavelength λ scaled to the plasma wavelength λ_P. The films were deposited onto glass. (From G. B. Smith et al., *Proc. SPIE* 6197 [2006] 6197OT 1–9. With permission.)

the NIR and can be up to 6. This implies that it is very inaccurate to use the "standard" Drude model with ε_∞^* equal to one rather than to ~6. However, it is usually acceptable to neglect all explicit interband terms at wavelengths corresponding to thermal radiation.

The frequency for transition to plasmonic behavior, that is, negative ε_1, is found by setting $\varepsilon_1(\omega) = 0$ in Equation 3.10 and solving for ω. For several important conductors, such as titanium nitride, gold, and silver, this can be simplified to solving

$$\varepsilon(\omega) = \varepsilon_\infty - \frac{\omega_P^2}{\omega^2 + i\omega_\tau\omega} \tag{3.12}$$

Now, putting $\varepsilon_1(\omega) = 0$ gives the transition to a plasmonic state at a "shielded plasmonic frequency" ω_P^* given by

$$\omega_P^* = \frac{\omega_P}{\sqrt{\varepsilon_\infty}} \tag{3.13}$$

which is well below ω_P. For example, gold has $\omega_P = 8.6$ eV and $\omega_P^* = 4.1$ eV.

One can roughly estimate the frequency for the onset of plasmonic behavior from experimental data, since reflectance starts to rise strongly for increasing wavelengths once a material becomes moderately plasmonic. Thus, the yellow color of gold and TiN tells us that plasmonic effects must start to have an influence in the middle of the visible spectrum, while in silver and aluminum they must do so in the violet or UV. The best electron-based conductors all make excellent infrared reflectors,

and some are also the best visible reflectors. They give the lowest emittance, down to a value as low as 0.015.

How strongly plasmonic can a material get? This is a key issue for
applications to solar heating and for energy efficient windows. In fact,
$\varepsilon_1(\omega)$ can drop far below –100 in the best plasmonic materials, such as
aluminum, gold, and silver. These metals are also preferred choices for
visible wavelengths, but then there is another recently developed alternative: multilayer polymer mirrors which utilize nanophotonics (cf. Sec.
4.1). Polymer-based mirrors do not have low thermal emittance, though,
so they cannot combine useful solar reflectance with thermal insulation.
We conclude from this discussion that—as a consequence of their electrical conductivity—the best metals have unique properties for energy-
related applications and are able to combine high solar reflectance with
low thermal emittance.

Other conductors with lower carrier density can still conduct usefully and may have a key advantage over the best metals; their usefulness for nanotechnology emerges from the fact that they have much
lower plasma frequencies than the best metals. This down-shift in ω_P,
and hence in conductivity, means that conductors such as ITO, ZnO:Al,
and SnO_2:F are transparent to visible light because reflection does not
rise until the NIR range. Their onset of plasmonic behavior, however, is
early enough to yield low thermal emittance, and this makes them useful
for thermal insulation in windows. A lower ω_P means that $\varepsilon_1(\omega)$ never
gets as negative as in gold or silver, because the transition to plasmonics
is much closer to the relaxation frequency ω_τ. A lower ω_P also makes
nanoparticles of ITO, ZnO:Al, and SnO_2:F uniquely useful as discussed
in Section 4.5.

The ability to vary the plasma frequency over a wide range by
choosing different materials clearly is valuable for green nanotechnologies, as well as for various other applications. Table 3.2 lists some useful plasmonic materials in decreasing order of their plasma frequency.
The transparent oxide conductors with ω_Ps in the NIR underpin some
energy-saving windows, thin-film solar cells, and systems for water distillation or purification. Their ω_Ps depend on the doping level, and the
values shown should be regarded as indicative. The factor ω_P^*/ω_P is in
the range 0.42 to 0.55 for all of the listed materials, and a particularly
important practical feature of Table 3.2 is, then, that the associated ω_P^*
values span the full solar range. This wide spectral coverage of ω_P^* is
useful both for dense thin layers and for their corresponding nanoparticles. Gold and silver have similar ω_Ps, but they look very different; gold
is "golden" rather than the neutral color of silver because it has a different value of ω_P^*. This, in its turn, is due to differences in the interband
transition at ω_1 in Equation 3.10.

TABLE 3.2   Plasma Frequencies of Some
Important Electron Conductors for Use in
Energy Systems*

Conducting Material	Plasma Frequency, ω_P (eV)
Al	14.8
Ag	9.0
Au	8.7
TiN	7.0
LaB$_6$	4.3
ITO	~2.3
ZnO:Al	~2.0

* The data were acquired on dense high-quality
thin films or on bulk samples; values for doped
conductors are indicative only and depend on
doping level.

3.5 IONIC-BASED MATERIALS WITH NARROW-BAND INFRARED PROPERTIES

3.5.1 Plasmonics Once Again

Good electron conductors are strongly plasmonic, and thus excellent reflectors and low emitters, but their reflectance and transmittance change only marginally across the blackbody spectrum. This can be a problem for applications related to the infrared properties of the atmosphere, and there is a clear need for spectral selectivity within the blackbody range. Certain ionic compounds can display a negative ε_1 in a limited wavelength range, and this opens possibilities to have spectral selectivity in the thermal infrared as discussed next.

In the infrared most ionic materials absorb within limited wavelength ranges, and otherwise they transmit rather than reflect. Ions oscillate in well-defined patterns, or phonon modes, with characteristic frequencies much lower than those of the much lighter electrons; they thus oscillate at thermal infrared frequencies. The dielectric response of the ionic materials consists of two terms: the first due to residual high-frequency electronic effects, which contribute a background screening constant ε_∞, and the second from the vibrating ionic lattice in the presence of the applied electric field. The dielectric response can be written as a Lorentz oscillator according to References 25 and 26 according to

$$\varepsilon(\omega) = \varepsilon_\infty + \frac{\omega_T^2\left[\varepsilon(0) - \varepsilon_\infty\right]}{\omega_T^2 - \omega^2 - i\omega\omega_\tau} \tag{3.14}$$

where $\varepsilon(0)$ is the asymptotic low-frequency response, ω_T is the resonant frequency of the transverse lattice excitations, and ω_τ is their damping frequency. For induced electric polarization, the transverse vibrations are directed normal to the wave direction. Such waves are called optical phonons and result from net electrical polarization of the lattice at any instant.

It is easy to show that $\varepsilon_1(\omega)$ is negative in the frequency range $\omega_T < \omega < \omega_L$ and hence the system then has plasmonic response with

$$\frac{\varepsilon(0)}{\varepsilon_\infty} = \frac{\omega_L^2}{\omega_T^2} \qquad (3.15)$$

where ω_L is the frequency of longitudinal lattice waves in the long-wavelength limit. Equation 3.15 is known as the Lyddane–Sachs–Teller relation [27]. High reflectance is thus expected for the $\omega_T < \omega < \omega_L$ range, because the extinction coefficient is then large. This is called the *Reststrahlen band*.

The value of ω_τ determines how close the reflectance actually gets to 100%, and in many ionic compounds it can be well above 90%. Thus, uniform layers of some ionic compounds can be good IR reflectors, but unlike good metal mirrors, they reflect strongly only over a limited range of IR wavelengths and therefore may have a range of thermal emittance values. Just as for metals, the ionic compounds can change from being strong reflectors to very strong absorbers when they go from uniform or microscale to nanoscale, and this property can be used in applications. As in the electronic case, the value of ω_τ is very important for determining the practical usefulness of a particular compound.

Table 3.3 lists parameters needed to model $\varepsilon(\omega)$ for three interesting compounds using Equation 3.14. Their Reststrahlen bands cover, or are located within, the sky window from 8 to 13 μm. The ω_τ value of silicon carbide is exceptionally small and is much less than that in BeO and BN.

TABLE 3.3 Oscillator Parameters for Ionic Materials with Plasmonic Response over a Limited Band Near or Inside the Sky Window

Ionic Material	ω_L (eV)	ω_T (eV)	ω_τ $(10^{-2}\,\text{eV})$	$\varepsilon(0)$	$\varepsilon(\infty)$
SiC	0.119	0.098	0.0065	9.72	6.7
BN	0.165	0.132	0.50	7.10	4.5
BeO	0.133	0.087	0.149	7.13	2.93

Source: From C. G. Ribbing, E. Wäckelgård, *Thin Solid Films* 206 (1991) 312–317; C. G. Ribbing, *Appl. Opt.* 32 (1993) 5531–5534.

This property makes crystalline silicon carbide (c-SiC) very useful in nanostructured form. Nanostructured BeO and BN, on the other hand, have comparatively larger ω_τs, which lead to much broader resonances and much weaker peak absorption. We will see in Chapter 7 that the parameter ω_τ makes nanostructured SiC by far the superior option for very low-temperature, low-cost sky cooling and for some water-related technologies. SiO_2 nanoparticles will be of use even though they have two local narrow resonances, one of which is lying within the sky window and arising from surface phonons. Amorphous silicon carbide (a-SiC) does not quite attain a negative ε_1 but still has elevated absorption in the sky window.

3.5.2 Phonon Absorption

An ionic compound, or a molecule, does not necessarily have to be plasmonic in order to absorb and emit strongly in a narrow energy band. Instead, the absorption and emission is often due to local molecular bond vibrations, such as the vibrations of the Si-O and Si-N bonds in thin amorphous oxides and nitrides of silicon [28,29], and various bonds in ammonia, ethylene, and ethylene oxide [30]. The spectrally localized nature of these absorption bands can be of direct use in energy applications and for metal-backed thin films of SiO [27,31], and related materials such as silicon oxynitride [29,32] have been found interesting for sky-cooling applications.

The main requirement then is to have lightweight compounds, preferably based on Si but perhaps also including Mg and Al; details are in Chapter 7. Although only nanostructured SiC and SiO_2 are useful for absorbing or emitting in the sky window, most of these materials can be used in various forms including as uniform coatings, microparticles, and nanoparticles. Composites of plasmon resonant nanoparticles within an otherwise dense layer, which absorbs in parts of the sky window not covered by the nanoparticles, may make an ideal combination; an example could be SiC in SiO.

Multilayer films are of interest with regard to strongly reflecting materials such as c-SiC. Applying thin-film models, it becomes possible to make judicious choices of the materials in the layers in order to create an angular selective gradient in the absorption or reflection of IR radiation. It will be demonstrated in Chapter 7 that this is very useful for sky cooling because of the directional dependence of the atmosphere's infrared properties. Angular selectivity becomes possible in the desired wavelength range when the two layers are strongly different in one or both components of the refractive index (n and k), provided that at least one of the materials has a large value of k. We note that the Reststrahlen effect, just as intrinsic absorption, can yield a high k.

3.5.3 Infrared Transparency

Transparency at thermal radiation wavelengths is also of value for energy-related applications. Materials that have good transparency across the whole blackbody range, or just the sky window, can be used to protect coatings, minimize dew formation, and insulate from local convective inflow without too much reduction of radiant outflow. Some polymers are well suited to the task but we noted earlier the problems that can arise with polymers due to narrow-band absorption, even in the solar NIR where C-H bonds absorb. Polyethylene, if not too thick, is a good candidate for the blackbody range but needs to be stabilized against UV degradation, as well as mechanically for outdoor uses [33]. Polypropylene is also of interest; it can be used in sheet and foil form, and it can also be woven into fabrics. The microfibers used for weaving can be doped with nanoparticles such as c-SiC. Some fluoro-polymers may also be useful as IR transparent layers. Among inorganic materials, ZnS may be one of the best for IR transparency in the sky window.

3.6 GENERIC CLASSES OF SPECTRALLY SELECTIVE MATERIALS

Spectral selectivity is a key concept for green nanotechnologies to be implemented in the built environment, and a number of distinct material types can provide the desired properties, either when used in bulk form or as thin films. They include well-known fully dense materials as well as nanostructures, and the latter group can be further subdivided. We classify the selective materials as follows:

- *Semiconductors* such as silicon; they are transparent for wavelengths longer than hc/E_g, with E_g being the band gap, and are strongly absorbing at shorter wavelengths
- *Strongly plasmonic materials* such as gold and TiN as smooth thin films on glass or clear polymer; they are transparent at optical frequencies above the shielded plasma frequency but reflect strongly at lower frequencies
- *Nanostructured materials*

Ordinary bulk materials do not provide the functionality and spectrally selective performance needed for most energy-related applications, but materials that are nanostructured can give these additional capabilities. Though engineering at the nanoscale is involved, nearly all of the nano-materials we address are amenable to mass production at commodity scales, which is why they have so much to offer, as we shall see in later chapters. An example list of nanomaterials, including their current market status, includes

- *Multilayer metal-insulator thin film stacks* in which the metal layer is around 10 to 14 nm thick (produced at commodity scale for many years)
- *Metal/oxide composites* in which the metal is in nanoparticle form (commercial products include solar collector coatings)
- *Photonic crystals* including multilayers of different nanothin polymers (produced in large quantity)
- *Transparent conductor nanoparticles embedded in polymers* (used in some glazing products)
- *Transparent conductors with nanostructured surfaces* (promising for thin film solar cells)
- *Nanoporous TiO_2 films* (commercial products include "self-cleaning" windows)
- *Nanoporous conductors* (under development for ultracapacitors and catalysis)
- *Nanocavities* (under development for many applications including all-electronic refrigeration, ultracapacitors, and nanochemical factories)

Many more examples are expected in coming years. Composite structures are considered scientifically complex materials, but for the consumer, builder, or manufacturer they have, in general, normal formats—and superior functionality. A major drawback of such complexity is that empirical possibilities are almost endless, so a good grasp of the underlying electromagnetic science, and of its adaptation to the application in question, is important.

3.7 THIN FILMS FOR CONTROLLING SPECTRAL PROPERTIES AND LOCAL LIGHT INTENSITIES

Uniform thin-film layers are widely used in photonic and electronic devices and systems, for corrosion protection and wear resistance, in medicine, in decorative coatings, on clothing, and for thermal control. For glazing alone, this is a commodity scale industry with sales exceeding 50 million square meters per year. Thin-film stacks, in which one or two of the layers is a conductor, such as nanothick silver, and the

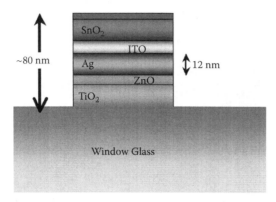

FIGURE 3.6   Schematic of a typical modern commercial thin-film stack for daylighting, solar control, and thermal insulation.

other layers are insulators, are commonly designed to transmit visible radiation and reflect both the NIR and thermal IR wavelengths. The insulator layers—typically consisting of TiO_2, ZnO, or SnO_2—have high refractive indices and are there to make the silver nearly transparent for wavelengths up to ~700 nm [34]. Figure 3.6 shows a schematic thin-film stack.

The silver layers in these stacks are as thin as 11 to 12 nm in most of today's products, and thickness must be tightly controlled during production. Induced transparency is possible because the refractive index of silver is small for visible wavelengths; in fact it is less than unity as seen from Figure 3.3. The thicknesses of the insulating layers have to be specially chosen using thin-film optical models to achieve the best results [35]. The relaxation time of carriers in the silver is very important for the optical performance, and the aim is for the lowest possible value of ω_τ because this leads to the most distinct spectral selectivity and the lowest possible thermal emittance, as is easy to show from the LD model in Equation 3.10 or 3.12.

These nanostacks have additional interesting features, which are relevant also for other energy applications such as thin film solar cells and quantum dot devices: the electric field, and hence the light intensity profiles within the stack, can be engineered so that regions of very high and very low intensity occur where they are desired [36,37]. Certain parts of the stack, hence, can "see" much more electromagnetic energy than others, and these parts then dominate the optical performance. This feature opens ways to achieve high-energy conversion efficiencies using minimal amounts of material. Furthermore, the absorption in stacks such as the one in Figure 3.6 can be lowered by keeping field penetration into the silver itself to a minimum.

Engineering of the local field distribution in materials also makes it possible to develop nanophotonics based on refractive indices that are negative for visible light [38,39], and possibly to achieve "cloaking" (making things appear to be invisible). All of these capabilities are frequency dependent, which may or may not be useful, depending on the nature of the incident radiation field and the application.

Even a sketchy presentation such as the one above clearly indicates that there are many possibilities ensuing from a capability to fine-tune optical field profiles at the nanoscale. Examples include the following:

- Large concentrations of energy can have a very low thermal impact compared to what a macroscale concentration would give.
- Charge carriers generated by light can be very efficiently collected before being lost by recombination.
- Chemical and photochemical reactions (including fluorescence) can be controlled, enhanced, and made high-yielding.

Nanoscale electric field control with light is in its infancy, and one has only recently begun to come to grips with the underlying science. However, nanoscale control is a key element in the way nanoscience is revolutionizing the world of technology, and it plays a role in even the simplest of nanosystems—the nanoparticle—which we address shortly. Local aspects of how some nanoparticles respond to light tended to get overlooked until recently since the focus was then on the overall response of nanosystems.

The discussion in this section has so far presumed "ideal" thin films, which are different from "real" films. The latter are rarely perfectly flat at the nanoscale and often contain some nanovoids. "Real" films can, under certain conditions, support surface charge density waves, which we will address specifically below. The assumption underlying the discussion so far in this section is that interface waves are not present; this is implicit, but rarely stated, in all standard thin-film optical modeling. This assumption is usually fine for well made and almost smooth films, even when there is a small level of nanoroughness. Thus, metal surfaces and films, and neighboring dielectrics, must be reasonably smooth and void-free at the nanoscale; otherwise surface waves can arise at interfaces, thereby introducing additional optical effects including energy loss [40].

This discussion of "ideal" versus "real" points at a possible negative impact of nanostructure, but it can be managed and diagnosed and sometimes even turned to advantage. New deposition techniques have evolved over the last 20 years, often involving energetic ion bombardment during coating, to avoid nanovoids and ensure sufficiently smooth surfaces. These aspects will be discussed in more detail in Section 4.3 and in Appendix 1.

3.8 NANOPARTICLE OPTICS

When photons strike particles, one of two outcomes normally occurs and the photons are either scattered or absorbed. Scattering means that the photons can go off in a variety of directions, and absorption usually means conversion to heat. The sum of scattering and absorption is called extinction and relates to the fraction of photons that are influenced. Remaining photons appear, in the far field, to go straight through and are unaffected by the particle. Cloaking is a special case when all incident photons appear to go through without any change; it involves complex structures and not just simple particles.

Light-particle interaction can be thought of crudely as incident light exciting tiny antennas, which then reradiate and also get warm if there is absorption. These outcomes are not just material dependent but they also depend strongly on particle size and wavelength, in particular the ratio of the two. The basic physics behind light-particle interaction is covered in a number of excellent classic books [41,42] to which the reader is referred for details.

3.8.1 Transparent and Translucent Materials

Particles are normally expected to scatter light to some extent as long as the particles and their embedding medium have refractive indices that are significantly different. However, even when the indices are close there is some scattering, which is special at large sizes and very useful in an energy context as will be discusses in Chapter 5. It is through the size dependence—via the particle size to wavelength ratio d/λ—that things get interesting for small enough particles. When this ratio is very small, below 0.02 to 0.1, scattering can get very weak. Whether or not useful absorption occurs depends on the k value of the material in the particle at the wavelengths of interest. But if $k(\lambda)$ is large, as it is near a resonance, scattering is enhanced relative to that for the same particle off resonance. Absorption still dominates for small enough nanoparticles, though, as wavelength to size and cross-sectional area are the key factors in scattering (relevant equations follow). Fortunately, these resonances lie outside the luminous range for many applications requiring materials with high k values, and then scattering hardly matters.

Figure 3.7 illustrates the qualitative difference in light scattering from small nanoparticles and light scattering from nanoparticles larger than ~100 nm or from microparticles. If scattering is weak, the transmitted light is undeviated while solar radiation still can be attenuated by absorption. Considering a window, this means no distortion of the

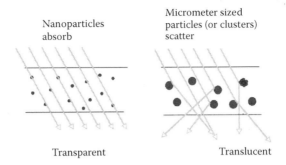

FIGURE 3.7   Schematic description of transparency and translucency obtained in nonabsorbing host media containing particles of different sizes.

outside view. An important aim for many glazings is to have negligible or weak absorption of visible light while NIR wavelengths are strongly blocked, and Section 3.10 discusses one approach to this. Scattering is usually strongest when particle absorption is large. This can be tolerated at nonvisible wavelengths, and it may even be beneficial to reject some NIR radiation.

Gold nanoparticles in glass can serve as a nice illustration of the importance of particle size, and Figure 3.1a, which appeared at the beginning of this chapter, showed the famous Lycurgus cup from Roman times. It contains gold particles that are large enough to back-scatter green light whereas they do not noticeably affect red light. The more modern vase shown in Figure 3.1b, on the other hand, has much smaller gold particles and appears red both in reflection and transmission. The more detailed reasons for this dichotomy will soon be explained.

The predominant scattering by weakly absorbing particles, and also for local defects in clear solids, was predicted by Rayleigh many years ago and is named after him. This scattering rises strongly and monotonically as wavelengths get shorter according to λ^{-4}. One consequence of Rayleigh scattering is that normally clear objects may appear slightly blue when viewed from the side or on reflection. A striking illustration of this phenomenon is given for silica aerogel in Figure 8.1.

Particles with high refractive indices, such as TiO_2 and ZnO, back-scatter strongly at all visible wavelengths as long as they are not too small, so polymers and paints containing these particles appear white. Another common particle used for making "whites" is $BaSO_4$. These kinds of particles are used for scattering in many skylights and roof glazings where neutral-colored transmitted light is required (particle size control is needed for the best results, though). Most white things we see, apart from sunlight and some phosphor emission, in fact get their appearance as a result of strong backscattering from high-index microparticles.

Small enough dielectric nanoparticles may optically be of no consequence at visible wavelengths. This is so, for example, for zirconia, which is very hard and hence useful in processes such as grinding and polishing of optical components. These particles can be difficult to remove after the treatment, but this may not be a problem as they can just be left behind with no effect on the optical properties.

As two final applications of scattering in energy technology, we note that multiple scattering from large particles can be used to harvest more solar energy in photovoltaic systems as well as to reflect solar energy while letting through thermal IR radiation to promote all-day sky cooling. ZnS particles are useful for these applications.

3.8.2 Some Basic Models

The color or solar control properties of a coating or polymer sheet can be altered by adding nanoparticles, which absorb or scatter in a narrow range of wavelengths. Such changes are based on absorption by the particles and sometimes on scattering as well. Each particle in an oscillating optical electric field $E(\omega)$ is polarized and forms an oscillating dipole moment of strength $p(\omega) = \alpha_{pol}(\omega)E(\omega)$, as illustrated schematically in Figure 3.8a. The dipole oscillates and hence it reradiates or scatters in the specific pattern whose cross section is shown. α_{pol} is called the polarizability and depends on the material in the particle, the host material in which it is embedded, and particle size and shape. A spherical particle of volume v and dielectric constant $\varepsilon_p(\omega)$, embedded in a host medium of dielectric constant $\varepsilon_h(\omega)$, has a polarizability given by

$$\alpha_{pol}(\omega) = 3v\varepsilon_h \frac{\varepsilon_p - \varepsilon_h}{\varepsilon_p + 2\varepsilon_h} \qquad (3.16)$$

Figure 3.8b indicates the scattering pattern for a simple dipole. Backscattering and forward scattered intensities are equal. This means that small nanoparticles, typically less than 20 to 50 nm in size, depending on the particle and host materials, often are best at sending solar radiation back by scattering, though one needs also consider the spectral band width. The + and – charges interchange at frequency ω.

The effective area of the incident beam scattered by a particle is called the scattering cross section C_{scatt}, while the absorption cross section, which determines the amount of incident radiation converted to heat, is C_{abs} according to

$$C_{scatt} = \frac{k^4}{6\pi} \left| \alpha_{pol} \right|^2 \qquad (3.17a)$$

$$C_{abs} = k \, \mathrm{Im}(\alpha_{pol}) \qquad (3.17b)$$

k is the wavevector in the host material and is given by $(2\pi / \lambda)(\varepsilon_h)^{1/2}$. If the host material is nonabsorbing, the absorption comes from the particle itself and can be enhanced at particular wavelengths by the influence of shape and surrounding medium.

One should notice that C_{scatt} is proportional to $(\pi r^2)(r/\lambda)^4$, that is, to the product of the particle's cross-sectional area and $(r/\lambda)^4$, where r is particle radius. Thus, scattering gets weak very quickly once r/λ drops below ~0.05 (~25 nm, say). This is Rayleigh scattering and explains why simple non-absorbing defects scatter blue light much more than red. The absorption strength, however, does not depend on projected area but on volume and is a parameter of central importance for several green nanotechnologies.

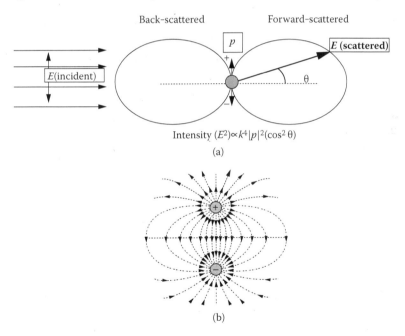

FIGURE 3.8   Part (a) shows a light intensity profile for scattering off a very small nanoparticle as a function of scattering angle θ to the incident axis. p is the oscillating dipole moment induced in the particle by the incident field E. Part (b) is a close-up view of the dipole and the electric fields it generates.

Equations 3.17a and b can be used not only for spheres but for other shapes such as ellipsoids, for which the polarizability can have up to three distinct components given by

$$\alpha_{pol,x.y,z} = v\varepsilon_h \left[\frac{\varepsilon_p - \varepsilon_h}{\varepsilon_h + L_{x,y,z}(\varepsilon_p - \varepsilon_h)} \right] \tag{3.18}$$

with L_x, L_y, and L_z being the depolarization factors for each axis. These factors are defined by shape and add up to unity. Only for the sphere, with $L_x = L_y = L_z = 1/3$, is the result independent of the applied field direction. Depolarization is due to the internal fields resulting from induced surface charge. The internal fields are constant inside small enough spheres and ellipsoids, and particle shape and field direction, then determine the dipole strength, which has the components p_x, p_y, p_z.

In practical experimental work it is often useful to test whether "nanoparticles" are of truly nano size. This can be done using scattering of laser light, as outlined in Box 3.2.

If particles are dilute, nonabsorbing and very small, they are in effect invisible and have no optical impact. A nanoparticle-host composite, however, in general displays interesting new optical properties. Such a composite behaves as a new homogeneous optical material with an "effective" refractive index $n^*(\lambda)$ determined by the number of particles and their arrangement. And if the particles absorb there can be significant changes for example in color or in solar reflection and absorption. An "effective" extinction coefficient $k^*(\lambda)$ can quantify these effects together with $n^*(\lambda)$.

For a wide range of nanoparticle concentrations, it is easy to estimate n^* and k^* from electrostatic models. This is so at least if the nanoparticles are randomly distributed, as the particles then behave quasi-statically. A good experimental indicator to tell whether such an approximation holds is to see if optical scattering is negligible. Quasi-static properties are also hinted at in the electric near-field pattern in Figure 3.8b which appears to be identical, at any instant, to that in electrostatics. However, the dielectric constant is still the one at optical frequencies and hence the designation "quasi."

Complex nanoparticles can be contemplated for solar energy applications. The aim would then be to tune the spectral properties continuously via internal structures of particles—rather than be restricted by the limited spectral bands provided by any specific material. For example, the wavelengths at which a particle is strongly absorbing can be altered through its structure. Local electric field intensities just outside the particle can also be engineered in complex nanoparticles, which may open

BOX 3.2   A TEST FOR NANO SIZE

Figure B3.2.1 outlines a useful technique to test whether the particles in a sample are of nano size or larger. Polarized incident laser light will remain polarized after scattering perpendicularly to the direction of the incident beam if the particles are small enough. This is a result of the dipole moment p, as illustrated in Figure 3.8, and there is no scattering along the direction of the dipole moment. The effect is well known and accounts for the fact that sky radiation is polarized (so that polarizing sunglasses can reduce glare). But large defects, including particles bigger than ~100 nm, typically scatter with outgoing waves forming increasingly complex patterns as the particle size goes up, thus resulting in perpendicular scattering being depolarized and also to more light going into the forward direction than backwards. This is due to additional component patterns of electric charge called multipoles which radiate in a variety of ways.

Thus, if side-scattered light is found to be unpolarized, defects are usually not less than ~100 nm in size. This test also enables one to quickly diagnose whether nanoparticles or some molecules form clusters, which is often an undesired property.

The next multipole component likely to add to the dipole performance is the quadrupole. Oscillating magnetic moments can also form at optical frequencies in some nanoscale metal systems, even though they are usually absent in normal bulk materials. A whole new area of physics, and possibly of technology, has opened up recently to deal with optically induced "magnetic" effects in plasmonic systems.

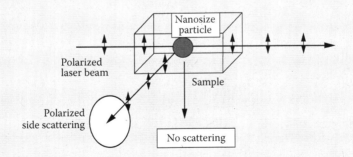

FIGURE B3.2.1   Optical set-up to test for nanosize particles. Incident polarized laser light will show polarization effects when detected from the side only if the particles are of nanosize. Larger particles, or clusters of nanoparticles, will instead depolarize the light.

up for applications such as enhanced sensing and photocatalytic conversion. A particularly interesting example is offered by core-shell nanoparticles whose central part is dielectric and to which a thin metallic shell is added. We come back to these structures when studying plasmonic nanoparticles. Practical production issues, including shell smoothness and small enough thickness, have limited their impact to date, but core-shell structures do have a potential for solar energy applications and for energy efficiency. Other interesting and useful core-shell systems involve fluorescent layers on a core which may itself be fluorescent.

We now turn to the basic science of what happens when many nanoparticles are embedded in another material or floating in a solution. These are the most basic nanocomposite structures and they already have widespread and growing commercial uses.

3.9 OPTICAL HOMOGENIZATION OF NANOCOMPOSITES

It is interesting to first take a quick look at the many ways to make nanocomposites. The ensuing nanostructures can be very different, and so are then their optical properties and the models to describe these properties. Thus, nanocomposites can be made by

- Codeposition of two materials that phase separate; certain polymer blends is one example [43]
- Deposition of atomic or molecular species under conditions so that limited diffusion leads to porosity; this is a major aspect of thin-film technology, which is discussed in Appendix 1
- Adding particles to a substrate by spraying or precipitation from a solution followed by self-assembly [44]
- Adding particles to a substrate by first nucleating them from a supersaturated gas [45,46]
- Early-stage thin-film growth in vacuum of materials, such as gold and silver, which initially form islands [24,47,48]
- Codeposition followed by etching of one component [49,50]

It is often desirable to keep the particles apart, especially for materials including metal particles, and then one can use particle self-assembly with binding agents that bridge particle surfaces [51] or precoat the particles with a dielectric [52]. There is a large literature on the production of nanocomposites, and the references above serve mainly as examples.

3.9.1 Models for Particle- and Inclusion-Based Composites

With N particles in a volume V, there is an additional dipole moment per unit volume ΔP to that induced in the host material by the applied field. Assuming that particles do not interact, one has $\Delta P = (N/V)p$. It is then easy to show from Equation 3.16 for a sphere that the total dipole moment gives, in the quasistatic limit, an effective dielectric constant ε^* for spheres according to

$$\varepsilon^* = \varepsilon_h + \left[N/V\right]\alpha_{pol} = \varepsilon_h\left[1 + 3f\,\frac{\varepsilon_p - \varepsilon_h}{\varepsilon_p + 2\varepsilon_h}\right] \tag{3.19}$$

where

$$f = (N\nu/V) \tag{3.20}$$

is the volume fraction of particles, also known as the "filling factor." But this works only in practice if (N/V) is small enough so that polarized particles do not interact with one another, that is, the particle density should be small.

At most densities, each particle "sees" a field E_p, which is the sum of the applied field and that due to induced dipole fields arising from all of the other particles. If the particles are randomly located in the composite, it has been possible to devise simple procedures [8,53–55] which lead, again for spheres, to a self-consistent expression for the combined field E_p at each particle according to

$$E_p = E_{appl} + \frac{P}{3\varepsilon_h} = E_{appl} + \frac{N}{V}\,\frac{\alpha_{pol}E_p}{3\varepsilon_h} \tag{3.21}$$

Solving this relation for E_p leads to the effective dielectric constant

$$\varepsilon^* = \varepsilon_{MG} = \varepsilon_h\left[1 + \frac{3f\left(\dfrac{\varepsilon_p - \varepsilon_h}{\varepsilon_p + 2\varepsilon_h}\right)}{1 - f\left(\dfrac{\varepsilon_p - \varepsilon_h}{\varepsilon_p + 2\varepsilon_h}\right)}\right] \tag{3.22}$$

This famous result is usually referred to in optics as the Maxwell–Garnett model (hence, the subscript MG), though the same equation can be used at lower frequencies and may then be labeled the Clausius–Mosotti or Lorentz–Lorenz model. It applies only when one medium, called the host, is continuous and the other is isolated [54,55]. For ordered structures, such as photonic crystals, it only works at low values of f, typically for $f < 0.4$ [56]. If particles get close but do not touch, the MG model can still give a good approximation at much higher densities provided the particles are randomly distributed [51].

An equation similar to Equation 3.22 can be developed for ellipsoids by starting from Equation 3.18. The resulting effective optics is then anisotropic, which means that the effective indices and dielectric constants depend on the field direction in the material. Various distinct vector or tensor components of the dielectric constant can then arise, for example $(\varepsilon_x, \varepsilon_y, \varepsilon_z)$ if the ellipsoid axes are aligned. In general a full tensor matrix is needed, containing elements such as $(\varepsilon_{xy}, \varepsilon_{xz}, \varepsilon_{yx})$, etc. Anisotropic materials are said to be uniaxial when two diagonal components are equal, say $\varepsilon_x = \varepsilon_y$, and biaxial if all are different. In general we will deal with isotropic media, but some energy-related applications do depend on anisotropic coatings. The optical models that are then needed will be presented briefly in that context.

When their density gets sufficiently large, the particles can join into chains. In such cases the particles may first become elongated, and the effective optical properties can then be modeled via ellipsoids as the system develops optical anisotropy. But if the particles join to form a linked network one needs a principally new type of effective medium model in which the two media may be topologically equivalent. If the particles that connect to each other are metallic, the system is said to percolate.

The class of effective or homogenized media described by Equation 3.22 has one material clearly serving as the "host" and the other as the "particle" or "inclusion." If such a distinction cannot be drawn, that is, if both types of components have equivalent topological status, one needs a different type of effective medium model. This is often simply called effective medium approximation (EMA), and in optics one speaks of the Bruggeman model, named after its initiator [57]. These notions are related to "mathematical" invisibility, which is discussed below.

3.9.2 Invisibility, Effective Medium Models, and Critical Percolation

A simple but extremely powerful idea for determining how the homogenized optical or electrical response of a nanocomposite relates to the properties and geometry of its components is based on invisibility [55]. If one embeds an entity in an otherwise homogeneous host and ensures that this entity gives a dielectric response identical to that of the host, then

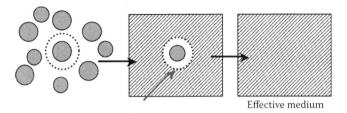

FIGURE 3.9 Development of an effective medium model in geometric terms. The random unit cell in the middle sketch is invisible.

the entity is "invisible" (formally meaning that it does not produce any net polarization or scattering). This entity, sometimes called a random unit cell or RUC, represents an average component from which the system is built, and the host is taken to be the final homogenized medium. The process is illustrated in Figure 3.9 for an array of particles.

If the host represents the homogenized material, its optical constants can be solved as long as it is possible to identify the random unit cell to embed in it. The basic example of this approach is its use to derive the Bruggeman or EMA model, where two distinct cells, one for each constituent, are needed to represent each local entity. If these cells vanish on average one retrieves the EMA model. The MG random isolated-particle model, treated in the previous section, has a single coated-particle RUC and is thus solvable using the "invisibility technique" shown in Figure 3.9. The core-shell particle is the average entity in place of the "orange" particle. Many other nanostructures can also be homogenized using the same approach but with different average entities. The reverse process can be employed for cloaking, as treated in Box. 3.3.

The standard EMA uses invisibility concepts for a two-component material with both species, labeled A and B, being topologically equivalent. However, it is possible to derive a more general EMA from a slight extension of the idea in Figure 3.9. To that end the average shape and size of each component is taken as an inclusion embedded in the final homogenized medium, which has a dielectric constant ε^*. Average invisibility again means that no additional (complex) polarization can arise, this time in the averaged sum of polarizations of each typical inclusion. The resulting equation for ε^* is based on Equation 3.18 for general shapes with ε_h replaced by ε^* for the two average inclusions for each species present; it reads

$$f_A\left[\frac{\varepsilon_A - \varepsilon^*}{\varepsilon^* + L_A\left(\varepsilon_A - \varepsilon^*\right)}\right] + (1 - f_A)\left[\frac{\varepsilon_B - \varepsilon^*}{\varepsilon^* + L_B\left(\varepsilon_B - \varepsilon^*\right)}\right] = 0 \qquad (3.23)$$

where L_A and L_B can be taken as effective depolarization factors for each component in the direction of interest. The most commonly used form of EMA has both Ls equal to 1/3, which avoids anisotropy.

When applied to metal-insulator composites, this model allows percolation transitions of the metal component, say A, and it can also include concentration regimes where both metal A and insulator B percolate. If one component in a percolating system is a metal, then the percolation threshold can lead to a very large change in both optical response and electrical conductivity. Recent work on energy efficient glazing by one of the authors [58] has shown that the optical and dc electrical percolation thresholds can be different because different length scales are involved and, more importantly, the optical transition is not as sharp as the electrical. Percolation also gives rise to a sharp transition from high to low thermal emittance for films on glass. Thus, the understanding of some interesting composites for energy applications requires insights into percolation issues.

BOX 3.3   CLOAKING AND ENERGY EFFICIENCY

It is interesting to consider a process that is the reverse of the one in optical homogenization, namely "designed invisibility" for the embedded entity (in jargon often called "cloaking"). It has been a hot topic of late and mainly theoretical, with practical devices so far operating at microwave frequencies and fixed wavelengths. The host medium is taken to be known, for example, water or air, and the aim is to design structures, with something inside, which appear to have no impact on the passing waves when illuminated from a reasonable distance. Light will always be distorted in the close vicinity, so the structure must ensure it bends light back into its original beam pattern. Furthermore, local attenuation should not be present. If cloaking is achievable, or nearly so, at the nanoscale it may be useful in some energy technologies, but not so much to "cloak" but rather not to impact too much on passing waves. For example cloaking could be used for protecting some chemical species, or in another approach to give "transparent" spectral selectivity.

Chapter 5 on lighting will give an example where "near invisibility" opens up new low-cost, low-energy lighting techniques. Nano- and microdesigns that "almost cloak"—that is, deliberately distort the emerging light just a little—also lead to interesting technology options. Thus, while theorists concentrate on a "perfect cloak," energy scientists might find use for one which is deliberately not quite perfect.

The percolation threshold of an insulator-conductor mixture is a phase transition from a conductor to an insulator and can involve conductivity changes of several orders of magnitude. Unique optical and electrical properties can arise near such a threshold and occur at a critical value of f which is labeled f_c. Phase transitions involve a change in state of a material (e.g., water to ice). They usually require energy and are then discussed in terms of a critical temperature τ_c. But in percolation one deals with a critical concentration f_c rather than a critical temperature. Some switchable materials, such as thermochromic VO_2 which can go from a metal to an insulator as the temperature is changed, have an intermediate mixed phase, or hysteresis regime, and thus can exhibit two transitions involving f_c as well as τ_c [59].

The model underlying Equation 3.23 is called a "mean field model." It can yield phase transitions but usually only describes the approach to them approximately. Composites in this case approach the threshold linearly in $(f - f_c)$ [47,60]. Properties at phase transitions tend to reach criticality as a power law function of either $(\tau - \tau_c)$ or $(f - f_c)$. From the conducting side, where $f > f_c$, this power law approach to percolation criticality for resistivity ρ is given by

$$\rho = \Gamma (f - f_c)^{-\alpha_+} \tag{3.24}$$

If this equation applies up to $f = 1$, then $\Gamma = \rho_m (1 - f_c)^{\alpha_+}$ where ρ_m is the resistivity for the bulk metal.

The effective conductivity σ^* of a mixture can also be estimated from an EMA equation such as Equation 3.23 but with conductivity σ in place of ε. If the insulator has $\sigma = 0$, as in the case of voids, then the basic EMA is

$$\sigma^* = \sigma \frac{f - L}{1 - L} \tag{3.25}$$

Clearly $f_c = L$ and is approached linearly. With regard to conductivity in a network of conducting "bonds," it is common to replace L by the quantity $1/D$ where D is the dimension of the bond network. D is determined by the number of bonds z_b emerging on average from each junction in the network [47,61] and $D = z_b/2$. This implies effectively that $L = 2/z_b$ in Equations 3.23 and 3.25. In two dimensions and with spherical particles distributed on a plane, it can thus be expected in this EMA approximation that $f_c = L = 0.5$, and for a random three-dimensional network of metal spheres one would have $f_c = L = 1/3$. Depending on growth processes and details of network structure, and also in more precise models than the EMA, percolation can occur at $f < 0.3$. Practical examples exist in growing thin films (examples follow in Section 4.3.2), when the metal network appears to be both hyperdimensional ($D > 3$) and anisotropic [47,62]. Hyperdimensionality is a relatively new idea for metal-based composites and has significant implications for critical behavior, which is fundamentally determined by dimension. This is a famous principle called the principle of universality [63].

Optical response can also display criticality but, as noted above, it has a different f_c and different critical parameters than for the dc response. An effective medium model which includes exponential critical behavior has been proposed for electrical conductivity [64] and works well for fixed-size metal particles embedded in an insulator matrix at all f. It can be extended to the optical regime and yields interesting results for plasmonic systems. The relevant phenomenological model for dc conductivity that replaces Equation 3.23 can be written [47,64]

$$f\left[\frac{\sigma^{1/\alpha_+} - \sigma^{*1/\alpha_+}}{\sigma^{1/\alpha_+} + (D-1)\sigma^{*1/\alpha_+}}\right] + (1-f)\left[\frac{\sigma_i^{1/\alpha_-} - \sigma^{*1/\alpha_-}}{\sigma_i^{1/\alpha_-} + (D-1)\sigma^{*1/\alpha_-}}\right] = 0 \quad (3.26)$$

where the possibility of different critical exponents on each side of the transition is included. It is easily seen that Equation 3.24 emerges from this EMA model when the insulator conductivity σ_i equals zero and $\alpha_+ = \alpha_- = 1.0$. For optical behavior, consistency requires that one replace D by $1/L$ and doing so one also recovers Equation 3.23 when $D = 3$, $L = 1/3$ (spheres) and $\alpha_+ = \alpha_- = 1.0$. Highly porous and hence hyperdimensional metals and other networks may have applications in thermal radiation control, supercapacitors for energy storage [49], chemical catalysis [65], and switching networks.

3.9.3 Core-Shell Random Unit Cells and Actual Core-Shell Particles

The average entity or RUC in Figure 3.9 is the average particle coated by an effective shell of host material. The shell shape and volume are determined by the average spacing and distribution of surrounding particles. For example, the outer surface of the shell is spherical if the surrounding particles are randomly distributed. The average entity for particle arrays is often a core-shell system. It is easy to show that the polarizability of a RUC consisting of a spherical core and spherical annulus embedded in the effective medium host simply is given by

$$\alpha_{pol,RUC}(\omega) = 3v\varepsilon^* \frac{\varepsilon_{MG} - \varepsilon^*}{\varepsilon_{MG} + 2\varepsilon^*} \quad (3.27)$$

and since we require $\alpha_{pol,RUC} = 0$ then $\varepsilon^* = \varepsilon_{MG}$ exactly as given in Equation 3.22. One should note that the particle fraction f is now equivalent to $(r/R)^3$ with r being the average nanoparticle radius and R the radius of the RUC.

Another way of looking at this is that the core-shell particle behaves externally like a particle made up of a material with $\varepsilon = \varepsilon_{MG}$. This mathematical result can also be of use for studying, in a very simple way, effective responses of real core-shell systems embedded in various media. In essence, in this three-phase system the real core-shell becomes the core of an RUC with the RUC shell made as usual of host material. The solution is now based on the polarizability of a core plus two shells. The problem reduces to that in the previous section, with the particle having effectively $\varepsilon_p = \varepsilon_{MG}(r/R)$, showing that the particle's effective relative permittivity now depends on (r/R), that is, the actual ratio of the core sphere to the outer radius. Thus, for a volume fraction f of core-shell spherical particles in a host material with $\varepsilon = \varepsilon_h$, the effective dielectric constant can be written immediately using Equation 3.22 as

$$\varepsilon^* = \varepsilon_h \left[1 + \frac{3f\left(\dfrac{\varepsilon_{MG} - \varepsilon_h}{\varepsilon_{MG} + 2\varepsilon_h} \right)}{1 - f\left(\dfrac{\varepsilon_{MG} - \varepsilon_h}{\varepsilon_{MG} + 2\varepsilon_h} \right)} \right] \tag{3.28}$$

The key feature of Equation 3.28 is that the constraint is no longer there to use simple nanoparticles for tuning responses and effective indices. Until now a variation of the spectral response has required changing particle materials, but with a core-shell particle the response can be tuned by changing the ratio (r/R), because ε_{MG} now depends not on f but on $(r/R)^3$. In the next section we shall see that spectral tuning based on varying particle nanostructure becomes dramatic and useful when the shell is a metal and the core a dielectric. Even more fascinating is the very recent discovery [58] that the geometrical counterpart—where all metal is replaced by dielectric and vice versa—is resonant and is in practice more widely tunable than the metal shell case.

The RUC and core-shell concepts can easily be developed to cover other nanostructures, including nonspherical particles, hemispherical and nonspherical nanocaps, and partial nanoshells, as well as a variety of nanoporous structures. They can also be modified to handle ordered lattice-type structures, such as those found in photonic crystals. Touching and near-touching particles, particle chains, and crystal lattices cause more complex responses including multiple resonances, which is one reason to keep particles apart using some of the production techniques sketched earlier. It should be emphasized that these rich geometry-based approaches to altering the optical properties of materials exemplify the power and potential of novel optical engineering based on modified nanostructures. Modern computer simulation methods may be needed to derive the responses in many of these structures.

The models outlined previously describe far-field properties, but associated near-field intensity changes very close to these nanostructures open up many additional technical opportunities, which we will mention in the context of their applications later. Very high local intensities are possible.

The scientific and technical literature is full of examples on the fitting of optical models to experimental data taken on inhomogeneous materials. Not all of there fittings have been executed with sufficient care, though, and in fact there are many ways to tread wrongly when it comes to modeling. Some of those are discussed in Box 3.4.

BOX 3.4   CHOOSING AND USING MODELS: PITFALLS, CHECKS, AND BALANCES

It should be noted that while the models covered explicitly in this chapter are the most widely used effective medium approximations, many more have been formulated for specific nanogeometries. The simple ones, however, tend to work more often than one might have expected because they are adequate for random structures, but "working" via fitting data does not necessarily guarantee "correctness."

Whenever systems are random and there is little evidence for multiple resonances, the simple models are a good starting point and may be adequate. Care is however needed because their underlying mathematical structures are fairly generic and thus may serve as approximations for other models over limited wavelength bands. Thus, apparent fitting parameters such as f, L, and (r/R) may, in fact, mean something else than filling factor, depolarization factor, and radius ratio.

The best test of validity is to have some independent checks on the geometric parameters from compositional analysis, imaging, surface profilometry, and other structural methods. It is also advisable to collect and model data over as wide a range of frequencies as possible. Powerful computer simulation may avoid these pitfalls and provide insights but serves very much as a trial-and-error exercise for limited structures and sizes. Good models to fit data, however, can often yield many useful parameters of the studied sample almost immediately.

3.10 SURFACE PLASMON RESONANCES IN FILMS, PARTICLES, AND "RECTENNAS"

We first introduce the least known of these concepts, the rectenna [66]. It is an optical antenna that resonates with light or solar energy and then turns the energy it picks up into lower-frequency electrical power before the oscillating fields in the antenna are lost as heat or reradiated. The antenna part is easy as we shall see here and can use metal or conductor nanoparticles or other special metal nanostructures, but the rectification is just an idea at present. If the complete package were achievable, solar power might take on a whole new meaning in terms of cost and efficiency. Energy loss in down-shifting the frequency no doubt will be a big hurdle to overcome. Optical antennas have other known uses, such as for focusing radiation and coupling it into solar cells by way of quantum dots, and for enhancing photochemical reactions.

The use of resonant absorption in embedded metal nanoparticles to get distinct and beautiful color dates back to Roman antiquity, as exemplified by the Lycurgus cup and the Victorian vase shown in Figure 3.1. Each contains gold nanoparticles. As noted earlier, the important difference between these two attractive glass vessels is that the Roman one is green in reflected light but red in transmission, while the Victorian vase remains red. There is a simple physical explanation for these observations, as explained shortly.

We first deal with the possible effects of mobile charge at the surface of a conducting material. As noted earlier, a material is plasmonic if $\varepsilon_1(\omega) < 0$. If this condition is fulfilled in a nanoparticle—or more generally on the surface of a metal, metal thin film, or metal grating—collective charge waves can form at the surfaces. These waves resonate with incoming radiation and give rise to novel and useful spectral properties. This phenomenon has been a real boon to various solar energy technologies for some years and is growing in importance.

Charge states that form on a conducting particle are called localized surface plasmons (LSPs), and charge states on extended conducting layers are called surface plasmon polaritons (SPPs). LSP charge polarization waves just slosh back and forth on the surface of a particle as the applied optical field switches direction. SPP waves, however, are charge polarization excitations which propagate along the interface between a conductor and an insulator. The associated electric fields of the SPPs are localized near the surface and decay exponentially into both the metal and the dielectric as indicated in Figure 3.10. These local fields are called evanescent waves and, importantly, can have wavelengths much smaller than those of the incoming radiation. The evanescent waves exist only in the near-field, which is a zone within one wavelength of the interface.

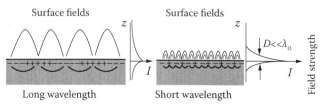

A continuum of SPP wavelengths exists
Two extreme examples shown

FIGURE 3.10   Surface plasmon polariton (SPP) charge density waves of different wavelengths on a flat metal surface and their interface electric field profiles. The field strength I is shown in the vertical (z) direction. The extent of the field (D) is much smaller than the wavelength λ_0.

Both LSP and SPP waves decay in two ways: either by turning into normal radiation or by conversion to heat by absorption. This absorption is different from the bulk absorption of the particle material given by the extinction coefficient k and can lead to an apparent new effective k^* for fine-enough nanostructures. The LSP in metal nanoparticles can be very sharp as a function of wavelength and hence very strong. An LSP is a very efficient means of harvesting incident radiation near its resonant wavelengths, for blocking some radiation, for heating, or as a rectenna. A large effect can be gained with very few particles at resonance, so in practice, large energy savings can be achieved at very low cost even if the particles themselves are moderately expensive. The strength, and hence width and height, of the resonance governs the quality of spectral selectivity for applications.

Consider a nanosphere. Equation 3.16 has a resonance when the real part of the denominator vanishes, that is, when $\varepsilon_{p1} = -2\varepsilon_h$. This condition can only be fulfilled if the particle material is plasmonic. For arbitrary spheroid shapes in which the internal field is constant, the resonance occurs from Equation 3.18 when

$$\varepsilon_{p1} = -\left[\frac{1}{L} - 1\right]\varepsilon_h \qquad (3.29)$$

Because ε_{p1} gets more and more negative as the wavelength increases, the resonance wavelength shifts to lower energies if either L drops below 1/3 or $\varepsilon_h = n_h^2$ gets larger. For example, there is a red-shift if the particle gets thinner in the field direction, or if one goes from a particle embedded in water ($n_h = 1.33$), to one in glass ($n_h = 1.58$), to one in titanium dioxide ($n_h = 2.4$). Controlled variation of the resonant wavelength is a key issue in green nanotechnology, which relies on LSPs on particles.

Other options to shift the resonance in the same host are to change the conductor or to use a core-shell system.

Silver and gold nanoparticles resonate in the visible, and if one wants to shift the resonance into the infrared one needs materials with much lower conductivity than Ag and Au. Why is this? The key parameter governing ε_{p1} is the plasma frequency, as apparent from Equation 3.7, and hence the conduction electron density N_e. Lowering N_e decreases both σ and ω_p. Equations 3.12 and 3.28 show with regard to nanospheres that the LSP resonance occurs at

$$\omega = \frac{\omega_p}{\sqrt{\varepsilon_\infty + 2\varepsilon_h}} \qquad (3.30)$$

Table 3.2 listed some key plasmonic materials in descending order of carrier density, and hence plasma frequency, and it is evident that nanoparticles of transparent conductors are very useful for absorbing in the NIR while letting through visible light.

The other option to shift resonances is to use core-shell particles with metallic shells. The single core-shell particle has a surface plasmon resonance at

$$\varepsilon_{MG} = -2\varepsilon_h \qquad (3.31)$$

as given by Equation 3.28. The resonant wavelength then depends critically on $(r/R)^3$. An ideal silver shell can produce a very sharp and deep resonance, as seen in Figure 3.11 [67], and one can in principle span the whole visible and NIR ranges by varying $(r/R)^3$. This would be great in practice, but it is difficult to make sufficiently ideal thin metal layers on a nanoparticle without large losses in the shell itself. But even so some useful core-shell systems have been developed [68], including ones with TiN thin shells [69].

Scattering increases at resonances but it can remain negligible if the particle is small enough, and conversion to heat then dominates. The latter depends on Im (α_{pol}) and, using Equation 3.17b, the absorption coefficient of a dilute set of nanoparticles becomes

$$\alpha_{p,h} = f \frac{2\pi\sqrt{\varepsilon_h}}{\lambda} \, \mathrm{Im}\left[\frac{\varepsilon_p - \varepsilon_h}{\varepsilon_h + L(\varepsilon_p - \varepsilon_h)} \right] \qquad (3.32)$$

At a resonance, $\alpha_{p,h}$ attains its maximum value which is easily shown to be

$$\alpha_{LSP,\max} = f \frac{2\pi\sqrt{\varepsilon_h}}{\lambda}\left(\frac{\varepsilon_p}{L\varepsilon_{p2}}\right) \tag{3.33}$$

Thus, a small value of ε_{p2} gives a strong resonance peak, as does a small L (which may be the result of a broad particle width in the field direction). This latter fact becomes important for the performance of angular selective thin films (cf. Section 4.7). The width of the resonance depends in a simple linear proportional way on ε_{p2}.

Carrier lifetime and energy loss are considered next. Equation 3.33 points at an important issue in nanoplasmonics and solar energy, namely the significant role of absorption loss as heat, which is ultimately determined by the lifetime τ_e or collision frequency ω_τ of conduction electrons excited by optical fields (cf. Equation 3.8). A small value of ω_τ leads to good spectral selectivity in most cases but, unfortunately, additional heat losses can occur in small enough nanoparticles when electrons collide with the particle surface, and this tends to broaden resonant peaks. This additional collision rate is determined by how fast the electrons are traveling—called the Fermi velocity υ_F—and the average distance r_{eff} they travel between each

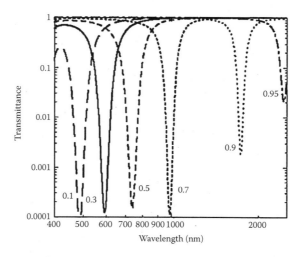

FIGURE 3.11 Calculated spectral transmittance of a laminated glass window with a 0.7-mm-thick layer of poly vinyl buteryl (PVB) doped with 0.05 vol.% silica-Ag core-shell nanoparticles having different silver thicknesses. The core volume fraction is indicated at the curves. Note that the vertical scale is logarithmic. (From G. B. Smith, Nanostructured thin films, in *Introduction to Complex Mediums for Optics and Electromagnetics*, edited by W. S. Weigelhofer and A. Lakhtakia, SPIE Press, Bellingham, WA, 2003, pp. 421–446. With permission.)

collision with the surface. In a sphere of radius r, it is often found experimentally that $r_{eff} = r/\zeta$ with $\zeta \sim 2$ and hence that the additional collision rate is $(\zeta \upsilon_F/r) \sim 2\upsilon_F/r$. For most conductors this effect does not become a major influence until $r < 50$ nm. And if the material has a high electrical resistance to start with, even lower sizes are needed before this is important.

Nanostructured materials in general have increased losses for the same reason, and thus ω_τ should be found from experimental results rather than taken from literature data for bulk materials. Even smooth thin films have enhanced losses if the crystal grain size is of nanoscale and small enough. Crystal alignment or relative orientation is important, too, and the losses can be small in one direction for a sample with aligned grains. For example, a silver film can have ω_τs between 0.025 and 0.06 eV, depending on deposition conditions.

If SPP waves are able to form on an extended interface between a metal m and a dielectric d, as shown in Figure 3.10, their wave vector along the surface is given by

$$k_x = k_0 \left(\frac{\varepsilon_m \varepsilon_d}{\varepsilon_m + \varepsilon_d} \right)^{1/2} \tag{3.34}$$

with $k_0 = 2\pi/\lambda$, and the combined wave has a wave vector (k_x, k_z) with the amplitude $k = \sqrt{k_x^2 + k_z^2}$. The real part of the wave vector corresponds to $2\pi/\lambda_{SPP}$. The SSP waves are only weakly damped and can travel up to hundreds of micrometers along the surface, depending on details of its structure and the specific metal. An important feature of λ_{SPP} is that it can be considerably shorter than λ, which leads to many new features and opportunities. The SPP wave component normal to the surface, on the other hand, is given by

$$k_{z,i} = k_0 \frac{\varepsilon_i}{\left(\varepsilon_m + \varepsilon_d \right)^{1/2}} \tag{3.35}$$

with i being m or d. This wave is heavily damped because ε_m is large and negative, as also illustrated in Figure 3.10.

3.11 TEMPORARY "STORAGE" OF LIGHT AT RESONANCES AND IN EVANESCENT FIELDS

3.11.1 General Considerations

Energy from an incident light wave can be stored for a significant time in an SPP or LSP resonant state at an interface. This is the plasmon lifetime before its energy is lost as heat or reradiated as light. Light can be slowed down in other ways, too; thus, modern photonic circuitry associated with fiber optic telecommunications often needs to delay the transit of photons, but current approaches for this are not applicable at the nanoscale.

What use has this extended storage time of light for energy systems? In fact, some of the most exciting new "green" ideas hinge upon this rarely highlighted issue, and if the evanescent energy locked into interface electric fields can be "tapped" before it decays as heat or light—and there are various interesting approaches to do this—then useful transformation of solar energy into power and light can be done much more efficiently and/or with much less material than with current practices.

The rectenna, introduced above, is an idea based on this possibility. Very high-speed switching materials may be needed to convert the light eventually to useful electric power frequencies, and this may need several down-conversion steps (viz. conversion of optical frequency electric charge oscillations to those at much lower frequencies). Conversion efficiency will be a challenge for energy applications, as much loss can occur in down-conversion, but this loss may not be unavoidable. We note that the conversion of light to low frequency radiation is currently the basis of most generators of terahertz (THz) radiation (used in security scanners).

The scientific concepts involved in evanescent waves are central to nanoscience and of growing importance in energy technology for areas apart from nanoplasmonics, such as conventional photonics, imaging, and computers. The latter is especially interesting since all-optical processing of data, and more use of all-light chips in signal and data processing, would save a lot of energy. The implication is that one could dramatically reduce the energy spent in today's "high technology" electronic computer chips, which populate not only computers but, for example, cars, airplanes, and ships, television and computer displays, manufacturing plants, energy grids, and various other energy-related systems. Energy evolution in chips plays a significant role in the growth of CO_2 emissions, urban heat island effects, and air conditioning needs. But all-optical chips will also eventually need to be nanoscale to be competitive. Communication systems will still rely on normal optical fibers, so coupling light or associated electric fields from them into nanophotonic systems will be an issue. This means dealing with evanescent fields which, as noted above, can have field profiles that change on a scale much smaller than the light wavelengths that excite them. For example, at every total internal reflection at the core-cladding interface of an optical fiber, some (evanescent) electric field penetrates into the cladding. The associated energies can be extracted and used (e.g., for sensing or coupling signals into nanophotonic systems). Thus evanescent fields are worth considering briefly from a generic point of view as, among other things, they allow a bridge between the normal spatial electric field profiles in the optical domain to the much smaller ones in the nanodomain.

3.11.2 Evanescent Optical Fields and How They Can Be Used

These are oscillating electromagnetic fields that do not propagate like normal waves but instead are localized at interfaces and die off within a wavelength or two. The projection of the interface wave in the transverse or z direction is, in simple terms, a standing wave, like that on a violin string. Such waves do not propagate. Fields around illuminated small particles have a localized and nonpropagating evanescent component which dominates the field structure out to distances of $\lambda/2\pi$ from the center of the particle. Beyond that, the propagating fields take over. These local particle fields can couple into neighboring systems such as optical waveguides, which also have a local field component.

The most commonly encountered situation where local evanescent fields arise is when total internal reflection (TIR) occurs at the boundary between a transparent material and one with lower refractive index. This is, of course, the basis of fiber optic technology and other light guides. Some electric field actually enters the cladding or outside medium in a TIR event, but this field does not propagate there. The internal and external wave vector component parallel to the interface has to be the same (i.e., Snell's law must be fulfilled) and is given by $k_{x,cl} = k_{x,c} = n_c k_0 \sin\theta$, with q exceeding a certain critical angle q_{crit} to ensure TIR. Subscripts c and cl refer to core and cladding, respectively, and $\dot{n}_c$ is the core material's refractive index. We further put $k_{z,cl}$ as the wave vector normal to the interface in the external medium or fiber cladding, whose refractive index is n_{cl}. Although $k_{z,cl}$ exists, it has to be purely imaginary in a TIR event because no external propagating waves can exist as they cannot have the small k_x values dictated by Snell's law. For $\theta > \theta_{crit}$, this imaginary wave vector component is given by

$$k_{z,cl} = \sqrt{k_{cl}^2 - k_x^2} = -ik_0\sqrt{n_c^2 \sin^2\theta - n_{cl}^2} \qquad (3.36)$$

since $k_x > k_{cl}$. The exponential decay rate of the external field is thus determined by $-ik_{z,cl}$.

If a thin metal film or metal nanoparticle is on the back of a glass or polymer prism, or sits on a fiber optic core or slab waveguide, then SPPs or LSPs can be excited by the evanescent fields arising through TIR. There are only two requirements: that sufficient energy is available in the evanescent field at the distance to the particle in order to provide coupling, and that $k_{x,SPP} = k_{x,cl}$ (in other words, that Snell's law applies). On a glass plate with refractive index n_g, for example, one requires

$$n_g k_0 \sin\theta = k_x = k_0 \left(\frac{\varepsilon_m \varepsilon_d}{\varepsilon_m + \varepsilon_d} \right)^{1/2} \tag{3.37}$$

using Equation 3.34.

The reverse situation is also possible, namely exciting SPPs or LSPs with external light and having their evanescent field couple into the waveguide or fiber optic modes. An interesting feature is then that the surface plasmon resonance peak can be shifted [70,71]; this follows because we are dealing with a coupled resonance system, just as two coupled pendulums which have new resonant modes different from those in each single pendulum. Thus, we also have a new way of adjusting the dominant surface plasmon resonant wavelength via coupled modes. These aspects are developed further in Box 3.5.

BOX 3.5   COUPLED PLASMON MODES

The idea of using evanescent fields for coupling resonant oscillators to systems at the nanoscale opens up a vast number of interesting possibilities, some of which have begun to be explored [72]. Evanescent fields can be built up and resonate between multiple nanothin metal layers, between or inside holes in thin "holey" metal layers, between closely spaced metal nanoparticles, and between metal nanoparticles, nanorings, and nanoshells, or metal nanogratings separated from extended thin metal layers by small distances through the use of an intermediate dielectric layer. Dramatic changes in optical spectra are then possible in systems, such as that in Figure B3.5.1, showing silver nanospheres or islands in the vicinity of a silver layer, provided that the oxide thickness is properly adjusted so that evanescent fields can bridge the gap. Clearly, the oxide layer must be very thin. The silver particles then resonate strongly with the silver layer due to SPPs excited in the latter. Optical modeling cannot be done with standard thin film equations, even if an effective medium is used for the nanolayer, because these models do not handle SPPs. Systems such as that in Figure B3.5.1 might be advantageous in thin film photovoltaics and solar thermal applications, as well as for windows and cooling systems.

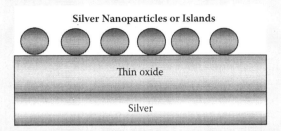

Silver Nanoparticles or Islands

Thin oxide

Silver

FIGURE B3.5.1   Silver islands on a nano-thin dielectric layer backed by a silver film. The particles resonate strongly with the silver layer due to SPPs excited in the latter if the gap is small enough.

New light coupling modes can arise between well-separated metal nanoparticles placed just above a thin dielectric coated metal plate as in Figure B3.5.2. Such modes have been studied for grids above a metal mirror, and the ensuing "gap modes" were found to exhibit unusual characteristics [73]. These gap couplings can occur via surface plasmons in the underlying metal layer. The plasmons can travel many micrometers, which means that it is possible to achieve significant coupling between the nanoparticles, even when they are separated by distances at which normal direct coupling would be negligible. Long-range plasmon–plasmon interactions are thus enabled. A local surface plasmon on a nanoparticle coupled to delocalized surface plasmons on a metal layer is called a "virtual bound plasmon resonance." This phenomenon is analogous to the well-known "virtual bound states" [74,75] that form when atoms of a magnetic material such as iron are located in a metal such as copper; it occurs because the conducting electrons in copper resonate with the localized magnetic electrons in the iron, with the mobile Cu s-electrons jumping on and off the iron, thereby temporarily occupying d-states.

Figure B3.5.2 illustrates plasmonic coupling between two well-separated nanoparticles via a metal layer. This situation also has a well-known analogy from magnetism and corresponds to the case of two dispersed iron atoms in copper that couple via induced magnetic oscillations in the conduction electrons of copper; the mechanism is known as the Rudermann–Kittel–Kasuya–Yoshida (RKKY) interaction [76].

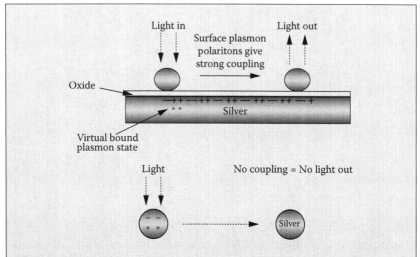

FIGURE B3.5.2   Conducting nanoparticles at resonance, but too far apart to interact normally, are shown to exchange energy via resonant interaction of their localized surface plasmons with the continua of surface plasmons on a nearby smooth silver layer. Light shining onto one particle can then lead to light emission from another. The state in the silver layer under each particle is a virtual bound one.

REFERENCES

1. U. Leonhardt, Optical metamaterials: Invisibility cup, *Nature Photonics* 1 (2007) 207–208; http://www.britishmuseum.org/explore/highlights/highlight_objects/pe_mla/t/the_lycurgus_cup.aspx.
2. S. Padovani, C. Sada, P. Mazzoldi, B. Brunetti, I. Borgia, A. Sgamellotti, A. Giulivi, F. D'Acapito, G.Battaglin, Copper in glazes of Renaissance luster pottery: Nanoparticles, ions and local environment, *J. Appl. Phys.* 93 (2003) 10058–10063.
3. C. M. Christensen, *The Innovator's Dilemma: When New Technologies Cause Great Firms to Fail*, Harvard Business School Press, Boston, MA, 1997.
4. C. M. Christensen, M. E. Raynor, *The Innovator's Solution: Creating and Sustaining Successful Growth*, Harvard Business School Press, Boston, MA, 2003.
5. G. B. Smith, unpublished work.
6. J. Rodríguez, M. Gómez, J. Ederth, G. A. Niklasson, C. G. Granqvist, Thickness dependence of the optical properties of sputter deposited Ti oxide films, *Thin Solid Films* 365 (2000) 119–125.

7. B. I. Bleaney, B. Bleaney, *Electricity and Magnetism*, 3rd edition, Oxford University Press, Oxford, U.K., 1976.

8. R. Guenther, *Modern Optics*, John Wiley & Sons, New York, 1990.

9. C. G. Granqvist, Far infrared absorption in ultrafine metallic particles: Calculations based on classical and quantum mechanical theories, *Z. Phys. B* 30 (1978) 29–46.

10. W. Cai, U. K. Chettiar, H.-K. Yuan, V. C. de Silva, A. V. Kildishev, V. P. Drachev, V. M. Shalaev, Metamagnetics with rainbow colors, *Opt. Express* 15 (2007) 3333–3341.

11. N. Engheta, Circuits with light at nanoscales: Optical nanocircuits inspired by metamaterials, *Science* 317 (2007) 1698–1702.

12. N. M. Litchinitser, I. R. Gabitov, A. I. Maimistov, V. M. Shalaev, Negative refractive index metamaterials in optics, in *Progress in Optics*, edited by E. Wolf, Elsevier, Amsterdam, the Netherlands, 2008, Vol. 51, pp. 1–68.

13. J. B. Pendry, Negative refraction makes a perfect lens, *Phys. Rev. Lett.* 85 (2000) 3966–3969.

14. E. D. Palik, *Handbook of Optical Constants of Solids*, Academic Press, Orlando, FL, 1985.

15. E. D. Palik, *Handbook of Optical Constants of Solids II*, Academic Press, San Diego, CA, 1991.

16. E. D. Palik, *Handbook of Optical Constants of Solids III*, Academic Press, San Diego, CA, 1998.

17. J. Ge, H. Lee, L. He, J. Kim, Z. Lu, H. Kim, J. Goebl, S. Kwon, Y. Yin, Magnetochromatic microspheres: Rotating photonic crystals, *J. Am. Chem. Soc.* 131 (2009) 15687–15694.

18. http://nanohub.org/resources/1748.

19. F. Wooten, *Optical Properties of Solids*, Academic Press, New York, 1972.

20. M. A. Ordal, L. L. Long, R. J. Bell, S. E. Bell, R. R. Bell, R. W. Alexander, Jr., C. A. Ward, Optical properties of the metals Al, Co, Cu, Au, Fe, Pb, Ni, Pd, Pt, Ag, Ti, and W in the infra red and far infrared, *Appl. Opt.* 22 (1983) 1099–1119.

21. A. I. Maroof, M. B. Cortie, G. B. Smith, Optical properties of mesoporous gold films, *J. Opt. A: Pure Appl. Opt.* 7 (2005) 303–309.

22. G. B. Smith, A. Maaroof, A. Dowd, A. Gentle, M. Cortie, Tuning plasma frequency for improved solar control glazing using mesoporous nanostructures, *Proc. SPIE* 6197 (2006) 6197OT 1–9.

23. C. G. Granqvist, O. Hunderi, Optical properties of ultrafine gold particles, *Phys. Rev. B* 16 (1977) 3513–3534.

24. G. B. Smith, G. A. Niklasson, J. S. E. M. Svensson, C. G. Granqvist, Noble-metal-based transparent infra-red reflectors: Experiments and theoretical analyses for very thin gold films, *J. Appl. Phys.* 59 (1986) 571–581.

25. C. G. Ribbing, E. Wäckelgård, Reststrahlen bands as property indicators for materials in dielectric coatings, *Thin Solid Films* 206 (1991) 312–317.

26. C. G. Ribbing, Reststrahlen material bi-layers: An option for tailoring in the infrared, *Appl. Opt.* 32 (1993) 5531–5534.

27. R. H. Lyddane, R. G. Sachs, E. Teller, On the polar vibrations of alkali halides, *Phys. Rev.* 59 (1941) 673–676.

28. C. G. Granqvist, A. Hjortsberg, Radiative cooling to low temperatures: General considerations and application to selectively emitting SiO films, *J. Appl. Phys.* 52 (1981) 4205–4220.

29. T. S. Eriksson, C. G. Granqvist, Infrared optical properties of silicon oxynitride films: Experimental data and theoretical interpretation, *J. Appl. Phys.* 60 (1986) 2081–2091.

30. E. M. Lushiku, C. G. Granqvist, Radiative cooling with selectively infrared-emitting gases, *Appl. Opt.* 23 (1984) 1835–1843.

31. M. Tazawa, H. Kakiuchida, G. Xu, P. Jin, H. Arwin, Optical constants of vacuum evaporated SiO film and an application, *J. Electoceram.* 16 (2006) 511–515.

32. C. G. Granqvist, T. S. Eriksson, Materials for radiative cooling to low temperatures, in *Materials Science for Solar Energy Conversion Systems*, edited by C. G. Granqvist, Pergamon, Oxford, U.K., 1991, chap. 6, pp. 168–203.

33. N. A. Nilsson, T. S. Eriksson, C. G. Granqvist, Infrared-transparent convection shields for radiative cooling: Initial results on corrugated polyethylene foils, *Solar Energy Mater.* 12 (1985) 327–333.

34. J. Ridealgh, Large area coatings for architectural glass: Design, manufacturing, processing faults, *MRS Proc.* 890 (2006) Y01–10.

35. M. Born, E. Wolf, *Principles of Optics*, 7th edition, Cambridge University Press, Cambridge, U.K., 1999.

36. R. Dragila, B. Luther-Davies, S. Vukovic, High transparency of classically opaque metallic films, *Phys. Rev. Lett.* 55 (1985) 1117–1120.

37. A. R. Gentle, G. B. Smith, Five layer narrow band position variable filters for sharp colours and ultra low emittance, *Appl. Phys. B* 92 (2008) 67–72.

38. M. D. Arnold, R. J. Blaikie, Using surface-plasmon effects to improve process latitude in near-field optical lithography, in *Proceedings of the 2006 International Conference on Nanoscience and Nanotechnology*, Gwanju, S. Korea, IEEE, New York, Vol. 1, pp. 209–212.

39. M. D. Arnold, R. J. Blaikie, Subwavelength optical imaging of evanescent fields using reflections from plasmonic slabs, *Opt. Express* 15 (2007) 11542–11552.

40. G. B. Smith, A. I. Maaroof, Optical response in nanostructured thin metal films with dielectric overlayers, *Opt. Commun.* 242 (2004) 383–392.

41. H. C. van de Hulst, *Light Scattering by Small Particles*, John Wiley & Sons, New York, 1957.

42. C. F. Bohren, D. R. Huffman, *Absorption and Scattering of Light by Small Particles*, John Wiley & Sons, New York, 1983.

43. C. Ton-That, A. G. Shard, R. Daley, R. H. Bradley, Effects of annealing on the surface composition and morphology of PS/PMMA blend, *Macromolecules* 33 (2000) 8453–8459.

44. P. Siepmann, C. P. Martin, I. Vancea, P. J. Moriarty, N. Krasnogor, A generic algorithm approach to probing the evolution of self-organized nanostructured systems, *Nano Lett.* 7 (2007) 1985–1990.

45. C. G. Granqvist, R. A. Buhrman, Ultrafine metal particles, *J. Appl. Phys.* 47 (1976) 2200–2219.

46. C. G. Granqvist, L. Kish, W. Marlow, Eds, *Gas Phase Nanoparticle Synthesis*, Kluwer Academic, Dordrecht, the Netherlands, 2004.

47. G. B. Smith, A. I. Maaroof, M. B. Cortie, Percolation in nanoporous gold and the principle of universality from two dimensional to hyperdimensional networks, *Phys. Rev B.* 78 (2008) 165418 1–11.

48. P. C. Lansåker, J. Backholm, G. A. Niklasson, C. G. Granqvist, TiO_2/Au/TiO_2 multilayer thin films: Novel metal-based transparent conductors for electrochromic devices, *Thin Solid Films* 518 (2009) 1225–1229.

49. M. B. Cortie, A. I. Maaroof, G. B. Smith, Electrochemical capacitance of mesoporous gold, *Gold Bull.* 38 (2005) 14–22.

50. A. I. Maaroof, A. Gentle, G. B. Smith, M. B. Cortie, Bulk and surface plasmons in highly nanoporous gold films, *J. Phys. D: Appl. Phys.* 40 (2007) 5675–5682.

51. S. Schelm, G. B. Smith, G. Wei, A. Vella, L. Wieczorek, K.-H. Muller, B. Raguse, Double effective medium model for the optical properties of self-assembled gold nanoparticle films cross-linked with dithiols, *Nano Lett.* 4 (2004) 335–339.

52. L. M. Liz-Marzán, P. Mulvaney, The assembly of coated nanocrystals, *J. Phys. Chem. B* 107 (2003) 7312–7326.

53. J. C. Maxwell Garnett, Colours in metal glasses and metallic films, *Phil. Trans. Roy. Soc. London*, Series A, 203 (1904) 385–420.

54. G. B. Smith, Dielectric constants for mixed media, *J. Phys. D: Appl. Phys.* 10 (1977) L39–L42.

55. G. A. Niklasson, C. G. Granqvist, O. Hunderi, Effective medium models for the optical properties of inhomogeneous materials, *Appl. Opt.* 20 (1981) 26–30.

56. R. C. McPhedran, D. R. McKenzie, The conductivity of lattices of spheres: I. The simple cubic lattice, *Proc. Roy. Soc. London A* 359 (1978) 45–63.

57. D. A. G. Bruggeman, Berechnung verschiedener physikalischer Konstanten von heterogenen Substanzen, I. Dielektrizitätskonstanten und Leitfähigkeiten der Mischkörper aus isotropen Substanzen, *Ann. Phys. Lpz.* 24 (1935) 636–679.

58. G. B. Smith, A. E. Earp, Metal-in-metal localized surface plasmon resonance, *Nanotechnology* 21 (2010) 015203 1–8.

59. A. R. Gentle, G. B. Smith, A. I. Maaroof, Frequency and percolation dependence of the observed phase transition in nanostructured and doped VO_2 thin films, *J. Nanophotonics* 3 (2009) 031505 1–15.

60. G. B. Smith, A. R. Gentle, A. I. Maaroof, Metal-insulator nanocomposites which act optically like homogeneous conductors, *J. Nanophotonics* 1 (2007) 013507 1–15.

61. J. P. Clerc, G. Giraud, J. M. Laugier, J. M. Luck, The electrical conductivity of binary disordered systems, percolation clusters, fractals and related models, *Adv. Phys.* 39 (1990) 191–309.

62. S. Shabtaie, C. R. Bentley, Unified theory of electrical-conduction in firn and ice: Site percolation and conduction in snow and firn, *J. Geophys. Res: Solid Earth* 99 (1994) 19757–19769.

63. P. M. Chaikin, T.C. Lubensky, *Principles of Condensed Matter Physics*, Cambridge University Press, Cambridge, U.K., 1995.

64. D. S. McLachlan, An equation for the conductivity of binary mixtures with anisotropic grain structures, *J. Phys. C* 20 (1987) 865–877.

65. C. Xu, J. Su, X. Xu, P. Liu, H. Zhao, F. Tian, Y. Ding, Low temperature CO oxidation over unsupported nanoporous gold, *J. Am. Chem. Soc.* 129 (2006) 42–43.

66. D. Y. Goswami, S. Vijayaraghavan, S. Lu, G. Tamm, New and emerging developments in solar energy, *Solar Energy* 76 (2004) 33–43.

67. G. B. Smith, Nanostructured thin films, in *Introduction to Complex Mediums for Optics and Electromagnetics*, edited by W. S. Weigelhofer and A. Lakhtakia, SPIE Press, Bellingham, WA, 2003, pp. 421–446.

68. S. J. Oldenburg, R. D. Averitt, S. L. Westcott, N. J. Halas, Nanoengineering of optical resonances, *Chem. Phys. Lett.* 288 (1998) 243–247.

69. M. R. Kuehnle, H. Statz, Encapsulated nanoparticles for the absorption of electromagnetic energy, U.S. patent publication No. US 2005/0074611 A1 (2005).

70. H. R. Stuart, D. G. Hall, Island size effects in nanoparticle-enhanced photodetectors, *Appl. Phys. Lett.* 73 (1998) 3815–3817.

71. S. Pillai, K. R. Catchpole, T. Trupke, G. Zhang, J. Zhao, M. A. Green, Enhanced emission from Si-based light-emitting diodes using surface plasmons, *Appl. Phys. Lett.*. 88 (2006) 161102 1–3.

72. F. Le, N. Z. Lwin, N. J. Halas, P. Nordlander, Plasmonic interactions between a metallic nanoshell and a thin metallic film, *Phys. Rev. B* 76 (2007) 165410 1–12.

73. M. D. Arnold, G. B. Smith, Comparisons of enhanced absorption in closely-coupled grating-mirror and random particle-mirror systems, *Proc. SPIE* 7404 (2009) 740407 1–8.

74. J. Friedel, Metallic alloys, *Nuovo Cimento* Suppl. 7 (1958) 287–310.

75. P. W. Anderson, Localized magnetic states in metals, *Phys. Rev.* 124 (1961) 41–53.

76. K. P. Sinha, N. Kumar, *Interactions in Magnetically Ordered Solids*, Oxford University Press, Oxford, U.K., 1980.

CHAPTER 4

Visual Indoors–Outdoors Contact and Daylighting
Windows

This chapter will bring us deep into green nanotechnology by considering a range of materials—often in the form of thin films with nanofeatures—for controlling the radiation going in or out through windows. "Windows" should be taken in a wide sense to include not only apertures in domestic buildings but also transparent façades such as those commonly used in modern commercial buildings. And the windows do not have to be integrated in walls; they can be skylights or any other opening in the envelope in a building. Figure 4.1 serves as a point of departure and shows one emerging type of windows, referred to as "smart" or "switching" windows. These are capable of changing the throughput of visible light and solar energy, thereby giving the user of the building a comfortable indoor climate, as well as minimizing energy for air conditioning while preserving visual indoors–outdoors contact.

This chapter discusses many aspects of the "view window" capable of giving both visual indoors–outdoors contact and daylighting. Chapter 5 on luminaires then continues the treatment of light in buildings, and covers artificial light and daylighting, primarily with systems that do not provide visual contact between the two sides.

In fact, (day)lighting and visual contact are two functions that are intertwined in a fascinating and poorly understood way. The dichotomy even appears in the very term for this multifaceted object. Thus, the English *window* literally means "wind eye" (from *vindauga* in Old Norse), and in this way is connected to vision, whereas the corresponding term is *Fenster* in German and *fenêtre* in French both going back to Latin *fenestra* meaning to provide light. Incidentally, there are other ways to represent the meaning of a window, such as the Greek παράθυρο, literally meaning "opening near a door."

FIGURE 4.1 Prototype of a "smart" window with four panes, two of which are dark and the other two transparent. The properties of the panes can be changed, gradually and reversibly, during the course of some tens of seconds. The smart functionality is obtained via a potentially low-cost foil that can serve as a lamination material and also provide mechanical ruggedness and acoustic damping.

4.1 GENERAL INTRODUCTION

Architectural windows and glass façades are problematic from an energy perspective, and the same is true for windows used in cars, trucks, and other vehicles. Nevertheless, the windows are absolutely essential for the users of a building and provide them with contact to the surrounding world. So diminishing window areas to the extreme may be good with regard for energy, but it is generally highly undesirable for the persons who spend their time in the building. Indeed, the primary function of fenestration—to provide unmitigated visual contact between indoors and outdoors, as well as daylighting—always must be kept in mind so that window areas are not made too small. Psychological well-being and health benefits associated with visual contact to the ambience and with daylight are issues whose importance is being increasingly recognized.

In a practical situation, there are normally unwanted energy flows with too much energy leaving or entering buildings via their windows, with concomitant demand for energy guzzling heating and cooling. Supplying daylight without glare can also save a lot of lighting energy and some thermal energy. Present architectural trends are to increase the window areas, and hence the energy issue may become even more pressing in the future.

This increase is especially marked with regard to roof glazing for homes, large buildings, and in atria attached to homes in cold climates. The average solar flux onto a roof is several times that onto a vertical façade, while heat loss via a glass roof during the winter is also much larger than for a wall. Glare can also be a major problem, though the daylighting benefits from roof windows are marvelous if these problems can be managed. Many building codes severely limit the area of roof glazing apertures unless special glazing is used. Glazings with variable optical properties—as illustrated in Figure 4.1 and discussed in the text following—is an ideal solution for roof windows since their problems vary with time and solar irradiation intensity, and such glazings can control heat flows as well as glare. Roof glazings are also combined increasingly with solar cells for power generation; these solar cells can reduce glare and solar heat gain while still admitting plenty of daylight. The solar cells, which are essentially opaque, may create illuminance patterns inside the building, which may be advantageous or problematic depending on the specific building; if such patterns are unwanted, they can be removed by having diffusing glass in the roof windows. Figure 6.1c shows a glazing with integrated solar cells.

4.1.1 Strategies for Energy Efficiency

Several strategies can be used to improve the windows' energy performance. All of them are based on the rational use of energy in our ambience, as discussed at length in Chapter 2. Three main strategies are that

- Spectral selectivity can be used to let in only the desired radiation whereas other radiation is removed or at least diminished.
- Angular properties can be taken advantage of, typically by maintaining good transparency along a desired line-of-sight while preventing radiation from other directions.
- The properties of windows can be varied in response to an external stimulus, so that visible light and solar energy are introduced or excluded depending on the need; materials capable of such dynamic performance are referred to as chromogenic.

The green nanotechnologies of particular importance for these strategies are discussed in detail below. But before we do so, it is important to have some basic knowledge about the properties of normal window glass and about plastic foils that are used in windows.

A note on terminology is useful for avoiding confusion. The thin sheet of plastic, typically a fraction of a millimeter in thickness, that can be used for upgrading windows are referred to here as "window foil," whereas the term window film is also commonly used. We prefer the former terminology and reserve the term film for the surface coatings—typically less than a micrometer thick—that can be applied to the surface of a glass pane or a foil, or any other substrate, in order to change its optical and electrical performance and, as a consequence, improve its properties.

4.1.2 Uncoated Glass and Plastic Foil

Most windows use glass panes. In principle, there are many different glass types, but float glass is employed in the great majority of all buildings—at least in the industrialized world—and is a highly standardized product [1,2].

Float glass is made by a process in which the glass melt is solidified on the surface of molten tin. As a result, this glass is characterized by uniformity and flatness almost on the atomic scale, which is important for its usefulness as a substrate for thin-film deposition. The middle curve in Figure 4.2 shows the spectral transmittance of 6-mm-thick standard float glass within the wavelength range relevant to solar radiation [3]. A characteristic absorption feature at $\lambda \approx 1$ µm is clearly seen and limits T_{sol}

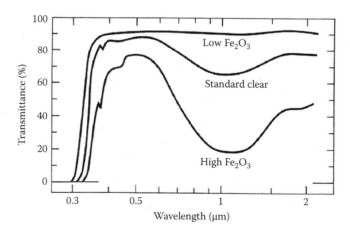

FIGURE 4.2 Spectral transmittance for float glass with three different amounts of Fe_2O_3. The upper two curves refer to 6-mm-thick panes and the lower curve to a 4-mm-thick pane. (From C. G. Granqvist, *Materials Science for Solar Energy Conversion Systems*, edited by C. G. Granqvist, Pergamon, Oxford, U.K., 1991; chap. 5, pp. 106–167. With permission.)

to some noticeable extent. This absorption is due to iron oxide, which is often added to the raw material of the glass in order to make the glass produced in one plant indistinguishable from that made in other plants.

Special float glass is available with varying amounts of Fe oxide, and Figure 4.2 shows that low Fe content leads to glass with very high values of T_{lum} and T_{sol}, and with substantial transmittance also in the UV. These properties are important for several different solar-energy-related applications such as for solar collectors for hot water production (cf. Figure 6.2). Glass with a high Fe content limits T_{sol}, while T_{lum} remains rather large; such glass has a greenish tint, which becomes increasingly distinct at large thickness. Other metal additions can provide different colors.

The reflectance of each interface between glass and air is about 4% in the $0.4 < \lambda < 2$ µm range, implying that the maximum transmittance for a glass pane is 92%. Glass is strongly absorbing for $\lambda > 3$ µm, and E_{therm} is as large as about 87%. In other words, *uncoated glass is almost a blackbody radiator.*

The production of float glass is enormous and currently (2009) amounts to more than 4 billion m^2 per year, with an overall annual growth of some 6%. This glass is used for different purposes, with 37% for residential buildings, 23% for commercial buildings, and 27% for vehicles. Some 1.8 billion m^2 is produced in Asia, whereas the numbers for the Americas and Europe are 1.0 and 0.9 billion m^2, respectively. The total value is expected to be about US\$ 8 billion in 2010. Considering these stunning numbers, and the large role of buildings in the world's use of primary energy, one conclusion to draw is that the potential for energy savings is huge for any technology that can significantly improve the properties of windows.

Polymers are also of great importance for windows, and transparent polymers can, in principle, replace glass. However, polymers degrade more easily than glass and hence have fewer applications. Polyethylene terephthalate (PET) foil deserves special mention since it can be coated and applied to windows in different ways: glued onto individual glass panes, suspended between glass panes, or integrated in laminates joining glass plates. Such foil can be retrofitted to existing windows or applied in new windows. An alternative to PET is polyethylene naphthalate (PEN) and some fluorocarbon plastics such as ethylene tetrafluoroethylene (ETFE). The latter plastics are less temperature sensitive and more UV resistant—but also more expensive—than PET. ETFE is employed in contemporary "membrane architecture," which we will touch upon in Chapter 9.

Foils comprising many alternating layers make it possible to achieve interesting optical effects. If the difference in refractive index among the layers, as well as the number of layers, is large it is possible to obtain high reflectance in a prescribed wavelength range, as can be understood from thin-film optics [4]; the principles for the optical functionality are discussed briefly in Section 5.3. Figure 4.3 shows data for two commercially available foils with different values of T_{lum} constructed in this way and backed by glass [5]. Obviously, the reflectance covers a significant part of the infrared solar radiation, specifically the $0.85 < \lambda < 1.2$ µm range, where the reflectance from the foil side is ~50%, and hence such a foil can be used to limit the energy inflow through windows. In the thermal infrared, the multilayer foil behaves as a homogeneous one and E_{therm} is high. Multilayer foils can be manufactured by coextrusion—which in principle is a very low-cost technique—and have been commercialized recently.

Laminated glass may be used for safety and other reasons. Such glass normally has a layer of polyvinyl butyral (PVB) sandwiched between two glass panes and bonded to the glass under heat and pressure. Laminated glass is of growing importance since it has to be used for safety reasons in high-rise buildings, and such buildings are the only option for housing the world's growing population, more than half of which lives in big cities today (2010). Car windscreens always use PVB laminates in order to avoid glass splinters in the case of breakage. The PVB is an efficient absorber of UV light, and the transmittance at $\lambda < 0.38$ µm is almost zero for a layer thickness of the order of 1 mm.

4.2 SPECTRAL SELECTIVITY: THE ENERGY EFFICIENCY THAT IS POSSIBLE

Consider a window with two or more vertical panes separated by air gaps. Generally speaking, the heat transfer should be low so that a desired indoor temperature can be maintained irrespective of the outdoor temperature being higher or lower than this chosen "comfort temperature." Heat transfer takes place via three mechanisms: radiation between the glass surfaces, conduction across the air gap, and convection (moving gas) in the air gap. The heat transfer can be decreased by minimizing each of these transfer mechanisms. The one that is easiest to influence is the radiative part, which can be cut back by coating at least one of the glass panes so that it does not emit as much thermal radiation as the blackbody-like bare glass surface.

What is the energy savings one can accomplish by having glass with low thermal emittance? To answer this question we now discuss some computations of the heat transfer in a window with one, two, or three vertical panes separated by air gap(s) of 12.7 mm, which is a typical value for insulated glass units [3]. The glass surfaces are designated by consecutive numbers, with the outside surface labeled 1. One of the surfaces is taken to have an emittance in the zero to 85% range. The upper value is characteristic for uncoated glass, as pointed out above, so the

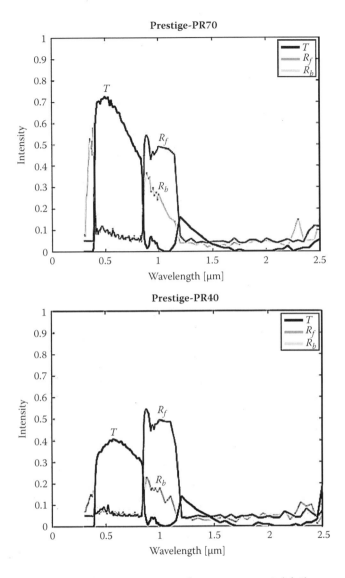

FIGURE 4.3 Spectral transmittance of two commercial foils comprising multilayer polymers and backed by glass. Data are given on transmittance T and reflectance from the front (foil) side R_f and back (glass) side R_b. Upper data correspond to T_{lum} = 69.3% and T_{sol} = 36.5%; lower data correspond to T_{lum} = 39.5% and T_{sol} = 21.9%. (From International Glazing Database (2009), Lawrence Berkeley National Laboratory, CA; http://windows.lbl.gov/software/window/window.html.)

computations illustrate the role of having one surface with a low value of E_{therm}. The calculations assumed a low outside temperature and that there were thermal losses due to wind. Data are shown in Figure 4.4. The upper two curves indicate that if the emittance of one of the surfaces in a single pane window is lowered there is a minor drop in the heat transfer (thermal conductance) from an initial value of about 6 Wm^{-2}K^{-1}, the effect being largest when the inner surface has a low emittance. For double glazing, the heat transfer can drop from about 2.8 to 1.4 Wm^{-2}K^{-1} when either of the surfaces facing the air gap has a low value of E_{therm}. For triple glazing, finally, the corresponding improvement is from 1.8 to 1.2 Wm^{-2}K^{-1}. The most significant result with regard to a typical double-pane window is that the *heat transfer can be cut to almost half by the use of a surface treatment giving a low thermal emittance.*

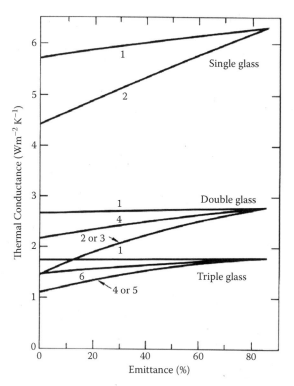

FIGURE 4.4 Computed thermal conductance for different window constructions with one glass surface having low emittance. The glass surfaces are numbered consecutively with the outermost surface designated 1. (From C. G. Granqvist, *Materials Science for Solar Energy Conversion Systems*, edited by C. G. Granqvist, Pergamon, Oxford, U.K., 1991; chap. 5, pp. 106–167. With permission.)

Detailed computations of the thermal performance of windows are technical, and rather complicated, issues which we do not wish to discuss here. Detailed information is available in the literature [6]. This literature also describes the role of gas fillings to diminish the thermal conductance still further, problems associated with thermal leaks at the frames in the windows, etc. Generally speaking, the effect of a low emittance treatment is accentuated when conductive and convective transfers are diminished. A judiciously chosen window design, with vacuum between the panes in a triple-layer construction and with low-emittance treatment, has given a thermal conductance coefficient as low as ~0.2 $Wm^{-2}K^{-1}$ [7]. That number probably gives a good estimate for what is feasible for the thermal insulation in practical fenestration. Vacuum insulation is discussed later in Section 8.3.

Controlling thermal conductance in a window is only one of the important tasks. Another, obviously, is related to the transmission of solar energy and visible light. As emphasized in Chapter 2, only about half of the solar energy falls in the spectral range for which the eye is sensitive. Since Figure 4.2 showed that regular float glass lets through most of the solar energy, the energy inflow through a window can be about twice as high as the one needed to provide full daylighting or, in other words, the radiative energy inflow through a window, in principle, can be cut in half without affecting the windows' primary functions of providing visual contact as well as daylighting.

We now return to the discussion of the ambient radiative properties in Chapter 2 and repeat, mainly for convenience, some of the essential facts. The upper part of Figure 4.5 shows the most relevant aspects of the ambient radiation: the solar irradiance after the radiation has gone through one air mass, the luminous efficiency of the eye, and blackbody spectra for some different temperatures [8]. As noted before, the blackbody spectra define the spectral ranges for thermal radiation.

The lower part of Figure 4.5 shows *ideal* properties of a coating that transmits in the solar range but does not emit thermal radiation. This is known as a "low emittance coating" ("low-E") and is characterized by

$$T(\lambda) = 1 \text{ for } 0.4 < \lambda < 3 \text{ μm} \tag{4.1a}$$

$$R(\lambda) = 1 \text{ for } 3 < \lambda < 50 \text{ μm} \tag{4.1b}$$

The other idealized coating specified in Figure 4.5, which transmits only visible light, defines a "solar control coating" characterized by

$$T(\lambda) = 1 \text{ for } 0.4 < \lambda < 0.7 \text{ μm} \tag{4.2a}$$

$$R(\lambda) = 1 \text{ for } 0.7 < \lambda < 50 \text{ μm} \tag{4.2b}$$

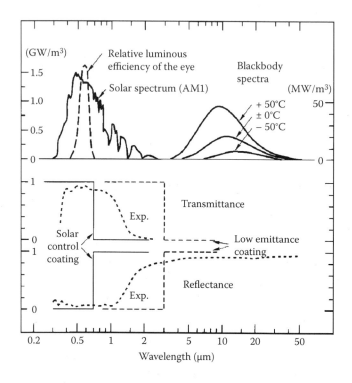

FIGURE 4.5   Upper part shows relative luminous efficiency for the eye, a solar spectrum for one air mass (AM1), and blackbody spectra for three temperatures. Lower parts show transmittance and reflectance for idealized window coatings designed for "solar control" and for "low emittance." Dotted curves refer to experimental data to be discussed in Section 4.4. (From I. Hamberg, C. G. Granqvist, *J. Appl. Phys.* 60 (1986) R123–R159. With permission.)

Both of the spectral profiles were given earlier in Figure 2.16. We emphasize that the spectral selectivity inherent in the relationships is idealized, and the dashed curves in the lower part of Figure 4.5 illustrate some typical experimental results. Practical ways to accomplish spectral optical data that approximate Equations 4.1 and 4.2 will be discussed at length next.

4.3 SPECTRAL SELECTIVITY OF NOBLE-METAL-BASED FILMS

Very thin films of free-electron-like ("noble") metals can combine luminous transmittance with infrared reflectance and electrical conductivity, as has been known for a long time. Thus, they can have properties that more or less resemble those in Equations 4.1 and 4.2 and, therefore, provide a good starting point for development of spectrally selective window coatings. Very thin layers of Cu, Ag, and Au are of particular interest; alternative materials are TiN and ZrN as well as Pt and Al. Whereas Ag-based films have unsurpassed optical properties (cf. Section 3.7), Au-based films may be of interest under corrosive or otherwise chemically demanding conditions. The nitrides are notable for their hardness and durability. A discussion of these matters is found elsewhere [9], and this earlier article can serve as a general reference to much of the material covered below.

Ag-based films are used on a very large scale in practical fenestration. Such films are delicate and are only suited for application on surfaces facing insulated gas-filled cavities in double-glazed windows. The films are sometimes referred to as "soft coats" and are normally applied by high-rate magnetron sputtering.

4.3.1 Thin Metal Films Are Not Bulk-Like

The *idealized* performance of a *uniform* metal-based thin film is readily computed using formulas for thin film optics. The metal is uniquely characterized by two parameters—known as the optical constants n and k or the complex dielectric function ε—pertinent to a uniform bulk-like material. Figure 4.6 shows computed results of T_{lum}, T_{sol}, R_{sol}, and E_{therm} as a function of thickness d for Ag [3,10]. It is evident that $d = 5$ nm yields an impressive set of data, specifically with $T_{lum} = 85\%$, $T_{sol} = 74\%$, and $E_{therm} = 8\%$. A glass pane with such a thin film would make an excellent low-E glazing. Increasing the film thickness predictably makes the transmittance drop and the reflectance go up.

But can a uniform film, such as the one just discussed, really be made as a window coating? The answer may be easy to guess, since we emphasized the idealized nature of the films considered in Figure 4.6. Before discussing this matter further, let us look at *experimental* data for thin silver films, such as those given in Figure 4.7 [11]. Here are shown measured reflectance and transmittance for conventionally evaporated Ag films on glass. At $d = 6$ nm one observes that R_{therm} is low, implying that E_{therm} is high, which is very much in opposition to the data in Figure 4.6. It is only for films with $d \approx 9$ nm or thicker that E_{therm} becomes small, but then T_{lum} and T_{sol} are rather low. Specifically demanding a high reflectance at long wavelengths—that is, low E_{therm}—evidently confines us to $T_{lum} < 50\%$ and $T_{sol} < 40\%$ for evaporated Ag films. Somewhat superior properties, with $T_{lum} \approx T_{sol} < 60\%$ are possible for other types of film deposition. If a much higher transmittance is required—as is frequently the case for architectural windows and glass façades—one must use multilayer coatings as considered in Section 4.3.4.

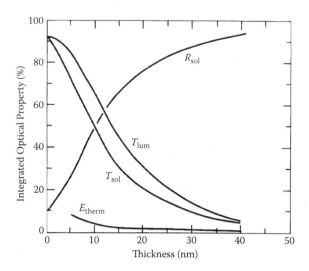

FIGURE 4.6 Integrated optical properties as a function of thickness computed from an ideal plane-parallel Ag film. (From C. G. Granqvist, *Materials Science for Solar Energy Conversion Systems*, edited by C. G. Granqvist, Pergamon, Oxford, U.K., 1991; chap. 5, pp. 106–167. With permission.)

4.3.2 Very Thin Metal Films Are Nanomaterials

A thin metal film does not grow uniformly on a glass pane or a polymer foil. Rather the growth resembles what happens when water vapor condenses on a fatty surface—though with the difference that the metal "droplets" are on the scale of nanometers to micrometers. This effect of nonuniform "wetting" has been demonstrated many times.

Information on thin-film growth and evolution has usually been gained from transmission electron microscopy applied to films deposited onto electron-transparent substrates such as thin carbon films, that is, onto substrates that are different from the ones actually used in fenestration. Figure 4.8, however, shows recent scanning electron micrographs for Au films made by sputter deposition onto glass and the images hence are appropriate for assessing what really happens for window coatings [12]. The thicknesses of the films are "equivalent" ones, that is, the thicknesses hypothetical uniform films would have; the term "mass thickness" is sometimes used. The initial deposition clearly leads to tiny metallic nuclei formed at certain sites on the substrate, and continued deposition makes these nuclei grow, which is expected to occur via diffusion of atoms or molecules over the substrate surface as well as by direct impingement of atoms or molecules. The metal "islands" that are

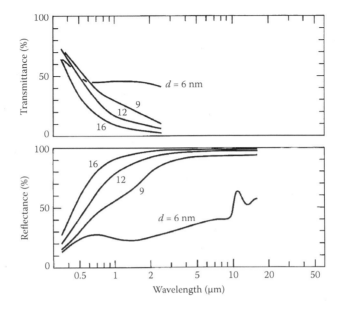

FIGURE 4.7 Spectral normal transmittance and near-normal reflectance measured for Ag films on glass. The film thickness is denoted d. (Replotted from E. Valkonen et al., *Solar Energy* 32 [1984] 211–222. With permission.)

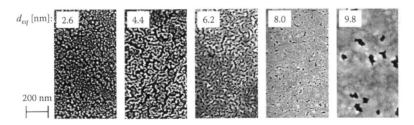

FIGURE 4.8 Scanning electron micrographs for Au films made by sputtering onto glass to the shown equivalent thicknesses d_{eq}. The gold appears bright and the uncoated parts of the substrate look dark. (From P. C. Lansåker et al., *Thin Solid Films* 518 [2009] 1225–1229. With permission.)

then formed can have shapes somewhat resembling ellipses, thus implying that the "islands" can be represented in geometric terms by a more or less well-defined eccentricity. Progressing deposition makes some of the "islands" touch and reorganize into larger and more irregular objects; this is normally referred to as "coalescence growth." The growing film then passes through what can be called "large-scale coalescence," signifying that a meandering metallic network of macroscopic extent is formed. This point in the growth sequence is referred to as the "percolation threshold." It is only then that metallic conduction can be measured across the film. Optically, the film begins to display plasmonic features [13] with increased IR reflection and an "effective" plasma frequency ω_p^* due to the still low average density of carriers and their difficulty in getting through the tortuous metallic network (so the effective mass m^* is high). Subsequent deposition makes the voids between the metallic paths become smaller and increasingly regular so ω_p^* rises. Again, an eccentricity may be useful for representing the geometry. The filling-in of the voids may not be a continuous process, though, since crystallization occurs concurrently in the Au film, as can be realized by comparing the films with thicknesses being 8.0 and 9.8 nm in Figure 4.8. A close-to-uniform layer with atomically flat grain tops may ultimately be formed.

Figure 4.9 shows the structural evolution of a metal film in a pictorial way and also points out the important fact that the thickness range gone through toward the growth of a continuous film depends on the method for making the films [14]. Thus, a gold film made by conventional evaporation reaches large-scale coalescence at a "critical" thickness d_c of ~10 nm, whereas a film made by evaporation under simultaneous bombardment with energetic ions (known as ion-assisted evaporation) has $d_c \approx 4$ nm. For the sputter-deposited Au film reported on in Figure 4.8, d_c was ~4 nm, too, which indicates the important effect of ion bombardment for a film undergoing growth via sputter deposition.

It is worth noting that the percolation thresholds in sputter-deposited Au and Ag films are specific and nonidentical despite the fact that both types of films go through the same structural evolution [13]. Thus,

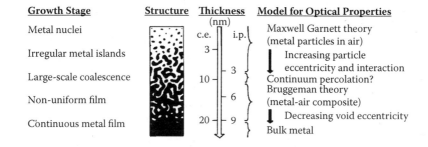

FIGURE 4.9   Survey over growth stages, structures, and thickness scales for thin Au films deposited onto glass by conventional evaporation (c.e.) and ion-assisted evaporation (i.e.). Optical models for representing the optical properties are shown. The Maxwell Garnett and Bruggeman theories for optical homogenization were discussed in Section 3.9. (From G. B. Smith et al., *J. Appl. Phys.* 59 [1986] 571–581. With permission.)

$f_c \sim 0.3$ for Au and $f_c \sim 0.5$ for Ag, with f being metal volume fraction or area coverage and c denoting the "critical" value. This difference ensues because highly mobile gold atoms grow together anomalously near the percolation transition. It implies one can get an Au film to behave as a metal with less material than in the case of Ag.

4.3.3 Toward a Quantitative Theoretical Model for the Optical Properties

We now endeavor to go one step beyond the purely empirical description of the optical data and see to what extent these data can be understood from models for nanomaterials, that is, we want to compare experimental data with predictions from the optical homogenization theories introduced in Chapter 3. Specifically, we will discuss some results from an investigation by the authors [14]. Figure 4.10 reports spectral transmittance and reflectance for Au films made by ion-assisted deposition. The transmittance has a plateau-like appearance except for

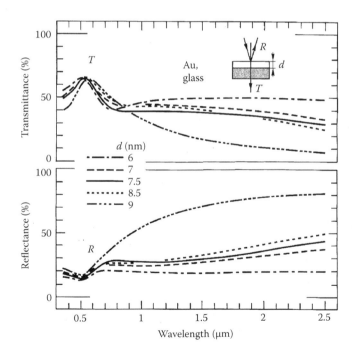

FIGURE 4.10   Spectral transmittance T and reflectance R for Au films made by ion plating and having the shown thicknesses d. (From G. B. Smith et al., *J. Appl. Phys.* 59 [1986] 571–581. With permission.)

the thickest film with $d \approx 9$ nm, which has a distinctively metallic-like performance with a transmittance that drops and a reflectance that increases monotonically with increasing wavelength in the IR. The value of R_{therm} exceeds 75% for $d > 7.5$ nm. But why are there plateaus in the optical data?

The optical homogenization theories can be modified to include the role of nonspherical voids represented by a depolarization factor L. The calculations used Equation 3.23 with $L_A = L$ and $L_B = 1/3$ and led to the data in Figure 4.11, indicating that the transmittance develops a plateau for the case of highly irregular voids specified by a large depolarization factor [15]. The analysis is not expected to be quantitative for the simple reason that the nanostructure of the film is highly irregular and not possible to capture accurately by one single parameter. Nevertheless, the theoretical model leads to an intuitive understanding of the relationship between the nanostructure and the characteristic optical data for the films. Further analysis leads us to expect that different homogenization theories are appropriate to the various growth stages, as illustrated schematically in the right-hand part of Figure 4.9. Recent work has made it possible to complement and extend the theoretical analysis, as discussed in Box. 4.1, and the last word on the optical properties of noble-metal-based films may not have been spoken.

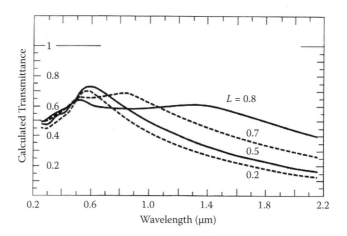

FIGURE 4.11 Computed spectral transmittance for an Au film with thickness $d = 9$ nm, filling factor $f = 0.2$, and the averaged optical properties represented by the Bruggeman theory. The structural entities (voids) are represented by the values shown of one depolarization factor L. (From G. A. Niklasson, C. G. Granqvist, *Appl. Phys. Lett.* 46 [1985] 713–715. With permission.)

BOX 4.1   A SPECIAL LOCAL "METAL-IN-METAL" PLASMONIC RESONANCE, FIRST IDENTIFIED IN GROWING THIN METAL FILMS

The "anomalous" spectral region at $\lambda \sim 1 \mu m$ in Figure 4.11 is due to absorption. An analogous anomaly has also recently been seen in nanoporous gold made by etching [16]. The large values of L used in Figure 4.11 showed that there was a special resonance, but its physical origins were unraveled only recently [17]. The optical effect is related to residual isolated islands that have not joined up with the surrounding percolating network; examples of this are circled in Figure B4.1.1.

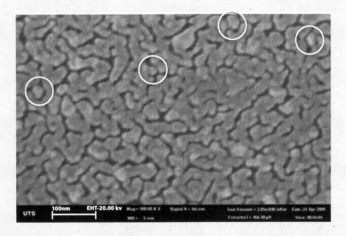

FIGURE B4.1.1   Percolating nanoporous Ag network containing a small number of nanoislands which are not electrically connected to the network (encircled). (From G. B. Smith, A. A. Earp, *Nanotechnology* 21 [2010] 015203 1–8. With permission.)

This new class of local surface plasmon resonance is extraordinarily strong because the matrix is a metal, not an insulator. To understand and model the optical response of this nanostructure, we can go back to the random unit cell approach in Section 3.9, modified as in Figure B4.1.2. The entity of interest is a metal nanoparticle with a very thin insulating shell, embedded in a conducting matrix. The equations for the core-shell structure introduced in Section 3.9 apply here if suitably modified. The conducting matrix surrounding the core-shell particle is porous, but it can be a fully dense metal, which brings forth particularly interesting results.

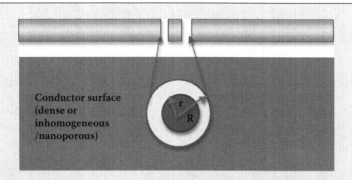

FIGURE B4.1.2   Schematic cross-sectional and plan views of a resonant entity being a core-shell structure. (From G. B. Smith, A. A. Earp, *Nanotechnology* 21 [2010] 015203 1–8. With permission.)

The structure shown is two-dimensional and applicable to thin films, including those in Figure B4.1.1. However, the principle can be extended to three-dimensional core-shell systems wherein the resonating entity is totally buried [17]. In this latter case one can speak of a "Babinet's principle structure" [4], and the terminology indicates that metal is replaced by insulator and vice versa. Babinet's structures have begun to attract attention in several areas of metamaterials science, in part because they can give enhanced responses to external conditions, in part because they are easy to engineer in thin metal films and metal foils, including with new patterns which lead to novel resonant and plasmonic transport properties. To amplify and shift the resonance, and hence show it very clearly, one can overcoat to fill the voids with oxide. The core-shell equations introduced in Section 3.9, modified for a Babinet's structure, fit the anomalous resonance very well with a distribution of shell thicknesses from 1.5 to 3.5 nm [17]. The resonance wavelength is very sensitive to small changes in shell thickness.

These metal-in-metal systems make it possible to engineer strong and very narrow absorption features within a reflector by rather simple means. Figure B4.1.3 shows data for a 40-nm-thick continuous Au film on glass and for such a film on top of a discontinuous Au layer with 12-nm-diameter nanoparticles coated with 2 nm of alumina. The measurements are done from the substrate side and show a prominent reflectance minimum in the NIR for the metal-in-metal system. These data indicate a new way for tuning solar properties and color in metal thin films.

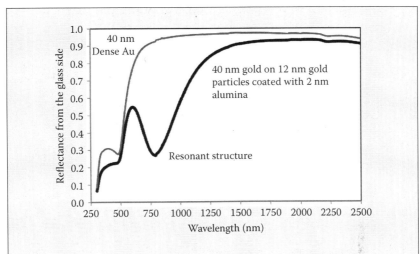

FIGURE B4.1.3 Spectral reflectance of an Au-based metal-in-metal system and a corresponding continuous Au film, as described in the main text. (From G. B. Smith, A. A. Earp, *Nanotechnology* 21 [2010] 015203 1–8. With permission.)

4.3.4 Multilayer Films for Spectrally Selective Windows

The low values of T_{lum} and T_{sol} in contiguous noble metal films having low values of E_{therm} are largely due to reflectance at their surfaces. Consequently, it is possible to increase the transmittance by adding layers that anti-reflect the metal. One is then led to dielectric/metal and dielectric/metal/dielectric multilayers. Dielectrics with high refractive indices—such as Bi_2O_3, In_2O_3, SnO_2, TiO_2, ZnO, and ZnS—give the largest enhancement of the transmittance. It is possible to optimize for low-E or for solar control trough proper choices of film thicknesses.

Figure 4.12 shows measured transmittance and reflectance spectra for TiO_2/Ag/TiO_2 films on glass [3]. These data are characteristic for what one can accomplish by using the three-layer design. It is concluded that $T_{lum} > 80\%$ and $E_{therm} \ll 20\%$ are obtainable for the two types of coatings. The solid curves pertain to the coating with maximum IR reflectance; it yields $T_{sol} \approx 50\%$ and $R_{sol} \approx 42\%$. The dotted curves correspond to $T_{sol} \approx 67\%$ and $R_{sol} \approx 26\%$.

Joint development of electrical and optical properties in three-layer films is of interest for a number of applications, and Figure 4.13 shows some recent data for ZnO/Ag/ZnO films with each of the ZnO films being 20 nm thick [18]. The sheet resistance $R_\square$ is given; it is defined by $R_\square = \rho/d$ where ρ is the resistivity. The stated transmittance corresponds

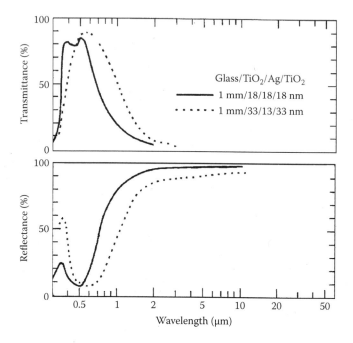

FIGURE 4.12   Spectral normal transmittance and near-normal reflectance measured for $TiO_2/Ag/TiO_2$ coatings on glass. Film thicknesses are shown. (From C. G. Granqvist, in *Materials Science for Solar Energy Conversion Systems*, edited by C. G. Granqvist, Pergamon, Oxford, U.K., 1991; chap. 5, pp. 106–167. With permission.)

to its maximum value. Clearly $d < 4$ nm yields poor electrical properties while $d > 8$ nm produces poor optical properties. The optimum is at $d \approx 6$ nm, where one finds $R_\square \approx 3\ \Omega$. The results are in basic agreement with those discussed above.

The optimum Ag thickness in large-scale manufactured multilayer coated windows is typically 10 to 11 nm, as pointed out in Section 3.7, which gives a value of T_{lum} approaching 90%. Two features are important for achieving the highest quality: to keep the Drude relaxation rate (Equation 3.8) in the Ag as low as possible and to minimize surface nanostructure. Doing this leads to sharper spectral transitions, more distinct colors, and lower values of E_{therm}. Silver grown on TiO_2 has a lower relaxation rate than on glass, and on ZnO it is lower still. This is so because the silver tends to grow more epitaxially on some substrates than on others, which allows electrons to pass more easily between neighboring Ag grains. Surface nanostructure can introduce losses by interface plasmons, as discussed above in Section 3.10, which reduces transmittance. These plasmons are exacerbated by adjacent thicker high-index

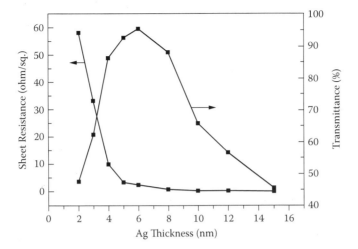

FIGURE 4.13 Sheet resistance and maximum transmittance as a function of Ag thickness in a three-layer structure comprising ZnO(20 nm)/Ag/ZnO(20 nm). (From D. R. Sahu et al., *Appl. Surf. Sci.* 252 [2006] 7509–7514. With permission.)

films [19,20], which localize the plasmon fields closer to the Ag. This is the likely reason the coatings in Figure 4.13 perform worse than those of optimized thin-film stacks in commercial products, such as those illustrated in Figure 4.14 below. The adjacent high-index layers in the latter coatings are either thinner or slightly separated from the Ag, and the Ag itself is very flat.

The examples above were thus selected to illustrate scientific principles rather than today's (2010) commercial products. In fact, Ag-based thin films for energy efficient fenestration are now highly optimized, and a very large number of products are available on the market. The production is of the order of 60 million m² per year, with specified thermal, solar, and luminous (including color) properties. The current technologies are proprietary, and the reader is directed to vendors' technical publications or comprehensive databases [5] for specific details to the extent such are given. Typical coatings go far beyond the three-layer design, and a "simple" low-E coating can have six layers comprising glass/TiO_2(15–20 nm)/ZnO(5–10 nm)/Ag(9–12 nm)/ITO(2–5 nm)/SnO_2(35–40 nm)/TiO_2(1–5 nm) [21]. An advanced solar control coating can have as many as nine layers, including two Ag films. The layer sequences around the silver must ensure impedance matching, just as in the simple case of a three-layer coating.

Figure 4.14 illustrates two sets of data for commercially available glass with coatings of these general types. The curves designate transmittance, reflectance from the film side of the glass, and reflectance from

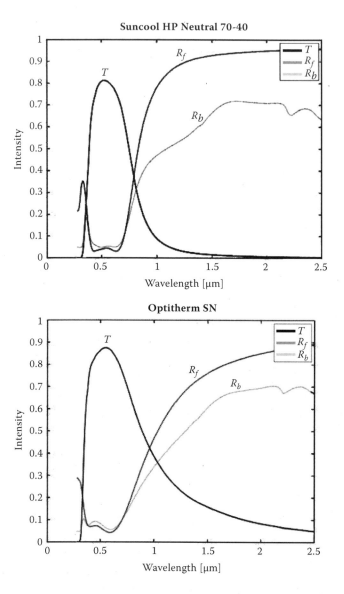

FIGURE 4.14 Spectral transmittance and reflectance from the front (film) side (R_f) and the back side (R_b) for two commercial window coatings. The upper set of data corresponds to $T_{lum} = 79.8\%$ and $T_{sol} = 40.9\%$; lower set of data corresponds to $T_{lum} = 86.7\%$ and $T_{sol} = 57.5\%$. (From International Glazing Database [2009], Lawrence Berkeley National Laboratory; http://windows.lbl.gov/software/window/window.html.)

the back side (so that the light has passed through the glass before being reflected). It is remarkable that these multilayer window coatings are prepared with thickness control approaching atomic precision in high-performance production environments handling glass sheets up to 30 m² in size. Figure A1.4 in Appendix 1 shows such a machine. Extremely high cleanliness of the starting glass is essential, which means multiple washes and rinses.

Section 4.1.2 contained a brief discussion of multilayer polymer foils with strong spectral selectivity, and at this point one may wonder why a similar approach is not used to create solar-control coatings based on dielectric multilayer thin films with alternating high and low refractive indices. The reason basically is cost, and coating large surfaces—such as window panes—with a very large number of layers is time consuming and, hence, costly. Furthermore, window coatings normally should be able to provide thermal insulation and have high reflectance for $3 < \lambda < 50$ µm which is far too wide a range to be accomplished by dielectric multilayers. Thus, one cannot avoid a metallic-type coating, either of the kind described above or based on a heavily doped semiconductor as discussed next.

4.4 SPECTRAL SELECTIVITY OF OXIDE-SEMICONDUCTOR-BASED FILMS

There are a number of oxides with band gaps which are wide enough so that they are transparent for luminous and solar radiation and which also allow doping to sufficiently high levels, in practice up to several percent, so that they can achieve good electrical conductivity and hence IR reflectivity. They are often referred to as "transparent conducting oxides" or TCOs. In practice, these are oxides based on In, Zn, Sn, and combinations of these. Particularly good properties have been obtained with In_2O_3:Sn, ZnO:Al, ZnO:Ga, and SnO_2:F. The first of these is often referred to as indium tin oxide or ITO and can serve as a model example for this class of materials. A thorough discussion of these oxides and of their achievable properties has been given recently [22], and a review of their performance with regard to solar energy–related applications is available as well [9]. A recently discovered alternative to the oxides mentioned is TiO_2:Nb [23,24].

Why are these oxides conducting? Figure 4.15 illustrates the origin of the conductivity in a highly schematic way for the case of ITO [3]. The undoped In_2O_3 has a regular arrangement of oxygen (O^{2-}) and indium (In^{3+}) ions. When dopants are added, such as tin (Sn^{4+}), they can replace some of the indium. If the tin density is sufficient, each dopant atom can be singly ionized by giving off a free electron which, together with

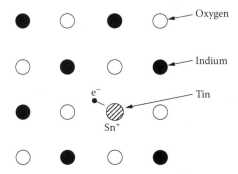

FIGURE 4.15 Oversimplified crystal structure and doping model for ITO. The actual crystal structure of In_2O_3 is complicated. (From C. G. Granqvist, in *Materials Science for Solar Energy Conversion Systems*, edited by C. G. Granqvist, Pergamon, Oxford, U.K., 1991; chap. 5, pp. 106–167. With permission.)

other free electrons with the same origin, renders the material metallic. The onset of metallic properties occurs at a doping level of ~10^{-19} cm^{-3}. Electron densities up to ~10^{21} cm^{-3}, or somewhat above this value, can be obtained by maximum doping. If one tries to push the doping to even higher levels, the result will be that the dopant atoms start to form strongly absorbing metallic clusters so that transparency is lost. The minimum resistivity is ~10^{-4} Ωcm, or slightly below, and similar data have been recorded for ZnO-based films. We emphasize the simplified nature of the doping model outlined here. In fact, In_2O_3 has a complicated crystal structure with 80 atoms per unit cell, and there are two nonequivalent In^{3+} positions. Also not every tin atom is active as a donor center [25].

A noble metal such as Ag or Au has an electron density of ~6×10^{22} cm^{-3}, that is, much higher than the electron density for the doped oxides, and their resistivity is ~2×10^{-6} Ωcm. These differences imply that a good low-E coating made of a doped semiconductor must be much thicker than a metal-based low-E coating with comparable properties. In fact, the oxide coating must have a thickness of ~200 nm or higher, as we will find below. This means that details of the initial film growth are not at all as important as they are for metal-based window coatings. Another consequence of the fact that the oxide films must have thicknesses comparable with the wavelengths of visible light is that minor variations of this thickness—which may be hard to avoid in practical large-area coating—can lead to optical effects such as iridescence (somewhat related to the exhibition of color in a soap bubble).

Another important difference between noble-metal-based and doped-semiconductor-based films is that the latter are much more durable and can be used even on the exterior surfaces of window panes. The term "hard coat" is sometimes used to distinguish them from the metal-based "soft coats." In particular, SnO_2:F films are used for practical fenestration; these coatings are normally made by spray pyrolysis directly onto the hot surface of float glass as it emerges from its solidification on the liquid tin bath. A similar technique is used to coat the windows used in cockpits of commercial aircraft (to allow electrical heating and maintain impact strength); this glass usually appears iridescent when viewed from the outside, which is a consequence of film thickness variations.

4.4.1 Some Characteristic Properties

Thin films of transparent and electrically conducting oxides require well-defined deposition conditions in order to have optimized properties. These features are illustrated next by showing some typical data for ITO made by reactive e-beam evaporation [8].

The left-hand panel of Figure 4.16 shows transmittance and reflectance of a 0.4-μm-thick ITO film produced on a substrate heated to ~300°C. The values of T_{lum} and R_{therm} are high. Strong interference effects show up as oscillations in the optical data. The onset of reflectance occurs rather steeply in the IR; it takes place at a wavelength that is somewhat too large for the coating to work well for solar control purposes. The

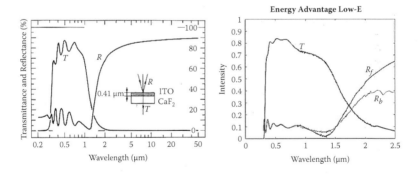

FIGURE 4.16 Spectral transmittance T and reflectance R for an ITO film made by e-beam evaporation onto a substrate at ~300°C according to the configuration shown in the inset (left-hand panel) and for float glass with a commercial SnO_2-based coating characterized by T_{lum} = 83.1% and T_{sol} = 71.3% (right-hand panel). The reflectance for the latter film was measured from the front (film) side (R_f) and from the back side (R_b). (Results for ITO from I. Hamberg, C. G. Granqvist, *J. Appl. Phys.* 60 (1986) R123–R159. With permission; results for SnO_2 from International Glazing Database [2009], Lawrence Berkeley National Laboratory; http://windows.lbl.gov/software/window/window.html.)

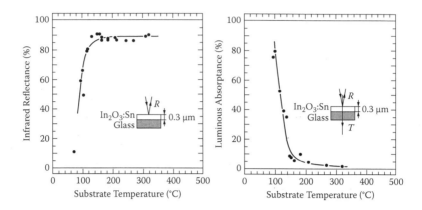

FIGURE 4.17 Infrared reflectance R at $\lambda = 10$ μm (left-hand panel) and luminous absorptance (right-hand panel) for an ITO film made by e-beam evaporation according to the configuration shown in the inset. (From I. Hamberg, C. G. Granqvist, *J. Appl. Phys.* 60 [1986] R123–R159. With permission.)

data are comparable with those for commercial low-E-coated float glass, with the coating based on doped SnO_2, as shown in the right-hand panel of Figure 4.16 where the presentation is analogous with the one for commercially available Ag-based products in Figure 4.14 [5].

The importance of a hot substrate for making good transparent conductors and IR reflectors can be inferred from Figure 4.17, where it is evident that a temperature of about 150°C is needed to produce ITO with a low visible absorptance and a high IR reflectance. It is noteworthy that A_{lum} can be as small as ~1% if the deposition takes place at a substrate temperature of ~300°C. The resistivity of those films was ~2 × 10^{-4} Ωcm.

The fact that A_{lum} is so low makes it of interest to explore the effect of antireflection treatments. This has been done by coating an ITO film with MgF_2; the top layer fulfills the requirements for a quarter-wave antireflection coating and, as expected, the transmittance is significantly increased. In fact, Figure 4.18 shows that the transmittance is higher than that of the uncoated glass over almost the entire luminous spectrum. Hence, there is no ground for believing, as is sometimes done, that a low-E coating necessarily degrades the luminous properties of a window. The MgF_2 layer enhances the thermal emittance by a few percent, that is, only marginally.

4.4.2 Typical Nanostructures of ITO Films

We first consider the structure of ITO films such as those discussed above. Figure 4.19 shows a transmission electron micrograph of a film

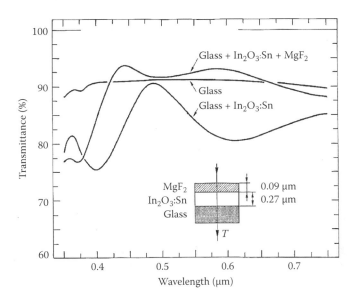

FIGURE 4.18   Spectral transmittance T for an ITO film made by e-beam evaporation as deposited onto glass and after coating with an antireflecting MgF$_2$ layer. (From I. Hamberg, C. G. Granqvist, *J. Appl. Phys.* 60 [1986] R123–R159. With permission.)

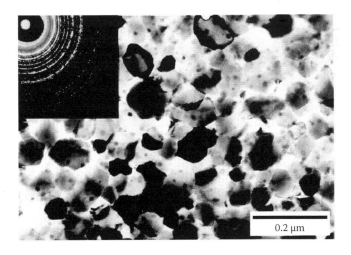

FIGURE 4.19   Transmission electron micrograph of a 40-nm-thick ITO film made by e-beam evaporation onto an electron-transparent amorphous carbon film at 230°C. Inset depicts part of an electron diffraction pattern. (From I. Hamberg, C. G. Granqvist, *J. Appl. Phys.* 60 [1986] R123–R159. With permission.)

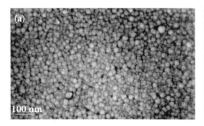

FIGURE 4.20 Scanning electron micrographs for an ITO film sputter deposited at ambient temperature and annealing posttreated at 200°C (left-hand panel) and for a similar film sputter deposited onto a substrate kept at 200°C (right-hand panel). (From U. Betz et al., *Surf. Coating Technol.* 200 [2006] 5751–5759. With permission.)

made at a substrate temperature of 230°C. It is seen to consist of crystallites with an average size of the order of 100 nm. This structure is typical for films with the highest conductivity and transmittance.

Generally speaking the nanostructure of ITO films is critically dependent on details of the deposition technology. Figure 4.20 illustrates this fact clearly for ITO films made by sputter deposition [26]. The left-hand panel shows a scanning electron micrograph for a film deposited onto a substrate at room temperature and subsequently annealing post-treated at 200°C, and the right-hand panel applies to a film that was sputtered onto a substrate at 200°C. A first assumption would be that the films should have similar nanostructures, but clearly this was not the case.

The granular nature apparent in Figure 4.20 can be even more pronounced for ITO films made by sol gel deposition using dispersions of nanoparticles. This kind of technique is of interest for some applications since it may open avenues toward simple and cheap deposition of patterned surfaces. As found from a recent study [27], the sol-gel-prepared films consisted of clusters of individually connected nanoparticles. Each of the nanoparticles had a resistivity of ~2 × 10^{-4} Ωcm, that is, as low as the resistivity of ITO films made by evaporation or sputtering under optimum conditions. However, clusters of such particles were weakly coupled to one another, so the overall resistivity of the nanoparticle-based films was not better than ~10^{-2} Ωcm. This clustering also led to a poor value of E_{therm}.

For some applications it is of interest to intentionally produce a nanostructure in order to create controlled light scattering. Such transparent conductors, sometimes referred to as "milky," can be applied as electrodes on thin-film solar cells and have the effect of increasing the optical path length of incident solar radiation and thereby the absorption of this radiation. Recent work has been made in particular on ZnO:Al prepared by sputtering and subsequent post-treatment in diluted HCl

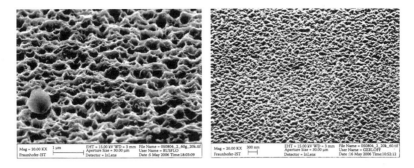

FIGURE 4.21   Scanning electron micrographs for a sputter-deposited ZnO:Al film before (left-hand panel) and after (right-hand panel) etching in 0.5% HCl. (From F. Ruske et al., *Thin Solid Films* 515 [2007] 8695–8698. With permission.)

[28]. Figure 4.21 illustrates the roughening that is then accomplished and that the treated film displays unevenness on a scale comparable to the wavelength of solar light. Films prepared under optimum conditions displayed a haze—defined as the diffuse transmittance divided by the total transmittance—of ~30% for λ = 550 nm. The haze then dropped off steeply toward longer wavelengths.

4.4.3 Theoretical Models for the Optical and Electrical Properties of ITO Films

Different deposition technologies lead to very different nanostructures, which makes it obvious that the optical and electrical properties depend on the deposition parameters. Hence, theoretical models of these properties are not universally true for all types of films. In order to develop a theoretical framework we first consider optimized e-beam-evaporated films, such as those reported on in Figures 4.16–4.19 above. The theory has been worked out in considerable detail [8] and the results in Box 4.2, which uses a somewhat simplified approach, show that theory and experiments are in excellent agreement [29].

Figure 4.22 illustrates the essentials of the physics that governs the optical and electrical properties of ITO films. At the shortest wavelengths, well beyond those of relevance for visible light and solar energy, the material is absorbing as a result of electron transitions across the fundamental band gap E_g. When the material is electrically conducting, the free electrons block the lowest states in the conduction band, implying that E_g becomes shifted toward larger wavelengths. The shape of the band gap is somewhat blurred (the band edge is logarithmic) by the

BOX 4.2   THEORETICAL MODELS FOR THE OPTICAL PROPERTIES OF ITO AND OTHER TRANSPARENT CONDUCTING OXIDES

The dielectric function of ITO is represented by additive parts from the undoped In_2O_3 lattice and the free electrons according to

$$\varepsilon(\omega) = \varepsilon^{In2O3} + i/\varepsilon_0\omega\rho(\omega) \qquad (B4.2.1)$$

Here $\rho(\omega)$ is the dynamical resistivity due to the free electrons and ε^{In2O3} contains contributions from valence electrons and polar phonons of In_2O_3. This particular formulation is convenient, as will become obvious later, since it clearly exhibits the role of electron scattering. The free-carrier part is of the greatest interest since it is responsible for the electrical conductivity and the IR reflectivity. We first use the Drude theory that was earlier introduced in Section 3.4. It can be written

$$\rho^{Drude} = 1/\varepsilon_0\omega^2_p\tau_e - i\omega/\varepsilon_0\omega^2_p \qquad (B4.2.2)$$

where ω_p is plasma frequency as given by Equation 3.7 and τ_e is a relaxation time.

One may get an approximate model for $\varepsilon(\omega)$ in this way, but the approach nevertheless is unsatisfactory, because τ_e is introduced as a constant fitting parameter without any connection to the actual mechanism for the electron scattering. However, it is possible to calculate $\tau_e(\omega)$ for a variety of scattering mechanisms in considerable detail [30]. The one of particular interest here concerns scattering of free electrons against ionized impurities; this type of scattering indeed is unavoidable for ITO and similar transparent conducting oxides, because the free electrons originate from the ionization of the tin atoms, so ionized impurity scattering represents a baseline giving the lowest electrical resistivity and the highest IR reflectivity (for frequencies below the plasma frequency). Other types of scattering, if important, can only degrade the properties.

Figure B4.2.1 shows that optimized ITO films come close to the theoretically predicted properties for a material having nothing but the unavoidable ionized impurity scattering. The upper panel reports experimental data of $\rho(\omega) = \rho_1(\omega) + i\rho_2(\omega)$ and the lower panel gives computed results using experimental parameters for N_e and m^*. The two sets of data are in good agreement, which shows that, indeed, ionized impurity scattering gives a good description of the experimental results. The data on $\rho_1(\omega)$ are particularly interesting because they are directly related to the electron scattering. Specifically, $\rho_1(\omega)$ is constant, and consistent with the dc resistivity, at energies below $\hbar\omega_p$ and falls off according to $\omega^{-3/2}$ above $\hbar\omega_p$. Other power dependencies are valid for other scattering mechanisms, and hence the assignment to ionized impurity scattering is unique. The drop of $\rho_1(\omega)$ at high energies causes other scattering mechanisms to become important and, in particular, grain boundary scattering can then start to influence the optical properties.

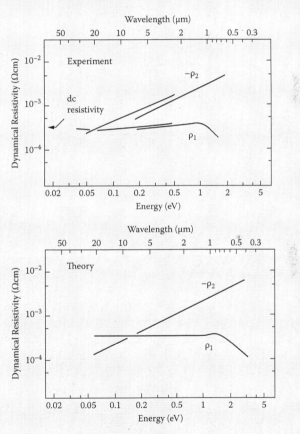

FIGURE B4.2.1   Experimental (upper panel) and calculated (lower panel) data on the real and imaginary contributions to the dynamical resistivity $\rho = \rho_1 + i\rho_2$ for e-beam-deposited ITO films as a function of photon energy and wavelength. The dc resistivity is shown. (From I. Hamberg, C. G. Granqvist, *Appl. Phys. Lett.* 44 [1984] 721–723. With permission.)

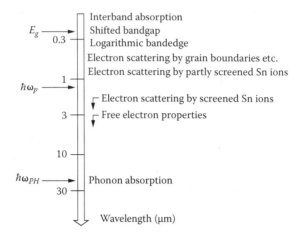

FIGURE 4.22 Overview of the fundamental physical properties and mechanisms for high-quality films of ITO and similar materials. (From I. Hamberg, C. G. Granqvist, Evaporated Sn-doped In_2O_3 films: Basic optical properties and applications to energy efficient windows, *J. Appl. Phys.* 60 [1986] R123–R159. With permission.)

electron scattering. Electron scattering by ionized impurities, screened by the free electron plasma, dominate the properties in most of the energy range from the plasma energy $\hbar\omega_p{}^*$ and downward, although grain boundary scattering can be of importance at the highest energies, including in the luminous part of the spectrum. Phonon absorption at the energy $\hbar\omega_{PH}$ lies far into the thermal IR; its influence tends to be negligible owing to the dominant influence of the free electrons.

The effects of the specific nanostructure are rather weak in ITO films of the highest quality, which is due to the fact that the individual crystallites are much larger than the distance between the ionized impurities. If the crystallites are smaller, as illustrated for specific films in Figure 4.20, the situation can be different and grain boundary scattering can be more dominant than ionized impurity scattering. Whether or not ITO, or similar heavily doped oxides, should be regarded as "nanomaterials" is somewhat of a matter of definition and—strictly speaking—depends on the quality of the samples.

4.4.4 Computed Optical Properties

Figure 4.22 outlined the backbone of a fully quantitative theory for ITO films that was also discussed in Box. 4.2. It is illustrative to use this theory to calculate $T(\lambda)$ and $R(\lambda)$ as a function of N_e. This is done in

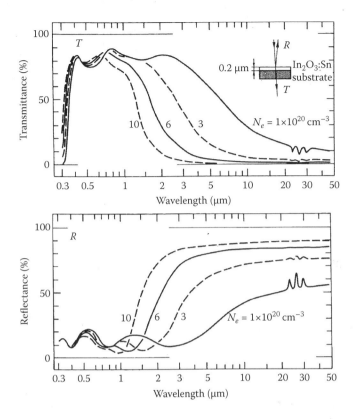

FIGURE 4.23   Spectral transmittance and reflectance computed from a quantitative theoretical model for ITO as a function of the electron density N_e. (From I. Hamberg, C. G. Granqvist, *J. Appl. Phys.* 60 [1986] R123–R159. With permission.)

Figure 4.23, where the largest value of N_e lies above the range of what is possible to achieve in practice. The main result is that the IR reflectance sets in at a wavelength that scales with N_e. Phonon effects, visible as three peaks in the thermal IR, become entirely swamped as N_e is increased.

The spectral data allow calculations of various integrated optical properties. Figure 4.24 shows results for T_{sol} as a function of film thickness for three values of E_{therm} with N_e as parameter. As discussed in Section 4.2, there are different heat transfer mechanisms in a window and, unless one deals with "vacuum windows," it does not make much sense to try to decrease E_{therm} to the extreme. Putting $E_{therm} = 0.2$ limits T_{sol} to 78% and this is obtained for a 200-nm-thick film. Setting $E_{therm} = 0.15$ limits T_{sol} to 70%, and allowing $E_{therm} = 0.25$ enhances T_{sol} marginally to 80%. The shaded region in Figure 4.24 pertains to

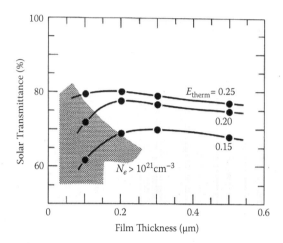

FIGURE 4.24 Solar transmittance as a function of film thickness computed from a quantitative theoretical model for the optical properties of ITO. Data are shown for three values of the thermal emittance E_{therm}. Shaded region corresponds to electron density N_e larger than 10^{21} cm^{-3}. (From I. Hamberg, C. G. Granqvist, *J. Appl. Phys.* 60 [1986] R123–R159. With permission.)

$N_e > 10^{21}$ cm^{-3}, which is not practically achievable. Antireflection treatments can be used to obtain a solar transmittance in excess of that in Figure 4.24 as already shown in Figure 4.18.

4.5 SPECTRAL SELECTIVITY: NOVEL DEVELOPMENTS FOR FILMS AND FOILS

There are many emerging techniques to achieve spectral selectivity. Next, we discuss nanomeshes and nanotubes based on metal and carbon, foils with conducting nanoparticles, possibilities to use metal films with designed nanoholes, and photonic crystals. There are still other ways. One of them uses films of poly(3,4-ethylenedioxytiophene):poly(styrene-sulfonate), known as PEDOT:PSS. Their electrical properties are not as good as for ITO but can be better than for carbon nanotubes. The transmittance can be large in the luminous range but drops in the NIR [31].

4.5.1 Silver-Based Nanowire Meshes

A new way to make transparent conducting window films that is radically different from those discussed above is to start from nanowires and

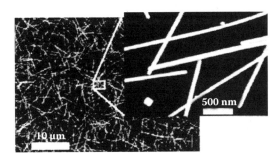

FIGURE 4.25   Scanning electron micrographs of Ag nanowire meshes. (From J.-Y. Lee et al., *Nano Lett.* 8 [2008] 689–692. With permission.)

deposit those onto a substrate so that an electrically conducting mesh is formed. This is a novel approach and a direct manifestation of the potential of nanotechnologies for "green" applications. Recent work has demonstrated that silver nanowires with lengths of ~10 μm and diameters of ~100 nm can be produced in large quantities via potentially inexpensive reduction of liquid Ag nitrate [32]. After deposition and annealing to promote electrical conduction between individual rods, a structure such as the one in Figure 4.25 can be observed.

Figure 4.26 shows the high spectral transmittance that can be obtained in Ag nanomesh films with a low resistance, specifically being

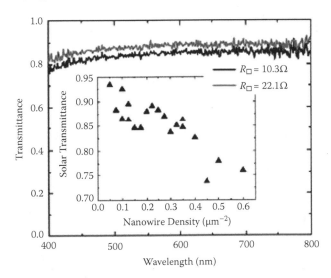

FIGURE 4.26   Spectral transmittance of Ag nanowire meshes with two values of the resistance. (From J.-Y. Lee et al., *Nano Lett.* 8 [2008] 689–692. With permission.)

10.3 and 22.1 $\Omega/\square$. The data compare very well with those for Ag-based multilayer films in Figure 4.13. The meshes consist of structural units that are large enough to cause significant scattering, and ~20% of the visible light is scattered at angles exceeding 10°. This haze is a severe limitation for most windows in buildings, whereas it is insignificant, or can even be advantageous, for applications related to solar cells and LEDs.

4.5.2 Carbon Nanotubes and Graphene

Carbon can form an infinite number of nanostructures, some of which are illustrated in Figure 4.27. Graphite has a layered structure such as the one to the right. If the individual layers are separated, one gets graphene. Carbon can also form nanotubes, as indicated in the middle, and nanoparticles consisting for example of 60 atoms in a C_{60} unit (also known as a buckminsterfullerene molecule or a "buckyball").

An emerging type of transparent conducting films, with possibly even larger potential than the Ag nanowires, uses meshes based on carbon nanotubes. Single-walled carbon nanotubes, which can have lengths up to several centimeters, consist of one layer of hexagonal graphite lattice rolled to form a seamless cylinder with a radius up to a few nanometers. The nanotubes are of two types: metallic and semiconducting

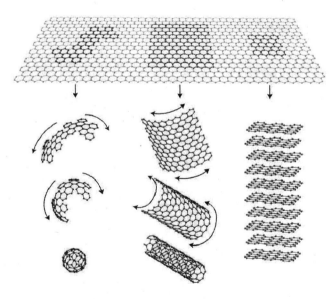

FIGURE 4.27 Schematic illustration of carbon-based nanostructures of different types and of their formation. (From A. K. Geim, K. S. Novoselov, *Nature Mater.* 6 [2007] 183–191. With permission.)

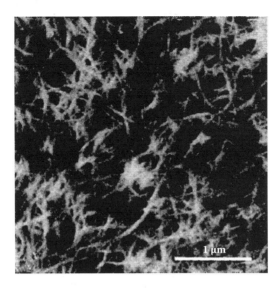

FIGURE 4.28   Transmission electron micrograph of a film comprised of carbon nanotubes. (From Y.-X. Zhou et al., *Appl. Phys. Lett.* 88 [2006] 123109 1–3. With permission.)

with conductivities comparable to those of Cu and Si, respectively. The individual nanotubes are electrically weakly coupled to one another. There are several different methods for preparing such nanotubes on substrates, including solution-based techniques such as dipping, spraying, spin coating, and printing. The ensuing layer is a mesh of nanotubes; a typical example is shown in Figure 4.28 [34]. Continuous and scalable manufacturing is feasible [35].

The optical transparency of carbon nanotube meshes can be high in the visible and IR, as apparent from Figure 4.29 pertaining to a film with a resistance of 120 $\Omega/\square$. It is apparent that the properties cannot yet fully match those of ITO and the best metal-based window films, but advantages regarding cost can outweigh this limitation depending on application. In the future, nanotube layers with a higher fraction of the metallic variety, and perhaps means to diminish the barriers between the individual tubes, conceivably will lead to better performance. A particular feature is the possibility to have large transparency even in the thermal infrared range [36].

Another possibility to make transparent conducting films is offered by graphenes, that is, two-dimensional graphite sheaths. These sheaths can be produced via mechanical exfoliation by repeated peeling of highly ordered graphite. The data obtained so far point at the feasibility of using graphenes [37], but the optical and electrical performance data are still far from those of the other coatings discussed above.

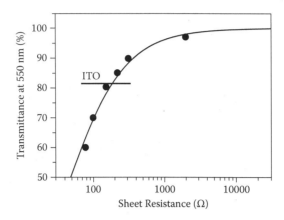

FIGURE 4.29   Transmittance at a mid-luminous wavelength versus sheet resistance for carbon nanotube films (dots). The horizontal bar denotes typical data for ITO films. (After Y.-X. Zhou et al., *Appl. Phys. Lett.* 88 [2006] 123109 1–3. With permission.)

4.5.3 Foils with Conducting Nanoparticles

Narrow-band surface plasmon resonances in conducting nanoparticles were introduced in Section 3.10. These resonances allow a simple and practical approach to solar control if they are used in polymer foils [38], and foils of this type—based on PVB as well as PET—have been applied commercially. Conducting nanoparticles, with little effect on T_{lum} while blocking in the NIR, are needed and the sizes of the particles must be below ~50 nm as they otherwise create haze and destroy the possibility of giving a clear view in transmittance. Figure 4.30 shows spectral absorption coefficient for a polymer with a refractive index of 1.5 and containing 0.3% by weight of a number of readily available conducting nanoparticles, and illustrates the sharp surface plasmon resonances that are then manifest. Clearly, the resonances can occur anywhere in the wavelength range for solar radiation, depending on the type of nanoparticles.

It is evident that metallic nanoparticles, such as TiN and ZrN, have too high plasma frequencies to yield a large value of T_{lum}, but they can produce distinct colors of glazings, as illustrated for Roman gold-doped glass in Figure 3.1 (and as will be seen also in heat reflecting paints; cf. Chapter 5). In stark contrast, commonly available transparent conductors such as ITO in nanoparticle form clearly have their absorption peaks well beyond the wavelengths for most of the IR in the solar radiation (i.e., 0.7 $< \lambda < 1.1$ μm). Foils with this type of nanoparticles look clear, and some weak damping of T_{sol} takes place only by the leading edge of the plasmon resonance. The best material so far for blocking NIR solar radiation in

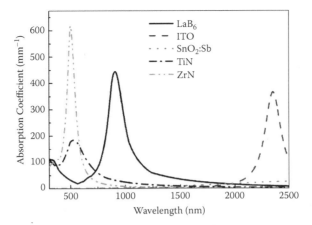

FIGURE 4.30   Spectral absorption coefficient for a polymer containing 0.3% by weight of the shown nanoparticles. Only a fraction of the initial part of the resonance due to SnO_2:Sb is shown.

transparent foils is LaB_6 [39] for which the resonance is centered at $\lambda \approx 1$ μm as shown in Figure 4.30. This material is excellent for glare reduction; it displays some enhancement of the coloration when the nanoparticle level is increased from zero to 0.03%. Figure 4.31 shows that very small amounts of LaB_6 are needed to significantly decrease T_{sol}, while T_{lum} is not affected to any major extent. ITO nanoparticles can serve as a replacement for LaB_6, but an order of magnitude higher nanoparticle contents must then be used to get the same blocking of the NIR solar radiation.

Transparent conducting oxide films can be made with doping levels from zero to a value exceeding 10^{21} cm^{-3}, as discussed at length in Section 4.4 above, and hence the plasma frequency can be tuned to a desired position for blocking NIR solar radiation in nanoparticle-containing foils. Calculations for 0.7-mm-thick PVB foils with 0.05 wt% of optimized ITO and ZnO:Al point at possibilities to achieve properties even better than those for LaB_6.

4.5.4 "Holey" Metal Films

An optically thick metal film with subwavelength holes can show "anomalous" optical transmission, that is, the transmission can be substantially larger than the area fraction that is occupied by the holes [40]. There is much interest in this novel phenomenon today, and we include some information here partly for completeness and partly because there may be a relationship to the optical data for thin metal films including holes discussed in Section 4.3.3 above.

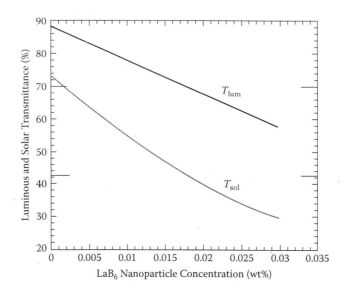

FIGURE 4.31   Luminous and solar transmittance as a function of LaB$_6$ nanoparticle concentration for 0.7-mm-thick doped PVB foils. (From S. Schelm, G. B. Smith, *Appl. Phys. Lett.* 82 [2003] 4346–4348. With permission.)

The effect is illustrated in Figure 4.32 showing transmittance through gold films with the indicated nanostructures; the holes are 0.25 μm in diameter and have a lattice constant of ~2 μm [41]. An "extraordinary" transmittance peak centered at λ ≈ 615 nm stands out clearly. Peak transmission levels up to 39% have been reported. Multiple transmission bands can be created with periodic arrays of nanohole apertures, and maxima and minima of transition spectra can be suppressed by adequate nanohole arrangements so that the transmittance can be tailored to different needs. The transmittance may not be sufficient for most energy-related applications, and the "holey" films may not allow visual indoors–outdoors contact, but the "anomalous" transmission nevertheless deserves some attention here because it can affect T_{lum} and T_{sol} for noncontinuous metal films.

4.5.5 Photonic Crystals

Also for completeness we now consider nanostructures based on arrays of identical oxide or polymer nanoparticles. They are known as "photonic crystals" and can be used for precise spectral control in either windows or mirrors. Particle arrays yield two- or three-dimensional

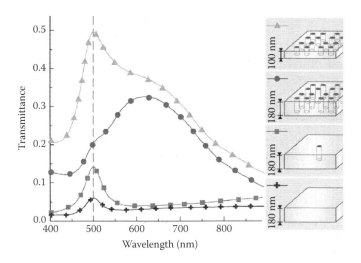

FIGURE 4.32   Spectral wide-angle transmittance through Au films with the shown configurations. The subwavelength hole arrays are seen to lead to extraordinary transmittance in a band centered at $\lambda \approx 615$ nm. (From H. Gao et al., *Nano Lett.* 6 [2006] 2104–2108. With permission.)

photonic crystals while multilayer films—such as the ones producing the data in Figure 4.3—can be described as one-dimensional photonic crystals. Lattice spacings or layer thicknesses must be repeatable to a high degree of accuracy, though in some applications a deliberate slight randomness in spacing may also be of use. These systems can act as near-perfect all-dielectric mirrors at wavelengths that cannot propagate inside the structure at the forbidden "band gaps" for photons. Such effects are well known for electrons in crystalline solids and result from destructive interference between electron waves. Figure 4.33 depicts a photonic band gap structure and illustrates optical interference leading to reflection depending on viewing angle [42,43]. Figure 4.34 shows viewing-angle-dependent colors and also indicates that there is a dependence on particle size, that is, on the lattice constant [42,43].

4.6 OPTIMIZED ANGULAR PROPERTIES: THE ENERGY EFFICIENCY THAT IS POSSIBLE

Spectral selectivity can give well-specified advantages with regard to energy efficiency for a window, as discussed in Section 4.2 above. The choice of appropriate angular performance for the fenestration is important, too, but the advantages cannot be as sharply defined.

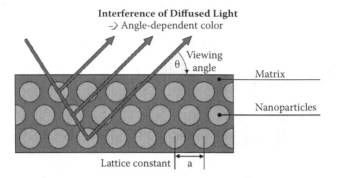

Interference of Diffused Light
-> Angle-dependent color

FIGURE 4.33   Interference of light in a photonic crystal consisting of an array of polymer nanoparticles embedded in a transparent crystal. (R. J. Leyrer, personal communication to one of the authors [GBS].)

FIGURE 4.34   *A color version of this figure follows page 200.* Color shifts, caused by a range of viewing angles, in reflected light from a photonic crystal such as the one depicted in Figure 4.33. Data are shown for different particle sizes. (R. J. Leyrer, personal communication to one of the authors [GBS]. Figure courtesy of BASF—The Chemical Company.)

Why are angular properties important for providing energy efficiency of windows? The reason is that the angles appropriate for viewing usually are not the same as the one from which the solar radiation emerges. Normal indoors–outdoors visual contact through a window is most often along a more or less horizontal line-of-sight. Of course, this is not always the case, and, for example, a person who is indoors may wish to look at the sky before taking a walk in order to decide whether or not to bring an umbrella. Or someone high up in a tall building may want to see what happens on the ground just outside the building. But these cases are not the most typical. Solar irradiation comes for most of the day from a small element of solid angle high up on the vault of heaven, and if this radiation penetrates freely into a building it may add significantly to the cooling load, and it would then be energetically advantageous to remove at least the direct impingement of solar radiation. This is what one can accomplish by the use of traditional techniques such as overhangs, awnings, and the like. But the problem is then that the basic function of the window—to give unimpeded indoors–outdoors contact—is severely eroded.

Daylight management is another bonus for glazings with properly chosen angular properties. There is usually plenty of daylight from the very large solid angle of sky not occupied by the sun and its circumsolar bright halo. Using this indirect radiation and blocking most of the direct solar rays can eliminate glare, which is a common reason for people to close blinds and turn on lights. Another type of glazing with carefully designed angular properties can be very useful when employed on a roof for collecting low-angle light in the morning or in the winter but rejecting much of the solar heat in the middle of the day and in mid-summer.

The fact that T_{lum} and T_{sol} often pertain to different angles leads naturally to the consideration of coatings with angular dependent properties. For the most common case of a vertical window, a simple coating such as a metal film will be most transparent at normal incidence simply because the thickness through which the optical radiation has to penetrate is then at a minimum. Hence, the desired angular properties are easily reached to some extent.

If the window is not vertical, one may want a higher transmittance at an off-normal angle of incidence (along a horizontal line-of-sight, say) than at normal incidence. This can be accomplished with multilayer structures having more than one metallic film and, essentially, depends on destructive and constructive interference of light reflected between the two metal films. As noted in Section 4.3.4, solar control films are often of this general type. In order to get an idea of the orders of magnitude that are involved, we can look at some computed data for a five-layer coating of the type discussed above, specifically incorporating two 12-nm-thick Ag films embedded between three SiO_2 films of equal thickness. It is straightforward to compute the optical properties of this structure using formulas for thin-film optics [4]; of course these properties are symmetrical with regard to the normal to the window pane. Figure 4.35 shows that T_{sol} can display very different angular variations, depending on the thickness of the SiO_2 layers [44]. If this thickness is 120 nm, T_{sol} falls off monotonically with increasing incidence angle θ, whereas the behavior is more complicated when the SiO_2 films are 170 nm thick. In the latter case, T_{sol} is about twice as high for $\theta = 45°$, and three times as high for $\theta = 60°$, compared to the case for normal incidence. It is clear that the use of proper angular properties will lead to energy efficiency, but the magnitude of this efficiency depends on many external param-

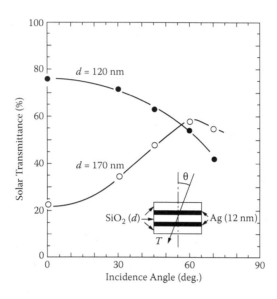

FIGURE 4.35   Solar transmittance for thin films of $SiO_2/Ag/SiO_2/Ag/SiO_2$ as a function of SiO_2 thickness d and incidence angle θ. (From G. Mbise et al., *Proc. Soc. Photo-Opt. Instrum. Engr.* 1149 [1989] 179–199. With permission.)

eters and cannot be generally quantified. What can be stated, though, is that the effect is far from insignificant.

An assessment of the energy efficiency for a practical case is hampered by the fact that commercially available coated window glass and foil are rarely specified with regard to their angular properties; however, empirical formulas can be used to estimate this dependence [45,46].

4.7 ANGULAR SELECTIVITY OF FILMS WITH INCLINED COLUMNAR NANOSTRUCTURES

Window coatings that have different optical properties at equal angles on the two side of the surface normal are of potential interest for energy-efficient windows, although they have not yet been implemented in practical fenestration. This angular selectivity cannot be accomplished with uniform films, such as the ones in Figure 4.35, whereas it is possible in films with a pronounced nanostructure comprised of inclined columnar features. The first issue to consider is a practical one: how to produce such nanostructured films.

Consider a film that is built up atom by atom or molecule by molecule using a process—such as vacuum evaporation or sputtering—wherein these atoms or molecules impinge from a more or less well-defined direction. Furthermore, assume that these impinging species have an energy that is not very high, meaning that they cannot diffuse widely over the surface where they hit and that they are incapable of causing major reconstruction of the growing film. Finally, assume that the species have an angle α with regard to the substrate normal. The film that is built up will then consist of columns separated by void spaces, and the columns will be oriented at an angle β toward the direction of the impinging species. The angles α and β do not coincide but are related to each other [47]. The relationship is often expressed in terms of a "tangent rule," $\tan \beta = \frac{1}{2} \tan \alpha$, which, however, cannot be taken as generally valid.

The evolution of the inclined columnar structure is made clear from simple model calculations illustrated in Figure 4.36 [48]. Here "atoms," depicted as two-dimensional hard discs, are allowed to travel with a well-defined direction but otherwise randomly toward a "substrate." The "atoms" stick where they hit this substrate or a previously deposited "atom" and relax into a stable position between other "atoms." Columns are then formed basically for the simple reason that a randomly formed protrusion in the film tends to shade whatever is behind it from further deposition, so there is a "self-shadowing" mechanism in place. The nature of the columns and the density of the nanostructure depend on the energy the "atoms" are taken to have in the simulation, that is, to the mobility of the deposition species. The panels in Figure 4.36 refer to three values of this mobility. As expected, an increase of the mobility leads to a denser "film." By moving the substrate during the deposition, one can accomplish an entire "zoo" of nanostructures [49], as shown in Figure A1.6.

Theoretical modeling of films with inclined columnar nanostructures is possible via the theories for optical homogenization presented in Section 3.9, essentially by representing the columnar features, in three dimensions, as elongated prolate spheroids (cigar-shaped ellipsoids) with their long axis coinciding with the columns' symmetry axis [50]. The modeling is technically rather complicated, though, and is not pursued here.

Particularly detailed data on angular selectivity are available for Cr films made by oblique angle deposition [51]. They are discussed in Box 4.3, where it is shown that empirical film structures can be strikingly similar to those in Figure 4.35, and that T_{lum} can differ by a factor of about two at equal angles on either side of the surface normal.

4.8 CHROMOGENICS: THE ENERGY EFFICIENCY THAT IS POSSIBLE

Chromogenics is a relatively new term which was introduced, in the sense it is used here, only in 1990 [52]. Chromogenic devices are able

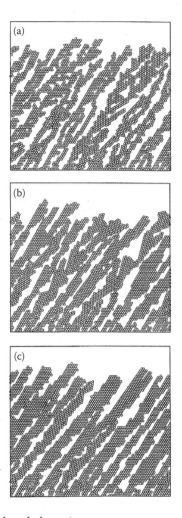

FIGURE 4.36   Simulated deposits grown at an angle of incidence α = 50° and at low (a), medium (b), and high (c) mobility of the impinging "atoms." (From M. J. Brett, *J. Mater. Sci.* 24 [1989] 623–626. With permission.)

BOX 4.3   ANGULAR SELECTIVITY IN OBLIQUELY EVAPORATED CHROME-BASED FILMS: A CASE STUDY

Figure B4.3.1 shows a scanning electron micrograph of the cross-section of a Cr-based film prepared by evaporation with $\alpha = 75°$. A pronounced columnar structure is clearly seen with $\beta = 40 \pm 5°$.

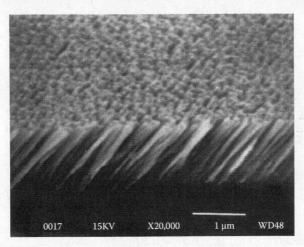

FIGURE B4.3.1   Scanning electron micrograph of a Cr film produced with oblique angle evaporation at $\alpha = 75°$. (From G. W. Mbise et al., *J. Phys. D: Appl. Phys.* 30 [1997] 2103–2122. With permission.)

Measurements of polarization-dependent optical transmittance, $T_s(\theta,\lambda)$ and $T_p(\theta,\lambda)$, were performed on the Cr-based films with the light beam incident at an angle θ to the film's normal and in the incidence plane (spanned by this normal, the direction of the evaporated flux, and the column direction). The geometry is explained in Figure B4.3.2, where $\phi = 0$ corresponds to the incidence plane. Data were recorded for $-64° < \theta < +64°$, with positive angles taken to be on the same side of the normal as the deposition direction and the column orientation. It is necessary to consider s and p polarized light separately; the transmittance appropriate for unpolarized light is then the arithmetic mean of the two values. Figure B4.3.3 shows data for $T_s(0,\lambda)$, $T_s(\pm55°,\lambda)$, $T_p(0,\lambda)$, and $T_p(\pm55°,\lambda)$ for a 45-nm-thick film evaporated at $\alpha = 85°$. The curves display pronounced angular selectivity with the transmittance increasing monotonically with wavelength in the displayed solar range.

It is found that $T_s(-55°,\lambda) = T_s(+55°,\lambda) < T_s(0,\lambda)$ and $T_p(-55°,\lambda) < T_p(0,\lambda) < T_s(+55°,\lambda)$, that is, selectivity is confined to the p polarized component, as expected from theory.

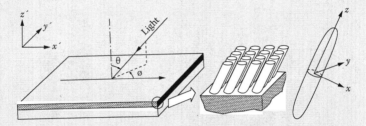

FIGURE 4.3.2 Geometry for describing angular selective optical properties of a film with inclined columnar nanostructure. Light is incident at a polar angle θ and an azimuthal angle ϕ onto a structure with nanofeatures represented by prolate spheroids. The theoretical modeling of the optical properties represents these spheroids in a coordinate system spanned by (x,y,z). The thin film is described in the coordinate system (x',y',z'). The theory accounts for the transformation between these coordinate systems.

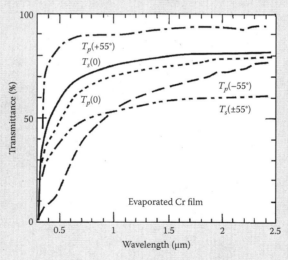

FIGURE 4.3.3 Spectral transmittance T for a 45-nm-thick Cr film made by oblique angle evaporation at $\alpha = 85°$. The curves refer to different incidence angles and states of polarization for the light. (From G. W. Mbise et al., *J. Phys. D: Appl. Phys.* 30 [1997] 2103–2122. With permission.)

Figure B4.3.4 illustrates $T_{s,lum}(\theta)$, $T_{p,lum}(\theta)$, and $T_{u,lum}(\theta)$, where u denotes unpolarized light, at $-64° < \theta < +64°$ for the same film as in Figure B4.3.3. Expectedly, $T_{s,lum}(\theta)$ is symmetrical around $\theta = 0$, whereas $T_{p,lum}(\theta)$ is strongly angular selective. $T_{u,lum}(\theta)$ varies from ~0.3 to ~0.7 when θ goes from $-64°$ to $+40°$. The maximum transmittance occurs at an angle θ_{max}. It was found that $\theta_{max} \approx \beta$, implying that the column orientation coincides with the direction for maximum transmittance.

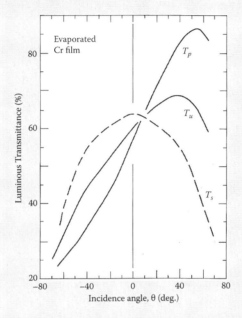

FIGURE B4.3.4 Luminous transmittance as a function of incidence angle for the same Cr film as in Figure B4.3.3. Data are given for s polarized, p polarized, and unpolarized light. (From G. W. Mbise et al., *J. Phys. D: Appl. Phys.* 30 [1997] 2103–2122. With permission.)

Data were recorded not only with regard to the incidence plane but also for other directions. A polar representation of $T(\theta,\phi,\lambda)$ is then appropriate; $\phi = 0$ and $\phi = 180°$ correspond to the incidence plane as shown in Figure B4.3.2. Measurements were taken with laser light at $\lambda = 543$ nm and a photodetector matched to the eye's sensitivity. Figure B4.3.5 shows data for the same film as above. The transmittance is illustrated by use of a gray scale with five 5%-wide intervals. The asymmetry between the left-hand and right-hand halves of the diagram provides striking evidence of angular selectivity for monochromatic as well as for luminous light, whereas the good agreement between the upper and lower halves of the diagrams indicates the symmetry expected from the geometrical features of the obliquely evaporated films. Data corresponding to $\phi = 0$ and $\phi = 180°$ can be directly compared with results in Figure B4.3.4.

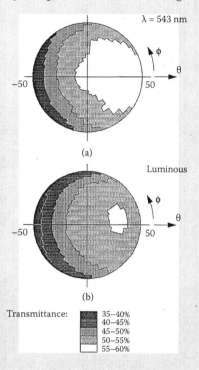

FIGURE B4.3.5 Polar plot of fixed wavelength and luminous transmittace through the same Cr film as in Figures B4.3.3 and B4.3.4. Data are given as a function of incidence angle θ and azimuthal angle ϕ. (From G. W. Mbise et al., *J. Phys. D: Appl. Phys.* 30 [1997] 2103–2122. With permission.)

to vary their optical properties, reversibly and persistently, in response to an external stimulus such as ultraviolet irradiation, temperature, or electrical charge or voltage. Variations of the optical properties are of obvious interest for achieving energy efficiency for the simple reason that the ambient properties vary during the day or season.

    The purpose of chromogenic fenestration can be to create energy efficiency in a building by admitting solar energy when there is a heating demand and rejecting solar energy if there is a cooling demand. Considering such fenestration, which is also referred to as "smart" or "switchable," there is also a human dimension related to comfort implying that there may be a need for good lighting, glare elimination, and prevention of thermal stress. Obviously the notion of "comfort" is a subjective one, even if quantification is possible to some extent, so chromogenics is also of interest for meeting individual demands.

How much energy can one save by using windows with variable properties? The answer to this question is complicated and multidimensional as will become apparent immediately below. In the rest of this section we then discuss three types of materials and their associated technologies in some detail. They are *photochromic* and change under ultraviolet irradiation, *thermochromic* and change when the temperature is altered, and *electrochromic* and can be changed electrically.

These are not the only chromogenic technologies, though, and another one that has attracted some attention recently is *gasochromics* [53]. It considers double-pane windows with facilities to introduce reducing (hydrogen containing), as well as oxidizing gases between the panes. One of the glass surfaces facing these gases has a coating of, for example, tungsten oxide and a very thin layer of a catalyst such as Pd. Tungsten oxide colors uniformly in the presence of a reducing gas and bleaches in an oxidizing gas. The physics underlying these phenomena are similar to that of the electrochromic materials to be discussed in Section 4.11. Gasochromic window prototypes have been built and have demonstrated good optical modulation. The necessity to invoke a gas handling system is an undeniable drawback, however, and gasochromics will not be discussed further here. Still another *magnetochromic* option was mentioned in Section 3.2.3.

How large is the possible energy savings with windows that are able to change their transparency? We first attempt to answer this question by a simple and intuitive line of reasoning and consider a building requiring cooling.

The solar energy falling onto a vertical surface, such as a normal window, each year is chosen here to be 1000 kWh/m^2. This energy can serve as a nominal value, whereas more correct numbers for south-facing/north-facing/horizontal surfaces are 850/350/920 kWh/m^2 for Stockholm (Sweden), 1400/450/1700 kWh/m^2 for Denver (Colorado), and 1100/560/1800 kWh/m^2 for Miami (Florida). Only about half of the solar energy corresponds to visible light (i.e., 500 kWh/m^2). This number is used in the analysis below because IR radiation can be reflected off, at least in principle, by use of a spectrally selective surface that does not exhibit variable transmittance (cf. Figure 2.16 or Figure 4.5). We now assume that the transparency of visible light can be altered between 7% and 75%, as has been demonstrated by use of electrochromic technology to be discussed further below. The difference between having the window constantly colored and constantly bleached then is 340 kWh/m². The next issue is to consider when the window should be colored and when it should be bleached. Assuming for simplicity that the window should be fully transparent when the room is in use, and that it is in its darkest state when there is nobody in the room, the relevant question regards the fraction of time that the room is "inhabited," or, more precisely, the fraction of the energy that comes into the room when it is not in use. Considering that a normal (office) room is empty during weekends, holidays, and vacations, early mornings and late afternoons (when the sun is near the horizon), etc., it is probably a conservative estimate that 50% of the energy in principle can enter the room when there is no one to look through the window. Hence, the estimate yields that 170 kWh/m² is the amount of energy saved annually by adopting the given control strategy.

Is this energy savings significant or not? To answer this question, one may note that 17% is a typical value for the photoelectric conversion efficiency of today's best solar cells and thin-film submodules devised for terrestrial applications (somewhat below the data for the research-type solar cells whose efficiencies are given in Figure 6.15). Thus, these solar cells would be able to generate 170 kWh/m² in the example given above. Of course, the analogy between energy *savings* in "smart" windows and energy *generation* in solar cells is not tied to the choice of the incident solar energy being 1000 kWh/m² but applies generally, irrespective of the orientation of the surface under consideration. The "smart" window saves thermal energy, but if a cooling machine—operating with an efficiency of 300%, say—runs on electricity generated with an efficiency of 33%, then the analogy becomes perfect.

We can summarize the reasoning above as follows: A "smart window" can save as much electricity for a cooling machine as a solar cell module, with the same position, could have generated.

This analogy between energy savings and energy generation is very simplistic, though, and more correct assessments of the potential of the "smart windows" technology must consider issues such as downsizing of cooling equipment, reductions in peak electricity demand, possibilities to achieve energy-efficient daylighting, and user aspects such as possibilities to avoid glare and thermal stress. Some studies for European climates have indicated that the energy for space cooling, on an annual basis, could be reduced by as much as 40% to 50% when using "smart" windows instead of conventional static solar control windows [54,55]. Other analyses, performed for California, also indicated very substantial energy savings [56]. A study from which Figure 4.37 was taken gives a very schematic illustration of the energies for cooling and for electric lighting with a number of fenestration types [57]. Not surprisingly, clear glass gives a comparatively small need for artificial lighting but is disadvantageous with regard to cooling energy. Going to tinted and

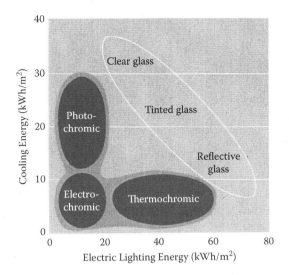

FIGURE 4.37 Schematic relationship between electrical lighting energy and cooling energy for a number of fenestration types. (After UNEP, Buildings and Climate Change: Status, Challenges and Opportunities, United Nations Environment Programme, Paris, 2007.)

reflecting glass diminishes the cooling energy but increases the demand for lighting. Chromogenic technologies—especially the one based on electrochromics—are found to have strong advantages both for cooling energy and electric lighting energy.

The detailed numbers in the mentioned studies should be regarded as tentative. However, there can be no doubt that chromogenic technologies are able to provide very significant energy savings especially in buildings that are cooled during a large part of the year. And these energy savings can be combined with increasing comfort [58]. It is expected that more analyses will be performed during coming years so that the energy savings potential can be set on firmer ground.

4.9 PHOTOCHROMICS

Photochromic glass darkens under UV irradiation and bleaches spontaneously in the absence of such irradiation. This kind of glass is well known and has been used in sunglasses for decades. In principle, there is nothing to prevent its use in windows, and test production of photochromic glass panes has been made in the past. There are also numerous thin films that show photochromism, but usually their ability to sustain many darkening and bleaching cycles is low.

Ordinary photochromic glass is based on photosensitive compounds that are randomly dispersed in the vitreous matrix. Metal halides, especially silver halide, are widely used as the light-absorbing substance. The glass that is most common also contains some chlorine and bromine ions as well as a small amount of copper ions. After appropriate heat treatment, the glass contains Cu-doped halide particles with sizes of 10 to 20 nm (i.e., the glass is a nanomaterial) [59]. Photo-induced reactions lead to the formation of clusters of metallic Ag, and therefore this material has some principle kinship to the selectively solar-absorbing coatings discussed in Section 6.1. When the UV irradiation ceases, the glass returns to its initial state. The details of the coloration and bleaching are still poorly understood.

The practical performance of photochromic glass is governed by the transmittance in dark and transparent states and the coloring and bleaching rates. The transmittance levels depend on the glass composition and values of T_{lum} up to 90% and down to 5% are possible. There is also some temperature dependence, and a temperature rise leads to enhanced dynamics. Figure 4.38 shows T_{lum} during darkening and clearing for two photochromic glass types measured at different temperatures [3].

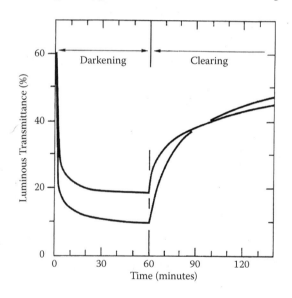

FIGURE 4.38 Luminous transmittance versus time during darkening and clearing of two photochromic glasses that had been cleared overnight prior to the measurements. Upper data were recorded at 25°C and lower data at 3°C. (From C. G. Granqvist, in *Materials Science for Solar Energy Conversion Systems*, edited by C. G. Granqvist, Pergamon, Oxford, U.K., 1991; chap. 5, pp. 106–167. With permission.)

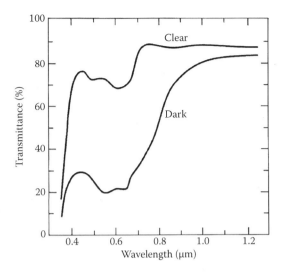

FIGURE 4.39   Spectral transmittance for a photochromic glass in clear and dark states. (From C. G. Granqvist, in *Materials Science for Solar Energy Conversion Systems*, edited by C. G. Granqvist, Pergamon, Oxford, U.K., 1991; chap. 5, pp. 106–167. With permission.)

Irradiation makes T_{lum} drop swiftly and, generally, darkening to 80% of the full range occurs during 1 min. However, some residual darkening is noticeable after times as long as 1 h. The clearing in the absence of irradiation is much slower and is incomplete even after 1 h.

Figure 4.39 shows typical spectral transmittance in the $0.35 < \lambda < 1.3$ μm range for photochromic glass in dark and clear states [3]. It is evident that the difference in the optical properties is confined mainly to the $0.35 < \lambda < 1$ μm range, implying that the modulation of T_{sol} is much smaller than that of T_{lum}, thus pointing at a limitation in the performance as a window coating.

Photochromic plastics have been known for many years, too, and have optical properties similar to those for photochromic glass. Photochromic spiropyran-silica mixtures can show both photochromic and thermochromic properties [60].

4.10 THERMOCHROMICS

Thermochromic materials are of interest for energy-efficient windows if they allow a substantially larger transmission below a "critical" temperature τ_c than above this temperature, and if τ_c is near room temperature. Under these conditions, the thermochromic material admits energy when there is a need for heating and rejects energy when heating is not required.

4.10.1 Metal-Insulator Transition and Its Nanofeatures in VO$_2$

There are several principles that can be used to create thermochromism. We first consider materials capable of undergoing reversible transitions between a high-temperature metallic-like, and hence reflecting, state and a low-temperature dielectric-like state that is transparent. Figure 4.40 displays temperature-dependent electrical conductivity for several metal

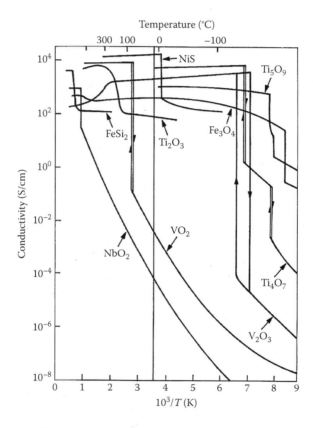

FIGURE 4.40 Electrical conductivity versus reciprocal temperature for several metal compounds. (From G. V. Jorgenson, J. C. Lee, in *Large-Area Chromogenics: Materials and Devices for Transmittance Control*, edited by C. M. Lampert, C. G. Granqvist, The International Society for Optical Engineering, Bellingham, WA, 1990; pp. 142–159. With permission.)

compounds [61]. Only two of them show transitions in the vicinity of room temperature, and only one has been investigated in detail, namely, vanadium dioxide, which has $\tau_c \approx 68$ °C. Its conductivity is shifted abruptly by several orders of magnitude at the transition.

Bulk crystals of VO_2 transform between a semiconducting, non-magnetic, and relatively IR-transparent state with monoclinic structure below τ_c to a metallic, paramagnetic, and IR-reflecting state with tetragonal structure above τ_c. The physics behind the structural transformation has been discussed for decades but is still not fully understood. The metal phase is most unusual and belongs to the class of materials known as "bad metals" because the relaxation rate ω_τ is so high that the carriers would travel less than one lattice spacing between collisions if they had the normal Fermi velocity. Properties in both phases depend on nano-structure and doping, but not in the expected way through increased relaxation rate but via changes in band structure [62]. Crystals of VO_2 tend to disintegrate after many structural transitions but this is not so for VO_2-based films.

Indirect evidence for percolation at the metal-insulator transition in VO_2 films has existed for many years. Recently, the notion of a gradual two-phase transition has been verified in direct measurements using scanning near-field infrared microscopy. Figure 4.41 shows such images for a series of temperatures between 68°C and 70°C [63] for films made by sol-gel technology. Clearly, the metallically conducting regions nucleate at different places and merge gradually as the temperature is increased (i.e., VO_2 is a clear case of a nanomaterial at the metal-insulator transition). There is some resemblance to the formation of a continuous metallic film on a glass substrate (cf. Figures 4.8 and 4.9). The main percolation transition appears to be unusual in that it occurs at different fractions of the metal phase under heating and cooling [62].

4.10.2 Thermochromism in VO_2-Based Films, and How to Adjust the Metal-Insulator Transition

Figure 4.42 shows typical data on the thermochromism in the IR for VO_2 films, in this case for 160-nm-thick films made by sputter deposition onto glass at 400°C [64]. The VO_2 film transforms between states with high and low transmittance at $\tau_c \approx 60$°C upon heating, and cooling brings back the original transmittance with ~10°C hysteresis (not shown). Clearly, the metal-insulator transition takes place at an undesirably high temperature for buildings-related applications, but τ_c can be lowered to room temperature by addition of a few atomic percent of tungsten, as also shown in Figure 4.42. The addition of W also tends to blur the

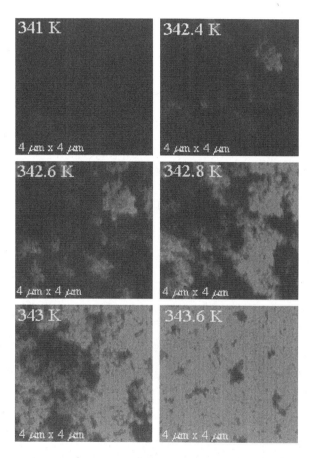

FIGURE 4.41   *A color version of this figure follows page 200.* Scanning near-field micrographs of 4 x 4 μm areas of VO₂ films as measured at the shown temperatures. Dark blue represent an insulating state and light blue, green, and red colors represent metallic conductivity. (From M. M. Quazilbash et al., *Science* 318 [2007] 1750–1753. With permission.)

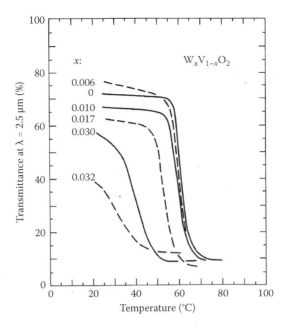

FIGURE 4.42   Temperature dependent transmittance at $\lambda = 2.5$ μm for sputter-deposited $W_xV_{1-x}O_2$ films having the shown magnitudes of x. (From M. A. Sobhan et al., *Solar Energy Mater. Solar Cells* 44 [1996] 451–455. With permission.)

metal-insulator transition. Figure 4.43 shows that τ_c drops linearly with the amount of tungsten doping [64]. The rate of this decrease seems to depend on the quality of the VO_2 films, and τ_c falls off by ~28°C/at %W in bulk-like samples [65].

A very sharp transition between the two states may not be desired for a practical application to windows. The reason is that a sharp transition in a narrow temperature range would tend to give rise to undesired optical effects with visible "color fronts" moving across the window under heating and cooling through τ_c.

Figure 4.44 shows spectral transmittance in the solar range for films of VO_2 and $W_{0.032}V_{0.968}O_2$ below and above τ_c [64]. It is evident that the transmittance is undesirably low and T_{lum} does not exceed 40%. Of course, a thinner film would display higher transmittance, but at the expense of degraded thermochromism (i.e., a less pronounced switching of the NIR transmittance and hence of T_{sol}). This points at a real problem of too high luminous absorptance in pure VO_2 as well as in W-doped VO_2. This feature has severely curtailed the applicability of thermochromic window coatings for many years. However, there are some recently discovered ways to boost the transmittance, as considered next.

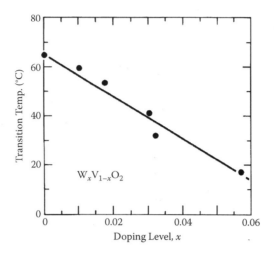

FIGURE 4.43 Thermochromic transition temperature τ_c versus tungsten content in sputter-deposited $W_xV_{1-x}O_2$ films. The line was drawn as a fit to the data points. (From M. A. Sobhan et al., *Solar Energy Mater. Solar Cells* 44 [1996] 451–455. With permission.)

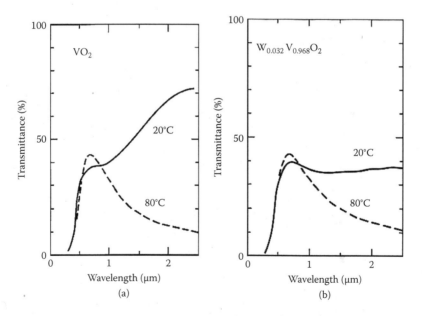

FIGURE 4.44 Spectral transmittance of films of VO_2 (panel a) and $W_{0.032}V_{0.968}O_2$ (panel b) at the shown temperatures. (From M. A. Sobhan et al., *Solar Energy Mater. Solar Cells* 44 [1996] 451–455. With permission.)

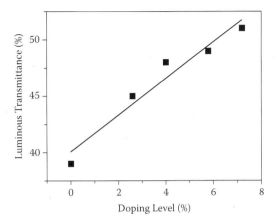

FIGURE 4.45 Luminous transmittance versus doping level in 50-nm thick $Mg_xV_{1-x}O_2$ films made by sputtering. (From N. R. Mlyuka et al., *Appl. Phys. Lett.* 95 [2009] 171909 1–3. With permission.)

4.10.3 How to Enhance the Luminous Transmittance in VO_2

There are several ways to enhance T_{lum} in thermochromic VO_2-based films. One recently discovered possibility is to dope with Mg, and Figure 4.45 shows that T_{lum} increases noticeably when the value of x in $Mg_xV_{1-x}O_2$ goes up [66]. Interestingly, τ_c drops upon Mg doping, though not as fast as in $W_xV_{1-x}O_2$.

Fluorination offers another possibility to boost T_{lum}, and Figure 4.46 illustrates the difference between films made in the presence and absence of a fluorine-containing gas in the sputter plasma [67]. Clearly, the fluorination can enhance T_{lum} by up to ~10%. Antireflection treatment is, of course, possible in order to increase the transmittance of VO_2-based films. This is illustrated in Figure 4.47, where the upper panel reports spectral transmittance for a $TiO_2/VO_2/TiO_2$ stack with a 50-nm-thick VO_2 film and 75-nm-thick dielectric TiO_2 films as measured at 20°C and 100°C, together with results for a single 50-nm-thick VO_2 film [68]. The maximum transmittance is strikingly higher in the three-layer structure and reaches 81.5% at $\lambda \approx 740$ nm, while the single-layer film peaks at 56.1% for $\lambda \approx 710$ nm. Integrated values of T_{lum} were 62.8 and 58% below and above τ_c, respectively, for the three-layer structure, whereas the single-layer film had 42 and 45% as corresponding measures. Considering T_{sol}, the three-layer structure had 63 and 57% below and above τ_c, respectively, and the single-layer film correspondingly displayed 47 and 42.5%.

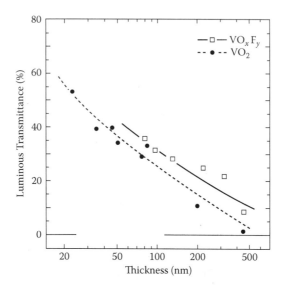

FIGURE 4.46 Luminous transmittance versus film thickness for sputter deposited films of VO_2 and VO_xF_y. Curves were drawn for convenience only. (From K. A. Khan, C. G. Granqvist, *Appl. Phys. Lett.* 55 [1989] 4–6. With permission.)

The lower panel of Figure 4.47 shows spectral transmittance for a $TiO_2/VO_2/TiO_2/VO_2/TiO_2$ stack with 50-nm-thick VO_2 films and 130-nm-thick TiO_2 films [68,69]. These data are compared with corresponding ones for a 100-nm-thick VO_2 film; the total amount of VO_2 hence was the same in the five-layer and single-layer configurations. Here T_{lum} was 45 and 42.3% below and above τ_c, respectively, for the five-layer structure, whereas the single-layer film had 41 and 40% as corresponding data. Considering T_{sol}, the five-layer structure had 52.1 and 40% below and above τ_c, respectively, and the single-layer film displayed 41 and 34.3%. The strong enhancement of the modulation of T_{sol}, as a result of the three dielectric layers, is noteworthy. The use of multiple coatings makes it possible to induce stress in the VO_2-type thin films. In this manner it is possible to influence τ_c to some extent too.

A different approach to high-transmittance thermochromic materials may be to incorporate VO_2-based pigments in polymer foils in the manner described for LaB_6 and other pigments in Section 4.5.3. Recent results on plasmonically resonant VO_2 pigments have given interesting results [70].

There are several other ways to create a thermochromic effect in windows. One of them uses cloud gels and is outlined in Box 4.4.

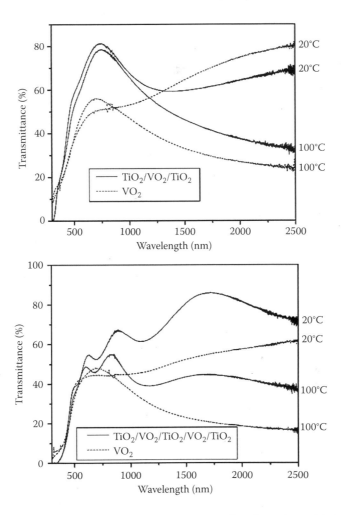

FIGURE 4.47   Spectral normal transmittance for three-layer $TiO_2/VO_2/$ TiO_2 films (upper panel) and five-layer $TiO_2/VO_2/TiO_2/VO_2/TiO_2$ films (lower panel), as specified in the main text, at two temperatures, one lying below the metal-insulator transition temperature and the other above this temperature. The data are compared with those for corresponding VO_2 films. (From N. R. Mlyuka et al., *Solar Energy Mater. Solar Cells* 93 [2009] 1685–1687. With permission.)

BOX 4.4   THERMOCHROMISM IN CLOUD GELS

Thermochromic windows can include a layer of a polymeric "cloud gel" [71,72]. Clouding—that is, transition to a diffusely scattering state—can set in above a temperature τ_c due to reversible thermochemical dissolution and a thermally induced modification of the length of the polymer molecules. This means, essentially, that the material changes between a "grainy" and light-scattering state and a homogeneous and nonscattering state, as indicated in Figure B4.4.1. The value of τ_c can lie anywhere within a wide range by proper choice of material. Figure B4.4.2 shows the total transmission (direct and diffuse) of radiation through a one-millimeter-thick cloud gel layer between two glass panes [3]. Both T_{lum} and T_{sol} can drop by ~50% when τ_c is exceeded. Apart from windows-related applications, "cloud gels" are of interest for overheat protection, for example in plastic solar collectors (cf. Section 6.1) [73].

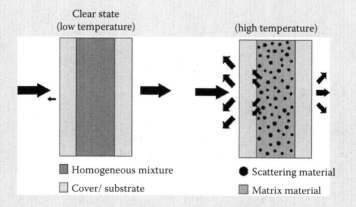

FIGURE B4.4.1   Schematics of the low temperature clear state and the high temperature backscattering state in a thermochromic glazing unit. (After P. Nitz, H. Hartwig, *Solar Energy* 79 [2005] 573–582. With permission.)

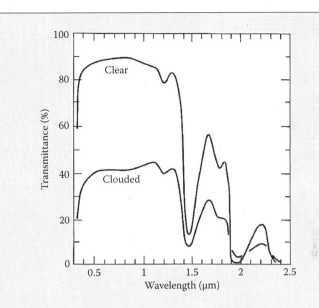

FIGURE B4.4.2   Spectral total transmittance through a cloud gel layer between glass panes. The data refer to a low-temperature clear state and a high-temperature clouded state. (From C. G. Granqvist, in *Materials Science for Solar Energy Conversion Systems*, edited by C. G. Granqvist, Pergamon, Oxford, U.K., 1991; chap. 5, pp. 106–167. With permission.)

4.11 ELECTROCHROMICS

Electrochromics is different from photochromics and thermochromics in that it allows the transmittance to be easily changed by an operator, and it is hence more flexible regarding applications than the other "chromics." It is also possible to construct devices with a large optical modulation, as illustrated in Figure 4.1. However, electrochromism is more complicated than photochromics and thermochromics, which rely on the change of a single material. In contrast to this, an electrochromic device includes several layers, and the optical modulation is connected to electrical charge being shuttled between different layers.

Electrochromism was discovered and made widely known in the late 1960s and early 1970s [74]. Early applications were sought for information displays, but electrochromic (EC) displays did not stand up to the then rapidly developing liquid crystal displays. The focus on EC research was changed to windows during the mid-1980s when it was realized that the technology had "green" attributes and could yield energy efficiency and user comfort in buildings. The concept of a "smart" window was then coined and captured interest both from researchers and the general public [75].

Electrochromism is well known both in a number of transition metal oxides and in several classes of organic compounds, but only the former are likely to be durable enough for long-time uses in architectural windows. Prior work on EC materials and devices has been summarized in considerable detail [76,77].

4.11.1 How Do Electrochromic Devices Work?

Figure 4.48 shows a standard design of an EC device [76]. There are five layers backed by one substrate or positioned between two substrates by use of lamination. The substrates are normally of glass, but plastic works, too, and flexible PET foil allows for interesting devices. The central part of the device conducts ions but not electrons. This can be an organic material, such as a transparent polymeric electrolyte or an ionic liquid, preferably with adhesive properties. It can also be a thin film, for example, a porous oxide incorporating ions. The ions should be small in order to be easily mobile in an electric field, and protons (H^+) and Li^+ are of concern in most cases.

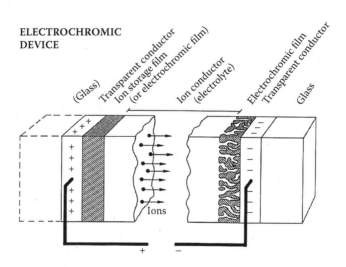

FIGURE 4.48   Basic design of an electrochromic device. The transport of positive ions in an electrical field is indicated. (From C. G. Granqvist, *Handbook of Inorganic Electrochromic Materials*, Elsevier, Amsterdam, the Netherlands, 1995. With permission.)

The ion conductor is in contact with an EC film capable of conducting both ions and electrons (i.e., a mixed conductor). Tungsten oxide is a typical example. On the other side of the ion conductor is an ion storage film, which also is a mixed conductor for ions and electrons. Ideally, this has EC properties which are complementary to those of the first EC film. This central three-layer stack is positioned between two transparent conducting films. ITO is often preferred because it has an unsurpassed combination of optical transparency and electrical conductivity, but SnO_2-based films can be used and offer cost benefits, especially if the SnO_2-based film is produced on glass by spray pyrolysis. Metal-based transparent conductors have not been tried to any large extent but should not be written off.

When a voltage of the order of 1 V is applied between the transparent conductors, ions can be transported between the EC film and the ion storage film. The charge of the ions is then balanced by electrons that are injected into or withdrawn from the EC film and ion storage films via the transparent conductors, and these electrons are the cause of the optical absorption as we will see shortly. Reversal of the voltage or, with suitable materials, short circuiting brings back the original properties. The coloration can be halted at any intermediate level, which means that the EC device has open circuit memory. The fact that power is needed only to change the optical properties is important for window-type devices designed for energy savings. The memory effect hinges on the fact that the ion conductor in the middle of the EC device does not conduct electrons, which is easier to accomplish with a laminate layer having a thickness on the order of several micrometers than with a thin ion-conducting film. The voltage level needed to move the ions is of the order of a few volt dc, which can be easily obtained by solar cells, and a number of EC devices with integral solar-cell-based powering have been researched [78].

It should now be clear that the EC device resembles an electrical battery with a charging state that corresponds to a degree of optical absorption. The analogy is useful and the two types of devices share many pros and cons. For example, both of them can easily degrade if they are mistreated by overcharging or overheating, but they also exhibit certain "self healing" features which are poorly understood. Also, both kinds of devices are unable to change their properties abruptly, and in the case of an EC device, the time for going from fully colored to fully bleached may vary from a few seconds in a device of the size of a few square centimeters to minutes, or even tens of minutes, for a window with a size of square meters.

What does it mean that the base EC film and the ion storage film should have complementary properties? The background is that there are oxides with two types of EC properties: those coloring under ion

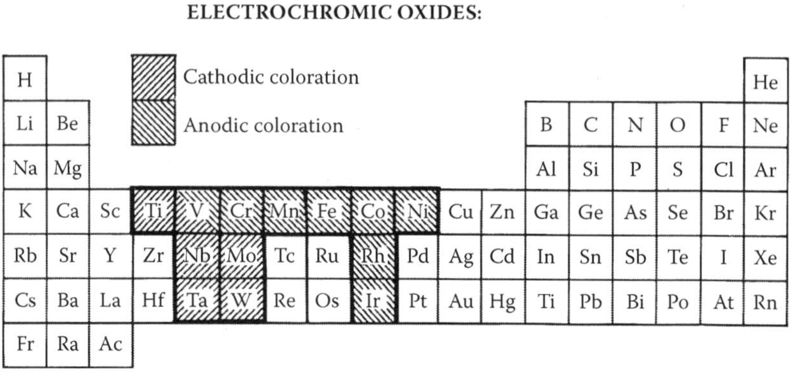

ELECTROCHROMIC OXIDES:

FIGURE 4.49 Periodic system of the elements (apart from the lanthanides and actinides). The differently shaded boxes indicate transition metal oxides whose oxides have clear "cathodic" and "anodic" electrochromism. (From C. G. Granqvist, *Handbook of Inorganic Electrochromic Materials*, Elsevier, Amsterdam, the Netherlands, 1995. With permission.)

insertion and known as "cathodic" and those coloring under ion extraction and known as "anodic." The terminology clearly points at the kinship between EC technology and battery technology. Figure 4.49 shows the metallic elements whose oxides exhibit the two kinds of coloration [76]. Among the cathodic oxides, most attention has been on oxides of W, Mo, and Nb. Among the anodic oxides, those based on Ir and Ni stand out as most interesting. It should be noted, though, that Ir is very rare and precious and hence ill suited for commodity-scale applications. The only metal with different properties is vanadium for which the V_2O_5 can exhibit cathodic and anodic features in different wavelength regions, but this is a special case which we return to briefly in the text to come. By combining, say, a cathodic EC film with an anodically coloring ion storage film, one can accomplish devices with both films becoming dark when charge is moved from one to the other and both films bleaching when the original charging state is brought back. This complementary feature also can be used to create color neutrality, provided that adequate combinations of "cathodic" and "anodic" films are used.

4.11.2 Facile Ion Movement Due to Favorable Nanostructures

Most of the EC oxides can be viewed as built from octagonal building blocks in different arrangements. The spaces between these octahedra are large enough to allow at least some ion transport. Furthermore, clusters of octahedra can aggregate into disordered and loosely packed

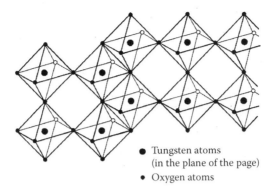

● Tungsten atoms
  (in the plane of the page)
● Oxygen atoms

FIGURE 4.50   Schematic illustration of corner-sharing and edge-sharing octahedra in a W oxide crystal. (From C. G. Granqvist, *Handbook of Inorganic Electrochromic Materials*, Elsevier, Amsterdam, the Netherlands, 1995. With permission.)

clusters with significant intergranular spaces. The nano features therefore enter on two levels as further elaborated shortly. Most of the discussion concerns W oxide, which is the most widely investigated EC material [76,79–81].

Figure 4.50 describes nanostructural features of W oxide and shows WO_6 octahedra with six oxygen atoms and a tungsten atom in the center [76]. Stoichiometric WO_3 corresponds to a structure with each octahedron sharing corners with adjacent octahedra. W oxide has a tendency to form substoichiometric phases in which some of the octahedra are not corner sharing but edge sharing, as is also illustrated in Figure 4.50. It is easily realized that a three-dimensional arrangement of octahedral "building blocks" leaves a three-dimensional network of "tunnels." They are wide enough to serve as conduits for small ions.

The schematic crystal structure in Figure 4.50 is, in fact, not entirely adequate because it refers to a cubic structure of W oxide, which does not form except under high pressure. The transition metal oxides are notorious for their large number of possible structures, and a tetragonal structure is appropriate for WO_3 crystals at normal temperature and pressure. It is shown in Figure 4.51 that the spaces between the octahedral "building blocks" are then larger than for the cubic structure, and an even more favorable structure with regard to the possibility of ionic movement is found in a hexagonal atomic arrangement [76]. The latter structure seems to be common in thin films of W oxide.

Nanostructures formed in thin films of W oxides have been investigated many times. Figure 4.52 reports results for films made by evaporation onto substrates at different temperatures τ_s and represent modeling based on x-ray scattering [76,82]. Clearly, the films exhibit cluster-type

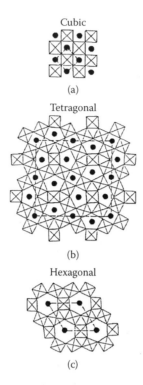

FIGURE 4.51   Atomic arrangements for W oxide with (a) cubic, (b) tetragonal, and (c) hexagonal structure. Solid dots indicate sites available for ions in the open spaces between the WO_6 octahedra. Dashed lines show the extents of the unit cells. (From C. G. Granqvist, *Handbook of Inorganic Electrochromic Materials*, Elsevier, Amsterdam, the Netherlands, 1995. With permission.)

features with clusters growing in size as τ_s rises. The individual clusters are believed to be linked by hydrogen bonds via water molecules. The cluster size is ~3 nm for $\tau_s = 150°C$, and further cluster growth has taken place at $\tau_s = 300°C$. The clusters start to interconnect at the latter temperature and a long-range ordered structure then starts to prevail. Two features in Figure 4.52 should be emphasized: the existence of large open spaces between the clusters, especially in the absence of substrate heating, and the hexagonal nature of the individual clusters (as apparent from a comparison with Figure 4.51c). Both of these features are advantageous for ionic mobility in the oxides.

The structural models discussed above are believed to be essentially correct, at least for the cathodic EC oxides. For the anodic ones, the situation is more complicated and the atomic distances are so small that any

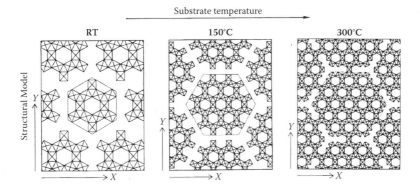

FIGURE 4.52   Structural models based on connected WO_6 octahedra for W oxide films made by evaporation onto substrates at room temperature (RT) and two elevated temperatures. Arrows in the x and y directions denote 2 nm. (From C. G. Granqvist, *Handbook of Inorganic Electrochromic Materials*, Elsevier, Amsterdam, the Netherlands; T. Nanba, I. Yasui, *J. Solid State. Chem.* 83 [1989] 304–315. With permission.)

direct insertion on ions is unlikely; therefore, it seems that film porosity and processes on the grains' surfaces are of overriding importance.

4.11.3 What Causes the Optical Absorption?

The origin of the optical absorption in electrochromic oxides has been the subject of much research. It is a complicated subject for various reasons, one being that oxides are poorly defined with regard to crystallinity and can incorporate mobile ions and water molecules to varying degrees. Another difficulty ensues from the fact that even the intrinsic electronic structure for materials such as NiO has been debated for decades without any consensus having been reached. We approach the optical absorption mechanisms in the EC materials step by step in Box 4.5 and focus on a particularly interesting combination of "cathodically" and "anodically" coloring EC oxides: those based on W and Ni [81].

4.11.4 Some Device Properties

Many different types of electrochromic devices have been studied over the years. Recently much interest has been focused on combinations of cathodic W oxide and anodic Ni oxide joined via with an electrolyte being an ion-containing inorganic thin film or an organic ion-conducting

BOX 4.5   ELECTROCHROMISM IN W OXIDE
AND Ni OXIDE: A DETAILED VIEW

We first consider insertion and extraction of protons (hydrogen ions) and electrons in WO_3 by the simple electrochemical reaction

$$[WO_3 + H^+ + e^-]_{bleached} \leftrightarrow [HWO_3]_{colored} \qquad (B4.5.1)$$

where it should be remembered that H^+ could be replaced by some other small ion such as Li^+, and that the reaction should only be partial in order to be reversible so that the colored compound should be written H_xWO_3 with $x < 0.5$. For the Ni-based oxide, the corresponding reaction, which is expected to be confined to hydrous grain boundaries, is

$$[Ni(OH)_2]_{bleached} \leftrightarrow [NiOOH + H^+ + e^-]_{colored} \qquad (B4.5.2)$$

An understanding of the optical properties under ion and electron exchange and of the principal differences between the cathodic and anodic can be developed by considering the electronic band structure appropriate for oxides comprised of octahedral "building blocks." The oxygen $2p$ bands are separated from the metal d levels, and octahedral symmetry leads to the splitting of the d levels into bands with the conventional notation e_g and t_{2g}. We refer to the literature for a detailed discussion of these matters [83].

Figure B4.5.1 illustrates schematically the cases believed to be relevant for typical cathodic and anodic oxides. The left-hand panel—for H_xWO_3, say—shows that the $O2p$ band is separated from the split d band by an energy gap. Pure WO_3 has a full $O2p$ band and an empty d band and hence is transparent as any semiconductor characterized by a wide-enough band gap. Insertion of small ions and accompanying electrons leads to a partial filling of the d band, accompanied by optical absorption as discussed in the text following. Incidentally, the filling of the lowest states in the band permits optical transitions across the band gap only with a larger energy than in the case of undoped WO_3, which is the same mechanism as the one leading to band gap widening, for example, by Sn doping an In_2O_3 to make ITO (cf. Figure 4.23).

The middle panel in Figure B4.5.1 applies to the anodic oxides. Here, the "pure" oxide has some unoccupied t_{2g} states, and insertion of ions and electrons may fill these states to the top of the band so that the material becomes characterized by a band gap between the e_g and t_{2g} states. Transparency then prevails provided that this gap is wide enough.

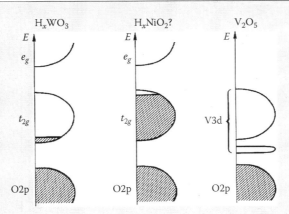

FIGURE B4.5.1 Schematic band structures for different classes of EC oxides. Shaded regions denote filled states, and E signifies energy. (From C. G. Granqvist, *Handbook of Inorganic Electrochromic Materials*, Elsevier, Amsterdam, the Netherlands, 1995. With permission.)

For V_2O_5, finally, the structure deviates sufficiently from the octahedra-based one that the d band displays a narrow low-energy portion lying in the band gap, as illustrated in the right-hand part of Figure B4.5.1. Low-level doping of V_2O_5 leads to filling of this split-off band and subsequent doping makes the split-off band fully occupied so that the "effective" band gap is widened. These features account for the fact that V_2O_5 is neither purely cathodic nor anodic [84].

The detailed mechanism for the optical absorption is considered next. It has been studied in greatest detail for W oxide. When ions and electrons are inserted, the electrons are localized on tungsten sites, so some of the W^{6+} sites are transformed to W^{5+} sites. By absorbing a photon, the inserted electrons can gain enough energy to be transferred to a neighboring site. Such transfer between sites i and j, say, can be written schematically as

$$W_i^{5+} + W_j^{6+} + photon \rightarrow W_i^{6+} + W_j^{5+} \qquad (B4.5.3)$$

This mechanism is effective only as long as transfer can take place from a state occupied by an electron to an empty one available to receive the electron. If the ion and electron insertion is large enough this is no longer the case and then not only transfer of the type $W^{5+} \leftrightarrow W^{6+}$ is important but also $W^{4+} \leftrightarrow W^{5+}$ and $W^{4+} \leftrightarrow W^{6+}$.

In a practical situation, such "site saturation" effects may not be so important, though, since the possibilities to have a highly reversible ion exchange tends to limit the insertion levels to those where $W^{5+} \leftrightarrow W^{6+}$ is dominant [85].

laminate [81]. The latter configuration has been implemented on flexible PET foil, which allows facile roll-to-roll manufacturing (cf. Figure A1.5) of products suitable for new windows as well as for retrofitting windows. The properties of such a foil are described in Box 4.6.

4.11.5 Alternative Electrochromic Devices

Several alternative EC device types have been studied over the years [9]. Thus, there are metal hydrides with some added catalysts that can change from a transparent, via an absorbing, to a reflecting state upon hydrogen exchange. The fact that a reflecting state can be reached is an asset since such a film does not heat up as much as an absorbing one in a typical windows-related application. However, the values of T_{lum} have so far been too low for general applications and long-term durability has not been demonstrated. These metal hydrides also exhibit gasochromic features [86].

Reversible electroplating of a metal onto a glass surface from an adjacent electrochemically active substance is another conceivable technique for reflectance modulation [87]. It has been researched for years, but no practical solution has yet been demonstrated for windows.

Other types of materials require constant electrical powering in order to remain dark or transparent and revert spontaneously to the other state in the absence of such powering. The drawbacks with regard to energy savings are obvious. These technologies include solution- and gel-based EC redox systems such as those used in today's antidazzling rear view mirrors for cars and trucks, and polymer-encapsulated liquid crystal devices that can switch between states of high and low scattering [88] and being somewhat analogous to the "cloud-gels" discussed in Box 4.4. Systems incorporating suspended rod-like particles can serve as "light valves" and can change their transparency, depending on whether or not the particles are aligned in an electrical field [89].

BOX 4.6   PROPERTIES OF AN ELECTROCHROMIC FOIL BASED ON FILMS OF W OXIDE AND Ni OXIDE

Figure B4.6.1 shows optical properties for a laminated device with W oxide and Ni oxide. The transmittance in the luminous and solar ranges can be modulated between widely separated extrema, which is due to a change of the absorption rather than of the reflection. The reflectance is not completely identical when the device is viewed from the two sides.

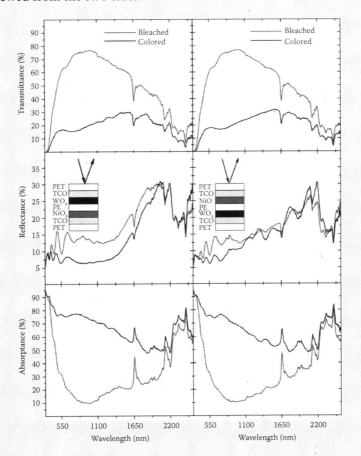

FIGURE B4.6.1   Spectral transmittance, reflectance and absorptance as measured for the EC device shown in the inset. It incorporates films based on W oxide, Ni oxide, and transparent conducting oxide (TCO), as well as a polymer electrolyte (PE) and PET foils.

The transmittance modulation during several color/bleach cycles for a similar EC foil device, 240 cm² in size, is shown in Figure B4.6.2. It is evident that the optical changes are not abrupt but take place during a time span of tens of seconds for devices of this size. The modulation range can be changed by having different thicknesses of the EC films. Using foil devices it is also feasible to make tandem foils in order to reach very low transmittance levels in the colored state. An example of this is shown in Figure B4.6.3.

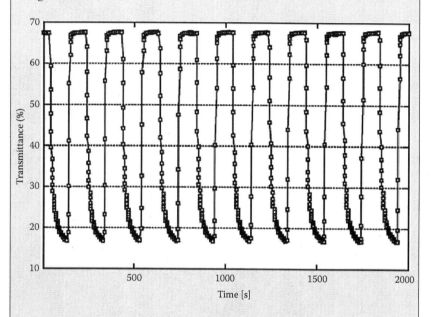

FIGURE B4.6.2 Transmittance versus time at λ = 550 nm for repeated coloring and bleaching of an electrochromic foil device of the type shown in Figure B4.6.1.

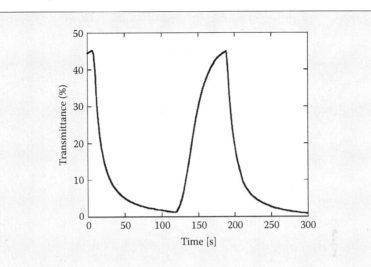

FIGURE B4.6.3   Transmittance versus time at λ = 550 nm for repeated coloring and bleaching of two superimposed electrochromic foil devices of the type shown in Figure B4.6.1.

REFERENCES

1. L. A. B. Pilkington, The float glass process, *Proc. Roy. Soc. London* 314 (1969) 1–25.
2. M. Wigginton, *Glass in Architecture*, Phaidon, London, 1996.
3. C. G. Granqvist, Energy efficient windows: Present and forthcoming technology, in *Materials Science for Solar Energy Conversion Systems*, edited by C. G. Granqvist, Pergamon, Oxford, U.K., 1991; chap. 5, pp. 106–167.
4. M. Born, E. Wolf, *Principles of Optics*, 7th edition, Cambridge University Press, Cambridge, U.K., 1999.
5. International Glazing Database (2009), Lawrence Berkeley National Laboratory, CA; http://windows.lbl.gov/software/window/window.html.
6. K. G. T. Hollands, J. L. Wright, C. G. Granqvist, Glazings and coatings, in *Solar Energy: The State of the Art*, edited by J. Gordon, James & James Sci. Publ., London, U.K., 2001; chap. 2, pp. 29–107.
7. H. Manz, S. Brunner, L. Wullschleger, Triple vacuum glazing: Heat transfer and basic mechanical design constraints, *Solar Energy* 80 (2006) 1632–1642.
8. I. Hamberg, C. G. Granqvist, Evaporated Sn-doped In_2O_3 films: Basic optical properties and applications to energy efficient windows, *J. Appl. Phys.* 60 (1986) R123–R159.

9. C. G. Granqvist, Transparent conductors as solar energy materials: A panoramic review, *Solar Energy Mater. Solar Cells* 91 (2007) 1529–1598.

10. E. Valkonen, B. Karlsson, Optimization of metal-based multilayers for transparent heat mirrors, *Int. J. Energy Res.* 11 (1987) 397–403.

11. E. Valkonen, B. Karlsson, C.-G. Ribbing, Solar optical properties of thin films of Cu, Ag, Au, Cr, Fe, Co, Ni, and Al, *Solar Energy* 32 (1984) 211–222.

12. P. C. Lansåker, J. Backholm, G. A. Niklasson, C. G. Granqvist, TiO_2/Au/TiO_2 multilayer thin films: Novel metal-based transparent conductors for electrochromic devices, *Thin Solid Films* 518 (2009) 1225–1229.

13. G. B. Smith, A. I. Maaroof, M. B. Cortie, Percolation in nanoporous gold and the principle of universality from two dimensions to hyperdimensions, *Phys. Rev. B* 78 (2008) 165418 1–11.

14. G. B. Smith, G. A. Niklasson, J. S. E. M. Svensson, C. G. Granqvist, Noble-metal-based transparent infrared reflectors: Experiments and theoretical analyses for very thin gold films, *J. Appl. Phys.* 59 (1986) 571–581.

15. G. A. Niklasson, C. G. Granqvist, Noble-metal-based transparent infrared reflectors: Improved performance caused by nonhomogeneous film structure, *Appl. Phys. Lett.* 46 (1985) 713–715.

16. A. I. Maaroof, A Gentle, G. B. Smith, M. B. Cortie, Bulk and surface plasmons in highly nanoporous gold films, *J. Phys. D: Appl. Phys.* 40 (2007) 5675–5682.

17. G. B. Smith, A. A. Earp, Metal-in-metal localized surface plasmon resonance, *Nanotechnology* 21 (2010) 015203 1–8.

18. D. R. Sahu, S.-Y. Lin, J.-L. Huang, ZnO/Ag/ZnO multilayer films for the application of a very low resistance transparent electrode, *Appl. Surf. Sci.* 252 (2006) 7509–7514.

19. Y. Tachibana, K. Kusunoki, T. Watanabe, K. Hashimoto, H. Ohsaki, Optical properties of multilayers composed of silver and dielctric materials, *Thin Solid Films* 442 (2003) 212–216.

20. G. B. Smith, A. Maaroof, Optical response in nanostructured thin metal films with dielectric overlayers, *Opt. Commun.* 242 (2004) 383–392.

21. J. Ridealgh, Large area coatings for architectural glass: Design, manufacturing, processing faults, *MRS Proc.* 890 (2006) Y01–10.

22. D. S. Ginley, H. Hosono, D. Paine, editors, *Handbook of Transparent Conductors,* Springer Science and Business Media, 2010.

23. Y. Furubayashi, T. Hitosugi, Y. Yamamoto, K. Inaba, G. Kinoda, Y. Hirose, T. Shimada, T. Hasegawa, A transparent metal: Nb-doped anatase TiO_2, *Appl. Phys. Lett.* 86 (2005) 252101 1–3.

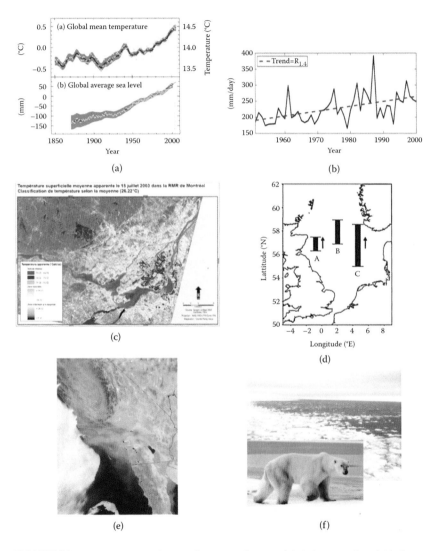

FIGURE P1 Panorama over climate changes and some of their impacts. Panel (a) shows increasing global mean temperature and associated rise of global average sea level since 1850. (See text for full caption and credits.)

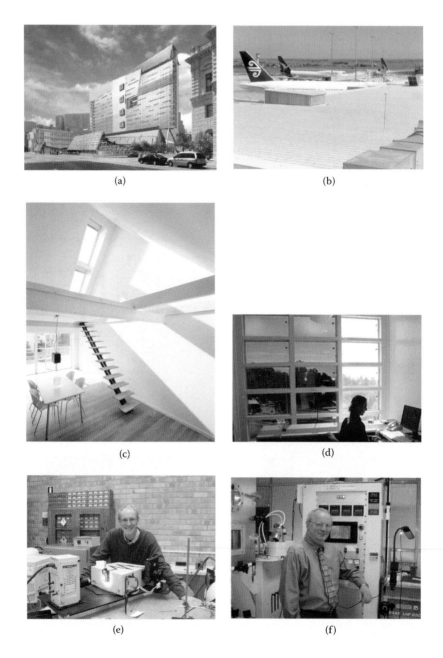

FIGURE P2 Panorama over some aspects of green nanotechnologies. Panel (a) shows a low-energy commercial building in San Francisco with energy-efficient glazing and natural ventilation. (See text for full caption and credits.)

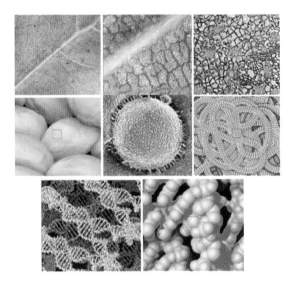

FIGURE 1.2   Images at successive magnifications of a green leaf. The linear scale in neighboring images falls by a factor of ten, starting at 1 cm and ×10 magnification and ending at 1 nm and ×100 million magnification. The last three images are in the domain of nanotechnology, with the final one showing molecular orbitals. (From http://micro. magnet.fsu.edu/primer/java/scienceopticsu/powersof10/index.html. Reprinted with permission from Dr. Mike Davidson, Florida State University National Magnet Laboratory.)

FIGURE 1.3   Examples of brilliantly colored nanoporous systems whose internal pattern of holes and solid matter leads to unique optical effects. The photos show the external color on the left and the associated nanostructure on the right; they refer to "peacock eye" butterfly wings (upper) and a famous opal gem, "the flame queen" (lower). (Images used with permission from the following sources: Upper left: The Peacock Eye butterfly © Copyright Lynne Kirton [image licensed under the Creative Commons Licence. http://creativecommons.org/licenses/by-sa/2.0/]; upper right: H. Ghiradella, *Appl. Opt.* 30, 3492–3500 (1991); lower panel: (c) en.wikipedia.org/wiki/Flame_Queen_Opal [image licensed under Creative Commons ShareAlike 3.0 en.wikipedia.org/wiki/Creative_Commons].)

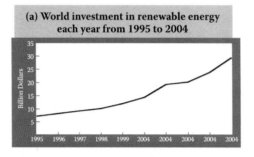

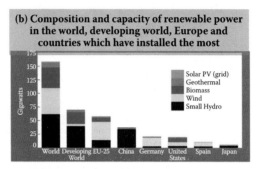

FIGURE 1.5 Panel (a) reports annual world investment (in billions of U.S. dollars) in renewable energy since 1995. Panel (b) shows renewable power capacities in GW for the world and for select groups of nations and individual countries in 2004. (Adapted from REN21 [Renewable Energy Policy Network for the 21st Century], Renewables 2005: Global Status Report, The Worldwatch Institute, Washington, DC, 2005; http://www.ren21.net.)

FIGURE 2.1 This oil painting on canvas by Prince Eugen of Sweden (1865–1947) dates from 1895 and is known as *The Cloud*. It is in the collection of the Gothenburg Museum of Art in Sweden. (From http://www.waldemarsudde.se/xsaml_molnet_g.html. Photo by Lars Engelhardt. Reprinted with permission.)

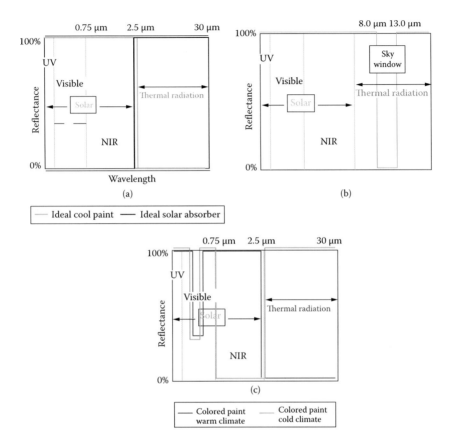

FIGURE 2.15 Ideal spectral reflectance for opaque surfaces. Panel (a) refers to a solar absorber and a surface (paint) for cooling applications; the sharp transition is at ~2.5 μm in wavelength. Panel (b) shows ideal spectral reflectance for maximizing radiation loss to the clear sky; the reflectance is 100% except where the atmosphere is transparent (between 8 and 13 μm). Panel (c) applies to colored surfaces (paints) that look the same but have widely different thermal performance; the properties are ideal for warm and cold climates. Ranges for ultraviolet (UV), visible, near-infrared (NIR), and thermal radiation are shown.

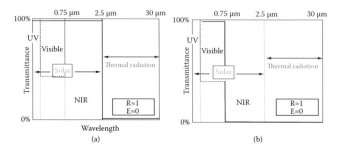

FIGURE 2.16 Ideal spectral transmittance for windows in (a) cold and (b) hot climates; the dotted option at visible wavelengths is for glare reduction. Ranges for ultraviolet (UV), visible, near-infrared (NIR), and thermal radiation are shown.

FIGURE 3.1 Two pieces of glassware, both doped with gold nanoparticles. Part (a) is the Roman Lycurgus cup from the British Museum and part (b) is an old Victorian vase owned by one of the authors (GBS). [Part (a) is from http://www.britishmuseum.org/explore/high-lights/highlight_objects/pe_mla/t/the_lycurgus_cup.aspx. Reproduced with permission.]

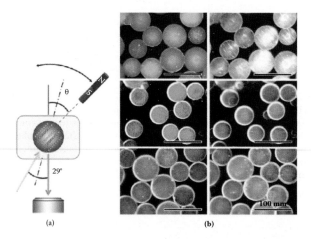

FIGURE B3.1.1 Magnetochromics based on nanoparticle arrays and magnetically rotatable photonic microspheres. Part (a) shows the experimental set-up, and part (b) illustrates the color changes that occur as the angle θ has different values. (From J. Ge et al., J. Am. Chem. Soc. 131 [2009] 15687–15694. With permission.)

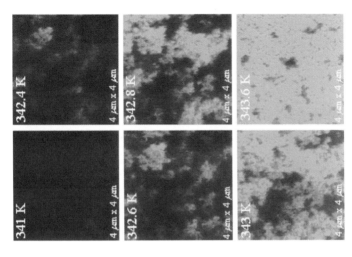

FIGURE 4.41 Scanning near-field micrographs of 4 × 4 μm areas of VO₂ films as measured at the shown temperatures. Dark blue represents an insulating state and light blue, green, and red colors represent metallic conductivity. (From M. M. Quazilbash et al., *Science* 318 [2007] 1750–1753. With permission.)

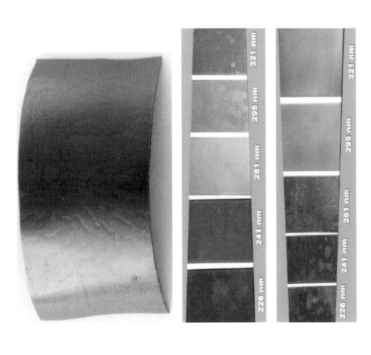

FIGURE 4.34 Color shifts, caused by a range of viewing angles, in reflected light from a photonic crystal such as the one depicted in Figure 4.33. Data are shown for different particle sizes. (R. J. Leyrer, personal communication to one of the authors [GBS]. Figure courtesy of BASF—The Chemical Company.)

FIGURE 5.4 Concentrated illuminance from clear skylights in a supermarket aisle (left-hand panel) and uniform illuminance in another supermarket with diffuse skylights (right-hand panel). The skylights are most easily seen for diffuse lighting. (After J. McHugh et al., Visible light transmittance of skylights, California Energy Commission, PIER, 2004; Contract Number 400-99-013. With permission.)

FIGURE 5.16 The left-hand panel shows a cross section of a one-dimensional photonic polymer crystal constructed from alternating layers; dark-colored layers are PMMA and light-colored layers are (birefringent) PET. The right-hand panel shows relative light transport along mirror light pipes of aspect ratio 17 including a foil of the shown material (a) and including specular reflecting aluminum (b) and silver (c) films. (From M. F. Weber et al., *Science* 287 [2000] 2451–2456. With permission.)

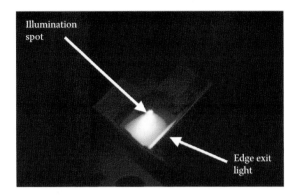

FIGURE 5.25   Light output from a TRIMM-doped PMMA sheet.

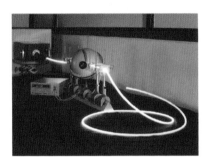

FIGURE 5.26   Flexible transparent polymer light guide containing transparent index matched microparticles. An integrating sphere (cf. Section 2.4.2) is used to measure the total light output at different segments along the guide.

FIGURE 5.28   Two applications of colored LEDs. (From K. Dowling, LED Essentials, Department of Energy, Webinar October 10 [2007]; http://apps1.eere.energy.gov/buildings/publications/pdfs/ssl/webinar_2007-10-11.pdf.)

(a)

(b)   (c)

FIGURE 6.1   Photos illustrating building-integrated devices for photothermal and photoelectric conversion of solar energy. Panel (a) shows a roof-mounted solar collector arrangement from Sweden, used to heat a number of private houses; panel (b) is a solar-cell-covered façade at Solar-Fabrik in Freiburg, Germany; and panel (c) is a solar cell installation forming an architectural element on a building belonging to Fraunhofer Institute of Solar Energy Systems in Freiburg, Germany.

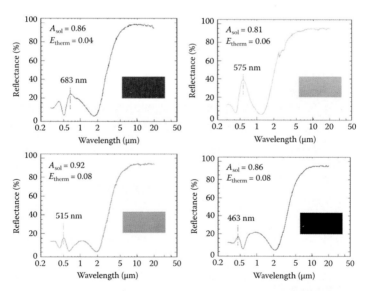

FIGURE 6.12   Spectral reflectance for $TiAlO_xN_y$ films made by sputter deposition onto Al. Different colors (shown) were obtained by selecting film thicknesses leading to reflectance maxima at the wavelengths shown in the luminous wavelength range. Corresponding values of A_{sol} and E_{therm} are stated. (From D. Zhu, S. Zhao, *Solar Energy Mater. Solar Cells* [2010] in press. With permission.)

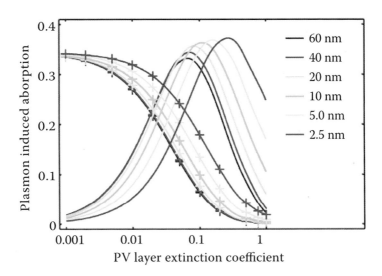

FIGURE 6.19 Absorption induced in various thicknesses of very thin Si by neighboring Ag nanoparticles (peaked curves to right). The x-axis is the absorption coefficient of the Si material. The particles themselves absorb according to the plots with crosses. (From C. Hägglund, B. Kasemo, *Opt. Express* 17 [2009] 11944–11957. With permission.)

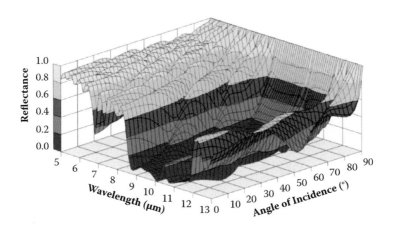

FIGURE 7.12 Spectral and angular-dependent reflectance for nanoparticles of SiO_2 and c-SiC embedded in polyethylene foil. (From A. R. Gentle, G. B. Smith, *Nano Lett.* 10 [2010] 373–379. With permission.)

FIGURE 8.1 Silica aerogel tile. (From http://www.airglass.se. With permission.)

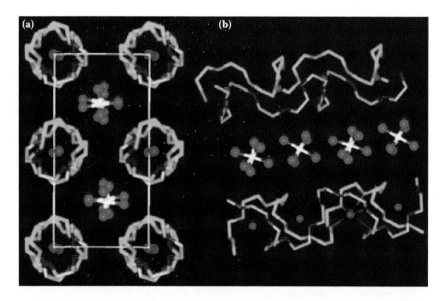

FIGURE 8.16 Structure of the nanostructured polymer ion conductor $PEO_6:LiAsF_6$ as seen looking into the chains and tunnels (a) and along them (b). The Li$^+$ (blue) sits inside the tunnels formed by the PEO. The atoms are fluorine (magenta), carbon (green), and oxygen (red). (From A. S. Aricò et al., *Nature Mater.* 4 [2005] 366–377. With permission.)

24. C. M. Maghanga, G. A. Niklasson, C. G. Granqvist, Optical properties of sputter deposited transparent and conducting TiO_2:Nb films, *Thin Solid Films* 518 (2009) 1254–1258.
25. Y. Shigesato, D. C. Paine, Study of the effect of Sn doping on the electronic transport properties of thin indium oxide, *Appl. Phys. Lett.* 62 (1993) 1268–1270.
26. U. Betz, M. Kharazzi Olsson, J. Marthy, M. F. Escolà, F. Atamny, Thin films engineering of indium tin oxide: Large area flat panel displays application, *Surf. Coating Technol.* 200 (2006) 5751–5759.
27. J. Ederth, P. Heszler, A. Hultåker, G. A. Niklasson, C. G. Granqvist, Indium tin oxide films made from nanoparticles: Models for the optical and electrical properties, *Thin Solid Films* 445 (2003) 199–206.
28. F. Ruske, C. Jacobs, V. Sittinger, B. Szyszka, W. Werner, Large area ZnO:Al films with tailored light scattering properties for photovoltaic applications, *Thin Solid Films* 515 (2007) 8695–8698.
29. I. Hamberg, C. G. Granqvist, Optical properties of transparent and heat-reflecting indium tin oxide films: The role of ionized impurity scattering, *Appl. Phys. Lett.* 44 (1984) 721–723.
30. E. Gerlach, Carrier scattering and transport in semiconductors treated by the energy-loss method, *J. Phys. C: Solid State Phys.* 19 (1986) 4585–4603.
31. C. S. S. Sangeth, M. Jaiswal, R. Menon, Charge transport in transparent conductors: A comparison, *J. Appl. Phys.* 105 (2009) 063713 1–6.
32. J.-Y. Lee, S. T. Connor, Y. Cui, P. Peumans, Solution-processed metal nanowire mesh transparent electrodes, *Nano Lett.* 8 (2008) 689–692.
33. A. K. Geim, K. S. Novoselov, The rise of graphene, *Nature Mater.* 6 (2007) 183–191.
34. Y.-X. Zhou, L.-B. Hu, G. Grüner, A method of printing carbon nanotube thin films, *Appl. Phys. Lett.* 88 (2006) 123109 1–3.
35. B. Dan, G. C. Irvin, M. Pasquali, Continuous and scalable fabrication of transparent conducting carbon nanotube films, *ACS Nano* 3 (2009) 835–843.
36. L. Hu, D. S. Hecht, G. Grüner, Infrared transparent carbon nanotube thin films, *Appl. Phys. Lett.* 94 (2009) 081103 1–3.
37. X. Wang, L. Zhi, K. Müllen, Transparent, conductive graphene electrodes for dye-sensitized solar cells, *Nano Lett.* 8 (2008) 323–327.
38. G. B. Smith, C. A. Deller, P. D. Swift, A. Gentle, P. D. Garrett, W. K. Fisher, Nanoparticle-doped polymer foils for use in solar control glazing, *J. Nanoparticle Res.* 4 (2002) 157–165.
39. S. Schelm, G. B. Smith, Dilute LaB_6 nanoparticles in polymer as optimized clear solar control glazing, *Appl. Phys. Lett.* 82 (2003) 4346–4348.

40. C. Genet, T.W. Ebbesen, Light in tiny holes, *Nature* 445 (2007) 39–46.

41. H. Gao, J. Henzie, T. W. Odom, Direct evidence for surface plasmon-mediated enhanced light transmission through metallic nanohole arrays, *Nano Lett.* 6 (2006) 2104–2108.

42. R. J. Leyrer, personal communication to one of the authors (GBS).

43. X. He, Y. Thomann, R. J. Leyrer, J. Rieger, Iridescent colors from films made of polymeric core-shell particles, *Polymer Bull.* 57 (2006) 785–796; W. Wohlleben, F. W. Bartels, S. Altmann, J. R. Leyrer, Mechano-optical octave-tunable elastic colloidal crystals made from core-shell polymer beads with self-assembly techniques, *Langmuir* 23 (2007) 2961–2969.

44. G. Mbise, G. B. Smith, G. A. Niklasson, C. G. Granqvist, Angular selective window coatings: Theory and experiment, *Proc. Soc. Photo-Opt. Instrum. Engr.* 1149 (1989) 179–199.

45. A. Roos, P. Polato, P. A. van Nijnatten, M. G. Hutchins, F. Olive, C. Anderson, Angular-dependent optical properties of low-E and solar control windows: Simulations versus measurements, *Solar Energy* 69 (supplement) (2000) 15–26.

46. I. R. Maestre, J. L. Molina, A. Roos, J. F. Coronel, A single-thin-film model for the angle dependent optical properties of coated glazings, *Solar Energy* 81 (2007) 969–976.

47. A. G. Dirks, H. J. Leamy, Columnar microstructures in vapor-deposited thin films, *Thin Solid Films* 47 (1977) 219–233.

48. M. J. Brett, Simulation of structural transitions in thin films, *J. Mater. Sci.* 24 (1989) 623–626.

49. J. J. Steele, M. J. Brett, Nanostructure engineering in porous columnar thin films: Recent advances, *J. Mater. Sci: Mater. Electron.* 18 (2007) 367–379.

50. G. B. Smith, Theory of angular selective transmittance in oblique columnar thin films containing metal and voids, *Appl. Opt.* 29 (1990) 3685–3693.

51. G. W. Mbise, D. Le Bellac, G. A. Niklasson, C. G. Granqvist, Angular selective window coatings: Theory and experiments, *J. Phys. D: Appl. Phys.* 30 (1997) 2103–2122.

52. C. M. Lampert, C. G. Granqvist, editors, *Large-Area Chromogenics: Materials and Devices for Transmittance Control*, The International Society for Optical Engineering, Bellingham, WA, 1990; Vol. IS4.

53. A. Georg, A. Georg, W. Graf, V. Wittwer, Switchable windows with tungsten oxide, *Vacuum* 82 (2008) 730–735.

54. A. Roos, M.-L. Persson, W. Platzer, M. Köhl, Energy efficiency of switchable glazing in office buildings, in *Proceedings Glass Processing Days*, Tampere, Finland, 2005; pp. 566–569.

55. A. Jonsson, Optical Characterization and Energy Simulation of Glazing for High-Performance Windows, Ph.D. Thesis, Department of Engineering Sciences, Uppsala University, Uppsala, Sweden, 2010.

56. E. S. Lee, S. E. Selkowitz, R. D. Clear, D. L. DiBartolomeo, J. H. Klems, L. L. Fernandes, G. J. Ward, V. Inkarojrit, M. Yazdanian, Advancement of electrochromic windows, California Energy Commission, PIER, 2006; CEC-500-2006-052.

57. UNEP, Buildings and Climate Change: Status, Challenges and Opportunities, United Nations Environment Programme, Paris, France, 2007.

58. R. D. Clear, V. Inkarojrit, E. S. Lee, Subject responses to electrochromic windows, *Energy Buildings* 38 (2006) 758–779.

59. H. J. Hoffman, in *Photochromic Materials and Systems*, edited by H. Dürr, H. Bouas-Laurent, Elsevier, Amsterdam, the Netherlands, 1990.

60. L. Malfatti, S. Costacurta, T. Kidchob, P. Innocenzi, M. Casula, H. Amenitsch, D. Dattilo, M. Maggini, Mesostructured self-assembled silica films with reversible thermo-photochromic properties, *Microporous Mesoporous Mater.* 120 (2009) 375–380.

61. G. V. Jorgenson, J. C. Lee, Thermochromic materials and devices: Inorganic systems, in *Large-Area Chromogenics: Materials and Devices for Transmittance Control*, edited by C. M. Lampert, C. G. Granqvist, The International Society for Optical Engineering, Bellingham, WA, 1990; pp. 142–159.

62. A. Gentle, G. B. Smith, Dual metal-insulator and insulator-insulator switching in nanoscale and Al-doped VO_2, *J. Phys. D: Appl. Phys.* 40 (2008) 1–5.

63. M. M. Quazilbash, M. Brehm, B.-G. Chae, P.-C. Ho, G. O. Andreev, B.-J. Kim, S. J. Yun, A. V. Balatsky, M. G. Maple, F. Keilmann, H.-T. Kim, D. N. Basov, Mott transition in VO_2 revealed by infrared spectroscopy and nano-imaging, *Science* 318 (2007) 1750–1753.

64. M. A. Sobhan, R. T. Kivaisi, B. Stjerna, C. G. Granqvist, Thermochromism of sputter deposited $W_xV_{1-x}O_2$ films, *Solar Energy Mater. Solar Cells* 44 (1996) 451–455.

65. J. B. Goodenough, The two components of the crystallographic transition in VO_2, *J. Solid State Chem.* 3 (1971) 490–500.

66. N. R. Mlyuka, G. A. Niklasson, C. G. Granqvist, Mg doping of thermochromic VO_2 films enhances the optical transmittance and decreases the metal-insulator transition temperature, *Appl. Phys. Lett.* 95 (2009) 171909 1–3.

67. K. A. Khan, C. G. Granqvist, Thermochromic sputter-deposited vanadium oxyfluoride coatings with low luminous absorptance, *Appl. Phys. Lett.* 55 (1989) 4–6.

68. N. R. Mlyuka, G. A. Niklasson, C. G. Granqvist, Thermochromic multilayer films of VO_2 and TiO_2 with enhanced transmittance, *Solar Energy Mater. Solar Cells* 93 (2009) 1685–1687.

69. N. R. Mlyuka, G. A. Niklasson, C. G. Granqvist, Thermochromic VO_2-based multilayer films with enhanced luminous transmittance and solar modulation, *Phys. Stat. Sol.* A 206 (2009) 2155–2160.

70. H. Bai, M. B. Cortie, A. I. Maaroof, A. Dowd, C. Kealley, G. B. Smith, The preparation of a plasmonically resonant VO_2 thermochromic pigment, *Nanotechnology* 20 (2009) 085607 1–9.

71. P. Nitz, H. Hartwig, Solar control with thermotropic layers, *Solar Energy* 79 (2005) 573–582.

72. O. Muehling, A. Seboth, T. Haeusler, R. Ruhmann, E. Potechius, R. Vetter, Variable solar control using thermotropic core/shell particles, *Solar Energy Mater. Solar Cells* 93 (2009) 1510–1517.

73. K. Resch, G. M. Wallner, R. Hausner, Phase separated thermotropic layers based on UV cured acrylate resins: Effect of material formulation on overheating protection properties and application in a solar collector, *Solar Energy* 83 (2009) 1689–1697.

74. S. K. Deb, Opportunities and challenges in science and technology of WO_3 for electrochromic and related applications, *Solar Energy Mater. Solar Cells* 92 (2008) 245–258.

75. J. S. E. M. Svensson, C. G. Granqvist, Electrochromic coatings for "smart windows," *Solar Energy Mater.* 12 (1985) 391–402.

76. C. G. Granqvist, *Handbook of Inorganic Electrochromic Materials*, Elsevier, Amsterdam, the Netherlands, 1995.

77. P. Monk, R. Mortimer, D. Rosseinsky, *Electrochromism and Electrochromic Devices*, Cambridge University Press, Cambridge, U.K., 2007.

78. C. M. Lampert, Large-area smart glass and integrated photovoltaics, *Solar Energy Mater. Solar Cells* 76 (2003) 489–499.

79. C. G. Granqvist, Electrochromic tungsten oxide films: Review of progress 1993–1998, *Solar Energy Mater. Solar Cells* 60 (2000) 201–262.

80. C. G. Granqvist, E. Avendaño, A. Azens, Electrochromic coatings and devices: Survey of some recent advances, *Thin Solid Films* 442 (2003) 201–211.

81. G. A. Niklasson, C. G. Granqvist, Electrochromics for smart windows: Thin films of tungsten oxide and nickel oxide, and devices based on these, *J. Mater. Chem.* 17 (2007) 127–156.

82. T. Nanba, I. Yasui, X-ray diffraction study of microstructure of amorphous tungsten trioxide films prepared by electron beam vacuum evaporation, *J. Solid State. Chem.* 83 (1989) 304–315.

83. J. B. Goodenough, Metallic oxides, *Prog. Solid State Chem.* 5 (1971) 145–399.

84. A. Talledo, C. G. Granqvist, Electrochromic vanadium-pentoxide-based films: Structural, electrochemical and optical properties, *J. Appl. Phys.* 77 (1995) 4655–4666.

85. L. Berggren, J. C. Jonsson, G. A. Niklasson, Optical absorption in lithiated tungsten oxide thin films: Experiment and theory, *J. Appl. Phys.* 102 (2007) 083538 1–7.

86. J. L. Slack, J. C. W. Locke, S.-W. Song, J. Ona, T. J. Richardson, Metal hydride switchable mirrors: Factors influencing dynamic range and stability, *Solar Energy Mater. Solar Cells* 90 (2006) 485–490.

87. J. P. Ziegler, Status of reversible electrodeposition electrochromic devices, *Solar Energy Mater. Solar Cells* 56 (1999) 477–493.

88. D. Cupelli, F. P. Nicoletta, S. Manfredi, M. Vivacqua, P. Formoso, G. De Filpo, G. Chidichimo, Self-adjusting smart windows based on polymer-dispersed liquid crystals, *Solar Energy Mater. Solar Cells* 93 (2009) 2008–2012.

89. R. Vergaz, J.-M. Sánchez-Pena, D. Barrios, C. Vásquez, P. Contreras-Lallana, Modelling and electro-optical testing of suspended particle devices, *Solar Energy Mater. Solar Cells* 92 (2008) 1483–1487.

CHAPTER 5

Electric Lighting and Daylighting
Luminaires

This chapter is devoted to luminaires, which are defined here simply as sources for artificial and natural light. Of course, the ordinary window can provide excellent light, but it is also capable of giving visual indoors–outdoors contact, and such windows were discussed at length in Chapter 4. Lighting is currently undergoing revolutionary changes, and nanoscience and nanotechnology are instrumental in these changes. Figure 5.1 illustrates one aspect and shows organic nanowires that serve as very efficient emitters of white light [1]. This type of lighting will be discussed in detail later in this chapter.

Lighting, thermal comfort, and air exchange are the three design aspects of a building with the largest influence on its energy consumption. They also determine the quality of life and work within a building. Light has two key roles: to enable activities whether for work or leisure, and to give visual form and appeal of interior spaces. The quality of the indoor environment is multifaceted and includes

- Daylight, glare, and thermal effects of light
- Access to views and windows
- User-operated and automatic control systems for lux levels and blinds
- Wall, floor, and ceiling materials and their colors
- Differentiation of task lighting and general lighting

These qualities must be compatible with low-energy lighting solutions. Much lighting is also used for TV screens and computers, in display and signage, and for a variety of outdoors applications. All of these lighting demands, whether indoors or outdoors, have considerable scope for energy savings and better quality. Finally, one must not forget that plants need

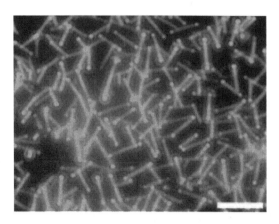

FIGURE 5.1   Organic nanowires emitting white light after color mixing.
The scale bar indicates 5 μm. (From Y. S. Zhao et al., *Adv. Mater.* 20
[2008] 79–83. With permission.)

lighting, and many are grown indoors in greenhouses, hydroponic farms
(relying on mineral nutrient solutions but not on soil), and in offices.

5.1 LIGHTING: PAST, PRESENT, AND FUTURE

Progress in lighting technology and in its energy efficiency have largely
mirrored the growth in standards of living within advanced economies.
Lighting is now more than 3000 times less expensive in real terms and
700 times more energy efficient than it was in 1800, and it is 80 times
less expensive and 50 times more efficient than in 1900 [2,3]. These
changes have led to the average home in industrialized countries having
a light-generating capacity of 200 times more lumens than its counterpart in 1800. Figure 5.2 illustrates the cost for lighting using gas, kerosene, and electricity [2].

Lighting use has outstripped lighting efficiency because costs have
fallen so much [2]. And the costs will soon fall even further, as indicated
in Figure 5.3 [4,5]. It can hardly be denied that advances in lighting
technology have underpinned the incredible advances in global living
standards for so many during the last 150 years. Good lighting simply
enables us to achieve more and perform better, and it is one essential
step toward creating broader economic opportunities in poorer nations.
The challenge now is to raise these already huge energy efficiency gains
by another factor of two to three (i.e., we need to become 1400 to 2000
times more efficient than in 1800). If lamp lifetime also goes up by a
factor of three to four, and the penalty in lamp costs for this is small,

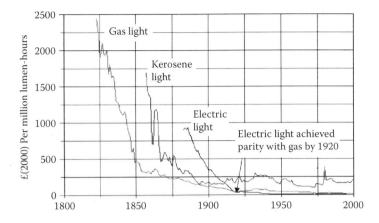

FIGURE 5.2   Evolution of lighting costs in the UK since 1800. (From R. Fouquet, P. J. G. Pearson, *Energy J.* 27 [1] [2006] 139–177. With permission.)

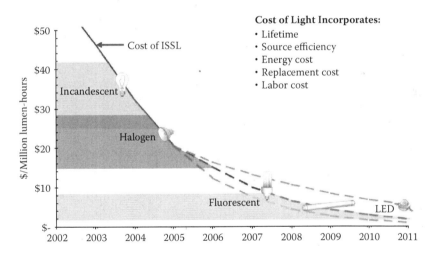

FIGURE 5.3   Overall cost of lighting with different lamps during a ten-year period. The cost of lighting and its division into various lamp segments are shown. The cost incorporates lifetime, source efficiency, energy cost, replacement cost, and labor cost. ISSL denotes intelligent solid state lighting. (From www.colorkinetics.com/support/whitepapers/CostofLight.pdf.; K. Dowling, LED Essentials, Department of Energy, Webinar October 10 [2007]; http://apps1.eere.energy.gov/buildings/publications/pdfs/ssl/webinar_2007-10-11.pdf.)

then we are aiming for lighting that is 12,000 to 30,000 times cheaper than it was in 1800 and 200 to 500 times cheaper in real terms than in 1900. It is interesting to speculate what such a gain might do for world economic growth, standards of living, and health, given the many millions of people who still rely on candles or kerosene. Add in controls and more daylight, as discussed later, and overall efficiency and costs improve even more.

A scientific approach to lighting requires understanding of the various aspects of the luminaire, including how it is powered, factors affecting the spatial distribution of light, the influence of windows and interior surfaces, required lux levels for different functions, and, most important, of combined eye/brain response and personal preferences. What we "see" and how we react are determined by the image processing that takes place in our brains when presented with the raw data taken in by each eye. Just as in the familiar and ubiquitous digital camera-computer processor link of today, human and animal vision—and any response that follows—is ultimately a computational process. This may also be why natural light is, in general, preferred to artificial light, provided it is delivered in a way that eliminates glare and minimizes solar heat gain in warm climates or heat loss in cold climates. Our brains may instinctively sense that this is the light in which humans evolved and thus enjoy. Hopefully, 21st century brains will increasingly note that this light is also free of cost and of carbon.

Energy savings and lifetime costs must be considered and are, of course, linked. By analogy with electricity costs in cents/kWh, lighting is costed in \$/[million lumen-hours]. A million lumen-hours (MLh) typically is achieved in running a lamp for one to two years of normal operation. The world currently uses 4×10^{10} MLh per year. Figure 5.3 compares this unit of cost for different lamps with consideration of maintenance and replacement costs, as well as power consumed and lamp life. The cost C in \$/MLh can be estimated from

$$C = (1000/Q)[(P + H)/L + RW] \qquad (5.1)$$

where Q is mean lumens, P is lamp price in \$, H is labor cost to replace a lamp in \$, L is rated lamp life in units of 1000 h, W is power required in watts, and R is cost of energy in \$/kWh.

This lifetime cost equation is usefully rewritten in terms of a widely quoted quantity, the lamp efficacy K in lumens per watt, which was introduced in Chapter 2. Using total energy E_L in kWh, consumed over the lamp life, the cost equation can be rewritten in terms of K according to

$$C = (1000/K)[(P + H)/E_L + R] \qquad (5.2)$$

This formulation, with cost inversely proportional to K, shows clearly the dominant influence of efficacy on costs. The cost per annum is $C \cdot \mathrm{MLh(pa)}$, with MLh(pa) the million lumen lamp hours used each year. The value of MLh(pa) can be reduced substantially by more use of daylight combined with good overall lighting design and room décor plus intelligent lighting control.

Despite their low purchase costs, standard incandescent bulbs are so inefficient and wear out so fast that they actually are more expensive to run than all of the other alternatives, as clearly seen in Figure 5.3. These rankings date from 2005 but are more or less on track in 2010. Of course, lamp prices vary widely with volume purchased. There is some dispute whether average LED products are as good as fluorescent ones in terms of lifetime costs, and confusion also arises because lamp lumens and lumens output into a room or space are not the same. All lamp types need to be downgraded for housing losses, sometimes substantially, and this should be included in the price of a million lumen-hours.

If one includes the influence of the luminaire design on the lumens provided, then lamp lumens Q and efficacy K should be downgraded to Q^* and K^*, respectively, in Equations 5.1 and 5.2. This, however, is a local design issue so Figure 5.3 and the two equations still serve as good guides for choosing lamps. The lifetime L is often contested because it depends on the electric power W used relative to the maximum rated value W_m. W can be set below W_m initially and ramped up over time to yield long-term stable lumens and longer life. L should achieve 50,000 h and can approach 100,000 h with such a strategy, which is equivalent to 10 to 20 years at 10 h per day. A full cost analysis also should add in the luminaire cost per annum, as averaged over its lifetime. Unless redecoration occurs often, the luminaire life generally exceeds lamp life by a large factor, but given the emerging values of L, there is a distinct possibility that luminaire replacement frequency soon could exceed or match that of lamps.

New classes of energy efficient lamps are emerging and challenge today's compact fluorescent lamps, LEDs and OLEDs. Some of these new lamps utilize nanostructures, as indicated in Figure 5.1 and discussed below. The way the lamp is powered is also an important consideration and can impact overall efficiency. Lamps that run off dc power, including LEDs, are of particular interest for solar-cell-driven lighting and can utilize solar charged batteries and possibly even super-capacitors (cf. Section 8.4). Thus, energy efficient lighting involves many issues in addition to the lamp. The best starting point—as with energy in buildings in general—is to minimize the amount of powered lighting that is needed. Lighting design, room décor, and controls all contribute, but daylighting is the first issue to address. And new materials and nanotechnologies play an important role in daylighting systems.

5.2 DAYLIGHTING TECHNOLOGY: THE "COOL" OPTION

Natural light is one of the coolest light sources available, but we tend to get fooled by its intensity and accompanying solar heat, the management of which is an integral aspect of all good daylighting technology. "Coolness" is measured by efficacy K, and outdoors daylight has $K \approx 100$ LW^{-1} because it is accompanied by much invisible NIR solar energy. Indoors, if daylight comes through a glazing that filters out a large proportion of the NIR, then K can approach 200 LW^{-1}, which is about as cool as light can get if it is color neutral. With 40,000 lumens per square meter (lux), say, there are potentially far too many watts in daylight, and one only needs between 150 and 500 lux for visual functions and comfort. Spreading the daylight and limiting its ingress hence are essential. "Coolness" is not the only bonus with daylight, though, and it usually comes with a view and always creates dynamic environments that are radically different from the static sterility of spaces lit entirely by lamps. Psychologically, it is good to stay in touch with at least part of the natural world.

According to a highly influential architect [6], daylight stands alone as "the most important design issue" for buildings. Why is this? Because human life is strongly influenced by architecture's functionality, form, and appearance, and today we have to add in the effect of energy use on the environment. The role of architecture might seem too strongly articulated, but it should be remembered that in modern society we spend some 80 to 90% of our time indoors, in buildings and vehicles [7]. As in the best works of art, architectural quality and visual impact depend critically on how lighting is rendered. The advent of the tubular fluorescent tube in the mid-1900s blinded people to the merits of natural lighting and led to changes in architectural form since sunlight was then perceived to be unnecessary. Many office towers of today are a sad legacy of that thinking, but new premises are gradually diversifying in form as needed to make the most of daylight and minimize heating and cooling demand in any specific location.

The light supplied by daylighting is often quantified using European conditions and hence overcast skies. Then the daylight factor (DF) is defined as DF $= I_{in}/I_{out}$ with I_{in} being the indoors horizontal illuminance and I_{out} the outdoors illuminance under an unobstructed overcast sky. DF is proportional to the solid angle Ω_{sky} that a window or skylight "sees" of the sky. For a skylight, with a 180° zenith angle swing, this solid angle can be up to 2π steradians; it is twice as large as the maximum solid angle for a vertical window which is π steradians near ground level where the window only has a 90° zenith swing. The lighting needs of a room under overcast conditions can thus be supplied by the daylight falling on a small fraction of the roof area or a larger area of a window. The same relative factors apply in places where clear skies predominate, but then the average illuminance is much higher so the required ratio between roof area and façade area is smaller. The illuminance can exceed 100,000 lux on a roof, while it ranges from a few thousand to 70,000 lux on vertical planes [8]. One should remember that daylight is beam-like rather than diffuse, so modified ray tracing analyses are often needed for detailed studies. And the ability to have controlled diffusion of the transmitted light is an important consideration when clear skies predominate.

5.2.1 Roof Glazing and Skylights

The illuminance on a roof can be very large, as noted above, and hence small apertures can supply all the lighting needs. A room with a floor area of 30 m^2 can be adequately lit by a roof aperture under 0.5 m^2 in size, with one important caveat: The daylight must be spread into the space as uniformly as possible because otherwise it ends up primarily just under the skylight, at least if the sky is clear. There are situations with clear glazing on a skylight for which illuminance levels in a room can range from 40,000 lux to below 150 lux. In this situation the low-lit parts of the room look exceptionally gloomy. Figure 5.4 shows an example for two supermarkets, one with clear and one with diffuse skylights [9]. Without the supporting lamp lighting, the duller sections in the building having the clear glazing skylight would barely be visible in this photo.

Technologies based on microstructured and nanostructured materials can achieve the desired spectral, directional, and intensity properties for transmitted solar radiation. Replacing clear skylight glazing with curved panels of PMMA or multiwall polycarbonate panels, in both cases pigmented with compounds such as $BaSO_4$ and $CaCO_3$, leads to more comfortable and uniform levels of illumination, ranging from 800 to 1000 lux. This lux level is about right, since people expect a larger illuminance when the sun is shining brightly. Backscattering from the pigments reduces the daylight throughput by around 50%, and this is accompanied by a parallel reduction in solar heat gain. Light diffusing materials are a core issue in this chapter for various luminaires, so a special section is devoted to them below.

Nano- and micromaterials that can switch from clear to translucent as they get hotter, which happens faster on a roof than on a vertical window, are of interest for controlling the illumination. These are called cloud gels or thermotropic materials, and were discussed in Section 4.10. Electrochromic glazing, discussed in Section 4.11, is also worth considering for skylights and roof glazing as it enables dynamic and seasonal management of glare and solar heat gain. It is also less of a cost impost to have smart glazing for a roof window than in a series of façade windows

FIGURE 5.4 *A color version of this figure follows page 200.* Concentrated illuminance from clear skylights in a supermarket aisle (left-hand panel) and uniform illuminance in another supermarket with diffuse skylights (right-hand panel). The skylights are most easily seen for diffuse lighting. (After J. McHugh et al., Visible light transmittance of skylights, California Energy Commission, PIER, 2004; Contract Number 400-99-013. With permission.)

capable of supplying the same lumens. Another good option for roofs and skylights is angular selective glazing which, as we will see below, can even out lumen throughput over a day and for different seasons as the sun's position in the sky varies.

Solar gains, glare, and winter heat loss can be problems for large-area roof glazings and skylights. Each of these factors has a bearing on cost savings, but the major benefits of glazings and skylights are in lighting costs, at least for large buildings such as shopping centers and airport terminals. Savings are worthwhile in homes, too. They may be small in each dwelling, but summed over a large fraction of the housing stock they can lead to very substantial reductions of emissions. And in addition to the actual cost savings, there are psychological factors that tend to add value.

Figure 5.5 illustrates calculated cost savings for skylights, consisting of diffusely scattering double acrylic sheets, versus skylight to floor area ratio (SFR) for the case of a 2,500 m² retail store in San Francisco [9,10]. A photocontrol system for dimming the electrical light was involved, too. The main finding is that the introduction of diffusing skylights leads

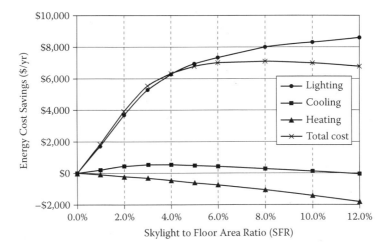

FIGURE 5.5 Energy cost savings due to skylights as a function of the ratio of its area to that of the floor. The data apply to a retail store in San Francisco. (From J. McHugh et al., Visible light transmittance of skylights, California Energy Commission, PIER, 2004; Contract Number 400-99-013. With permission.)

to large energy cost savings. These benefits level off when the SFR is about 6%, and larger SFRs make the savings drop. The heating cost savings goes down somewhat as SFR is increased, while the savings on cooling cost have a shallow maximum when SFR is about 4%.

Many national building standards and codes sensibly impose SFR limits for skylights consistent with the findings in Figure 5.5, unless special glazings are used. Better skylight design could lead to a smaller penalty resulting from the thermal conductance (U-value) than in the study reported on in Figure 5.5, and then a larger SFR would be optimum. Larger skylights may not be needed for lighting purposes, though. Switchable glazings could amplify these savings in the future.

The analysis so far has not accounted for any roof cavity. However, most installations of skylights use a light well to bridge the gap between roof and ceiling, and many wells are tapered outward from their base and hence are trapezoidal in shape. The well reduces the useable lumens and alters their distribution in a room. It adds a well efficiency (WE) factor which depends on the reflectance of the cavity walls, assumed to be diffuse, and the well aspect ratio (height/width). The standard reference in this field [11] uses a related property: the ratio of the cavity wall area to the exit aperture area, with wall area proportional to height. If this ratio is between 4 and 5, then WE is between 0.75 and 0.65 for a highly reflective white surface. If the well's wall reflectance drops to

80%, WE lies between 0.60 and 0.45. These losses are large, but they can be compounded if a light transmitting polymer or glass cover is used at the ceiling exit level. An increase in skylight area can also compensate. Some large-area, cheap-to-install "concertina devices" use thick, shiny aluminum folding foil on the well walls. Such wells are usually diffuse due to crinkles and hence lossy and need larger apertures than those with specular walls. The latter walls can cause severe glare unless highly diffuse glazings are used at ceiling level, but such diffusers generally have poor transmittance. Methods to resolve this glare problem are treated in the next section, and also in Section 5.4.

5.2.2 Mirror Light Pipes

Small-area cavity light pipe devices, which can be installed with minimum impact on the building, were introduced in the early 1990s and are now widely used for daylighting from a roof. The key parameters for cylindrical light pipes, which are the most common, are the aspect ratio p (i.e., length divided by diameter, and the specular reflectance R) [12]. The integer number ζ of individual reflectance events determines the transmittance. Since only two consecutive ζ values are possible for beam radiation, as seen in Figure 5.6, a weighted average of R^{ζ} and $R^{\zeta+1}$ comes into play at any angle of incidence. Apart from p, the angle of incidence of sunlight thus has the major effect on the lumens supplied, and the light throughput can drop dramatically as the sun's angle to the horizon falls so that ζ increases. Diffuse sunlight will also have a much reduced transmittance, and many rays with large ζ values will be present. The schematic in Figure 5.6 illustrates the fundamental issues for a two-dimensional slice and shows that one set of rays has $\zeta = 4$ while the remainder of the rays have $\zeta = 5$.

An exact theory of the transmittance through light pipes with circular and rectangular cross sections has been published [12,13] and

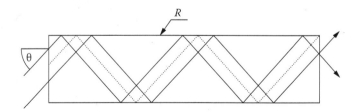

FIGURE 5.6   Schematic for two sets of rays, incident at an angle θ to the axis, traveling down a mirror light pipe. (From P. D. Swift, G. B. Smith, *Solar Energy Mater. Solar Cells* 36 [1995] 159–168. With permission.)

gives excellent fits to experimental data on devices with polymer walls coated with silver, as seen in Figure 5.7 for two different aspect ratios and a full set of incidence angles. Empirical data showed that significant color shifts can occur along with the multiple reflections unless the reflectance is essentially wavelength-independent, which is not the case for silver or aluminum.

Results such as those in Figure 5.7 are very sensitive to the specular reflectance, and if a higher R was used then the light transmission became much larger and the dependence on angle of incidence was much reduced. For example raising R from 95 to 98% and setting $\zeta =$

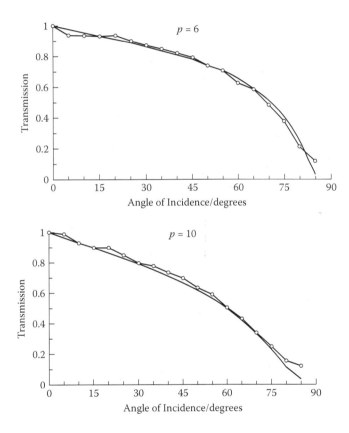

FIGURE 5.7 Theoretical (curves) and measured (dots joined by lines) light transmittance versus angle of incidence for cylindrical mirror light pipes. Their walls were specular mirrors with a measured reflectance of 79.2% at the wavelength used in the model. Data are shown for two values of the aspect ratio p. (From P. D. Swift, G. B. Smith, *Solar Energy Mater. Solar Cells* 36 [1995] 159–168. With permission.)

4.3, which is typical for an average incidence angle of ~35°, changes the transmittance from 80 to 92%.

Hollow cylinders lined with mirrors based on one-dimensional photonic crystals can do even better. They use multiple nano-thin polymer layers and can achieve R values as high as 99%. They can thus strongly outperform the best metal mirrors in this application or can enable smaller pipe cross sections for the same lumens. Supplying useful daylight to floors not directly under a roof then also becomes possible. Cost and large-area availability seem to have limited the use of dielectric mirrors in this field to date, though. An outline of nanostructured dielectric mirrors follows in Section 5.3, as it is relevant also to other aspects of lighting technology.

An alternative to having highly reflecting walls is to transmit inside solid clear polymers with smooth walls which act like fiber optic light guides. In common with fiber optics, they have a limited acceptance angle, but this can be overcome using some focusing or an angular selective device at the inlet. The main disadvantages of these devices are cost and heavy weight, plus the need for very smooth outer surfaces which must be protected. Having low-index tightly fitting outer layers means that even less daylight is collected, but protection arranged with a small air gap between the inner jacket and the low-index wall is possible and can improve the light throughput.

Hollow light pipes supply variable levels of light over a day and with change of season because of the angle-of-incidence effect shown in Figure 5.7. It is of much interest to flatten the temporal and seasonal response (i.e., get more light during the winter and less at noon). To this end a polymer sheet containing aligned cavities, to be positioned at the top of the light pipe, has been proposed and analyzed [14,15]. This device also diminishes the solar heating in the summer. Figure 5.8(a) illustrates the function of this device and compares it with a standard light pipe. Clearly the polymer sheet decreases the value of ζ, and hence enhances the light throughput for near-horizontal light. A photo of the device, and a rendition of its operation as a lighting device, are given in Figs. 5.8(b) and (c), respectively. Laser cut grooves are clearly visible in Figure 5.8(b).

Glare remains a problem for light pipes having specularly reflecting walls since various bright patterns form in the outlet light as the incidence angle varies. This effect is demonstrated in the lower row of photos in Figure 5.9 [16]; the patterns are called "caustics" because they can be harsh to view. Having a strong diffuser at the outlet of the light pipe can eliminate the problem but is not ideal. In fact these pipes supply only about 1,200 lumens as a maximum, and this value could be strongly reduced by backscattering. And weaker outlet diffusers do not eliminate glare. Having special diffusers which scatter into a narrow forward cone

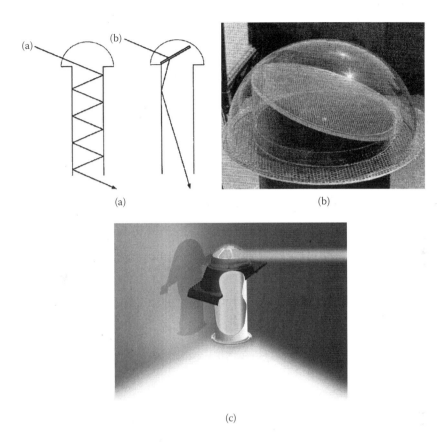

(a)   (b)

(c)

FIGURE 5.8   Panel (a) indicates the functioning of a standard light pipe (left) and a light pipe with a sheet having aligned cavities (right). Panels (b) and (c) show a photo of the device and its operation as a lighting device, respectively. (From I. R. Edmonds et al., *Lighting Res. Technol.* 27[1] [1995] 27–35. With permission; www.skydome.com.au.)

(cf. Section 5.5) in the top dome solves the problem, though, as shown by the images in the upper row of photos in Figure 5.9. It is obvious that the material in the diffuser must be selected with care, since too much diffuseness at the inlet will badly reduce the light throughput. For the same reason these light pipe systems do not work so well in places with low average hours per day of clear skies.

Apart from laser cut panels, such as those shown in Figure 5.8(b), internal cavities can be created by joining two very thin sheets of PMMA which incorporate mass-produced, micro-replicated prisms. When bonded together, the two prisms combine to create nearly flat microscopic air pockets or lamellae. Under illumination, they are able

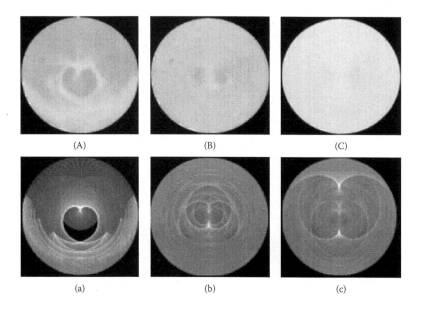

(A)   (B)   (C)

(a)   (b)   (c)

FIGURE 5.9   Caustics, or "hot-spot" light patterns, that form in perfect mirror light pipes at different angles of incidence (lower photos) and their experimentally found nearly complete elimination with increasing amounts of diffuser in the top dome (upper photos). (From P. D. Swift et al., *Lighting Res. Technol.* 38 [2006] 19–31. With permission.)

to split the throughput light and direct the two exit beams differently depending on the angle of incidence, as seen in Figure 5.10 [17]. Internal cavities have the advantages of staying clean and being easier to handle (though the laser cut ones are protected in Figure 5.8). These two systems for redirecting light also have applications in general angular selective heat control in glazings [18].

Ray tracing plots, such as those in Figure 5.10, do not tell the complete story since nano-optical effects often come into play in polymer-based macro- and micro-structures used in daylighting, as discussed below. These effects add new paths for the optical beams, thereby giving more complex patterns for the light output, and they also impair visual indoors–outdoors contact.

Finally, we note that mirror light pipes have also been studied and deployed as a means of transporting light away from the façade and deeply into buildings. Such pipes must have a horizontal axis, so an external optical device (examples are given in the next section) is needed to redirect daylight into the light pipe. Varying light output, as a function of distance from the wall, can be used to obtain uniform lighting deeply into a building [19]. Combinations of electrochromic "smart"

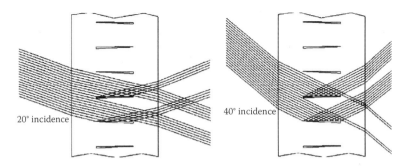

FIGURE 5.10 Internal cavities in PMMA sheet for redirecting solar radiation. (From http://www.bendinglight.co.uk/assets/pdf_downloads/How_Serraglaze_Works.pdf.)

windows and light pipes with controlled light output also allow even levels of illumination in deep rooms. This novel concept of "light balancing" enables energy-efficient lighting by optimally combining artificial and natural light [20,21].

5.2.3 Daylight Redirecting Structures for Façades

The internal groove structure, discussed above for mirror light pipes, is only one example of a variety of configurations which have been examined for applications on low-cost light redirecting systems. Most of them use structures on the surface of polymers or inside them. These structures can be produced by injection molding, casting or imprinting. Large areas are needed, and making the required masters for imprinting or molding with the necessary precision may be technically daunting but is manageable [22].

Figure 5.11 shows schematically the steps to produce large-area patterning by interference lithography [22]. The manufacturing scale requires detailed attention to optical control of the two beams in order to reduce net phase drift over the over the full area. Surface micro- and nanostructures can also be made using stamping from replicas prepared by holographic lithography [23].

Figure 5.12 illustrates some different surface structures that can be used for incident-angle-controlled light transport [24]. These are in effect sun-shading systems and allow much better use of daylight and better glare and heat control than blinds or curtains. They can also be used as part of a Venetian blind or louver system. Other light redirecting systems have also been applied for louvers [18].

Figure 5.13 elaborates on one of the microstructure types in Figure 5.12 and shows its effect on the total (hemispherical) transmittance [25]. The

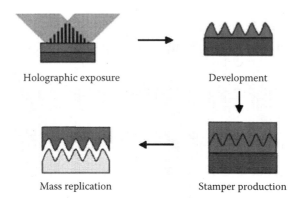

Holographic exposure   Development

Mass replication   Stamper production

FIGURE 5.11   Holographic lithographic steps for producing large-area stamps to mass produce daylight redirecting structures on glass or polymer. (From B. Bläsi et al., *Proceedings ISES Solar World Congress*, Gothenburg, Sweden, June 14–19, 2003; ISBN: 91-631-4740-8.)

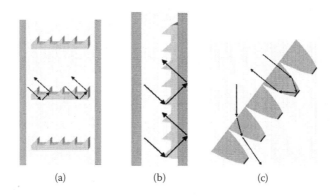

(a)   (b)   (c)

FIGURE 5.12   Solid microstructures for control of daylight and solar heat according to incident ray direction. The images are for louvers (a), windows (b) and compound parabolic structures (c). (From A. Gombert et al., *Opt. Engr.* 43 [2004] 2525–2533. With permission.)

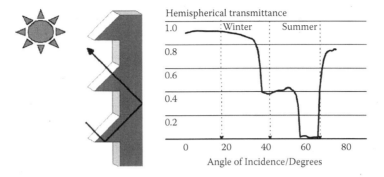

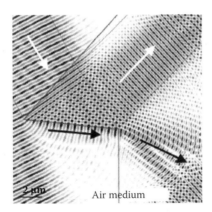

FIGURE 5.13   Schematic of an angular selective solar heat control system based on external polymer microstructures (left-hand part) and the impact of this structure on the solar heating during summer and winter (right-hand part). (From G. Walze et al., *Proc. SPIE* 6197 [2006] 61970Z 1–9. With permission.)

angular dependence makes the particular microstructure well suited for seasonal control of the throughput of solar energy.

The physical mechanism responsible for redirection of light rays to the exterior for a limited range of high-angle directions but allowing in others at lower angles is total internal reflection (TIR) at certain polymer–air interfaces, as illustrated in Figure 5.14 [25]. The distortion of the view in transmission is due to a number of features which include non-

FIGURE 5.14   Wave front perspective on the origins of the distortions of the external view caused by surface microstructures designed for angular selectivity, of the type shown in Figure 5.13. White arrows delineate directions of incident and reflected rays and black arrows show evanescent fields at TIR and the emergence of a new wave front. (From G. Walze et al., *Proc. SPIE* 6197 [2006] 61970Z 1–9. With permission.)

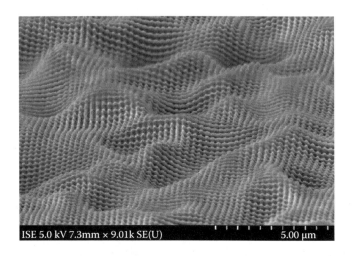

FIGURE 5.15   Microstructure overlaid with a nanostructure for anti-reflection without glare. (From A. Gombert et al., *Opt. Engr.* 43 [2004] 2525–2533. With permission.)

smooth laser cut surfaces and, for external surfaces, unavoidable wave- or diffraction effects occurring near the sharp edges in the construction. A TIR event is in fact not physically localized but a slight lateral shift occurs (called the Goos–Hänchen shift). The origin of this shift, and subsequent wave effects such as those shown in Figure 5.14, are connected to the evanescent fields present in TIR, as discussed in Sections 3.3 and 3.11. It is apparent from this figure that some evanescent fields can enter the polymer before the reflection event is complete, thus leading to some light in directions not predicted by ray optics [24–26].

Glare due to complex interference at some points might be an issue as a result of sharp corners. However, the problem can be made less disturbing by using a combination of micro- and nanostructure. Such a surface is illustrated in Figure 5.15 [24].

5.3 DIELECTRIC MIRRORS BASED ON NANOSTRUCTURE

Mirrors with very high reflectance over limited wavelength ranges can be made entirely from insulating materials by exploiting wave interference within highly ordered nanostructures. There are various names for these materials, but the underlying physics behind their properties are closely related. The names arose historically, and the three material categories discussed below are called photonic crystals, Bragg filters, and cholesteric or liquid crystals. They have periodicity or repeat units in their structure and can be constructed from properly chosen transmitting polymers, oxides, glasses and molecular liquids. They involve at least two materials with different refractive index and are thus a special category of nanocomposites. Photonic crystals may involve three different materials.

Wave interference depends on phase difference and varies with wavelength and angle of incidence. This makes it possible to devise nanomaterials which reflect strongly over a limited band of wavelengths and transmit strongly at others. Such systems can be very useful, for example, in combined solar cell and thermal collection systems. The interest for lighting, as noted earlier for mirror light pipes, is to maximize the throughput of daylight when multiple reflections occur. For applications in luminaires or LCD display system, the object is to maximize light output and hence reduce power needs. Some mobile phone displays already use dielectric mirrors.

The difference from normal thin films, which can also achieve high reflectance, is that the dielectric mirrors can have better bandwidth control, larger extremes such as reflectance values up to 99%, and in some cases angle of incidence control. The drawback is that a large number of layers or photonic crystal planes are needed to provide the necessary reflection. Accurate repeatability of the structure, as in a crystal lattice, is the key to such systems since this ensures that the phase relations between waves reflected at each interface or crystal planes are preserved.

We are mainly interested in planar or large radius-of-curvature mirrors, but the underlying physical effects can also be utilized on the outer surfaces and layers of microstructured or holey fiber optic light guides which are flexible. Light is then confined to a small area in the fiber core. Surfaces containing ordered arrays of conducting nanoparticles make it possible to guide light in the form of surface plasmons within nanoscale channels. Surface plasmons can even turn sharp corners in such guides, which light cannot do in fiber optics. Fiber optic guides play a role in daylighting, signage and display, and we will return to them in Section 5.4.

One-dimensional photonic crystals make excellent mirrors and can be fabricated entirely from plastic. The physical principles appropriate to those materials apply to the all-polymer-based spectrally selective foils for windows, discussed in Section 4.1, where the reflecting bands were tuned to the NIR rather than to the luminous range as for many other applications. Two or more different polymers, for example, PMMA and PET, form alternating layers with different refractive indices. Each layer of the same type has identical nano-thickness, as shown in Figure 5.16(a) [27]; the two kinds of polymers have different layer thicknesses in that system. The PET layer is optically anisotropic (birefringent), which

FIGURE 5.16 *A color version of this figure follows page 200.* The left-hand panel shows a cross section of a one-dimensional photonic polymer crystal constructed from alternating layers; dark-colored layers are PMMA and light-colored layers are (birefringent) PET. The right-hand panel shows relative light transport along mirror light pipes of aspect ratio 17 including a foil of the shown material (a) and including specular reflecting aluminum (b) and silver (c) films. (From M. F. Weber et al., *Science* 287 [2000] 2451–2456. With permission.)

means that the refractive index is different for light polarized perpendicular to the plane of the film and for polarization in the plane of the film. Interference effects in thin-film stacks and one-dimensional photonic crystals normally vary with angle of incidence, wavelength, and polarization of incident light as explained briefly in Chapter 3.

Phase changes, and hence interference, depend on both path length traveled and differences between refractive indices in adjacent layers. It is desirable that the dependence on angle of incidence and wavelength is reduced for lighting applications in order to avoid unwanted color shifting, and this can be accomplished if index components in adjoining layers have special relations [27]. One such special case is if the normal components of the refractive indices in adjoining layers, for example PET and PMMA in Figure 5.16, are nearly equal; another special case is if the difference between the normal and in-plane components of the refractive index has different signs in adjacent layers.

The system in Figure 5.16(a) exemplifies giant birefringent optics (GBO) and can be manufactured on a large scale. Two obvious advantages of such a system in mirror light pipes, relative to metal mirrors, are seen in Figure 5.16(b): the output is much brighter and the daylight's color is not shifted after multiple reflections. Clearly the aluminum mirror surface makes the transmitted light bluer while the silver mirror makes it redder.

The wave mechanism behind the high reflectance and photonic band gap in GBO is illustrated in Figure 5.17 [28]. The reflectance is the result of two different standing waves forming in the periodic structure when

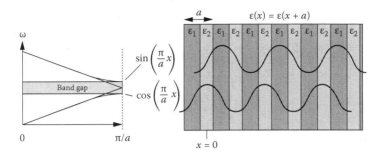

FIGURE 5.17   Standing waves and the frequency gap at which light cannot propagate in a one-dimensional photonic crystal. Note the periodicity in the dielectric constant ε with the repeat distance $2a$. (From http://ab-initio.mit.edu/photons/tutorial/L1-bloch.pdf; http://ab-initio.mit.edu/photon/tutuorial/spie-course-new.pdf. With permission.)

the wavelength equals $2a$, with a being the basic repeat distance. The two distinct standing waves form from adding two equal-amplitude waves traveling in opposite directions. The latter arises because the net back reflectance of each component traveling wave is very strong. Two outcomes are possible: either the reflected waves are all in phase with the opposite traveling wave, or they are 90° out of phase after reflectance. The resultant electric field amplitudes, and hence electric field intensities, peak at different locations, one in material 1 and the other in material 2 as indicated in Figure 5.17. The electromagnetic energy at any location depends on the product of local dielectric constant and intensity, and hence the two standing waves have different energies at this critical wavelength. Frequency gaps then form and prevent photons from propagating, so light of this frequency cannot enter the system. Ideally, light is thus totally reflected at these "gap" energies.

Liquid crystals are organic materials with internal layered structures, and some of them can also act as good mirrors in the same way as above. These are called cholesteric mirrors after the liquid crystal type on which they are based. When the spectrum is split by the mirror into reflected and transmitted spectra it is called a dichroic mirror, and such devices can have important applications in solar energy systems. These mirrors can be electrically tunable [29].

High-performance commercial Bragg filters, which also utilize wave interference, are available with sharp cut-offs between high reflectance and high transmittance bands. A Bragg filter is essentially a stack of partially reflecting interfaces whose spacing matches the normal component of a particular internal wavelength. This means that incident light at that wavelength is very strongly reflected. These filters have been used, as have the cholesteric mirrors, in various ways to enhance the performance of luminescent solar concentrators (LSCs) which will be addressed shortly below. Concentrators of the mentioned types have been employed with light pipes for daylighting and also for low-cost solar cell concentrators. We shall show, however, that some of these concentrators will not work well in real LSC systems of useful length. These spectrally selective mirrors can also be used in combined solar thermal and solar electric systems which are able to optimize their spectral properties for the solar cell response and use the remaining energy to generate heat. Another possible application is for splitting the solar spectrum to let in daylight but reflect the NIR component of solar energy in an awning or blind.

5.4 LUMINESCENT SOLAR CONCENTRATORS FOR DAYLIGHTING AND SOLAR POWER

When light strikes a sheet of PMMA doped with solar-stable fluorescent dyes, for example consisting of perylene, the dyes absorb light in a well defined range of wavelengths and emit light in a spectral band centered at a larger wavelength. The emitted light can be transported to one end of the sheet for lighting purposes or to its edges for irradiating solar cells. The systems for doing this are called "luminescent solar concentrators" or, alternatively, "fluorescent solar concentrators." Mirrors can be used to avoid losses in undesired directions so that the area used for light collection, or the area of the solar cells, is kept at a minimum. Dye-containing sheets of this kind were initially developed to provide low-cost concentration for solar cells attached to their edges and the interest in such applications remains [30–32].

5.4.1 Devices for Generating Daylight-Like Radiation

Figure 5.18 shows how LSCs can be used to generate radiation that is similar to daylight [33]. Three sheets, each with its specific dye, are stacked and produce light of different colors which give the desired spectrum when blended at the output. Dyes generating different colors cannot simply be mixed in one sheet since this would lead to light emission with only the longest wavelengths in addition to losses (such a mixture of pigments might be of use in some solar cell LSC systems, if it is desirable to shift light frequencies down to a common lower frequency in order to enhance efficiency). As also shown in Figure 5.18, the generated light can then be transported via clear optical light guides bonded to the LSCs and can be used for lighting in the interior of buildings [33]. The light guides must have the same cross-sectional area as the LSCs, or be

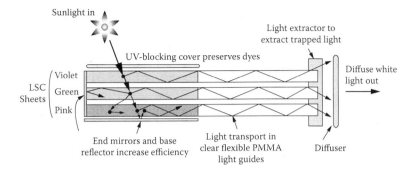

FIGURE 5.18 Schematic of a luminescent solar concentrator (LSC) stack attached to clear light guides and capable of producing white light. (From A. A. Earp et al., *Solar Energy Mater. Solar Cells* 84 (2004) 411–426. With permission.)

larger than those, since otherwise a lot of light guided to the end of the LSC sheets will not propagate inside the clear guides. Simple tapering of joints between the LSCs and light guides, or attempts to focus the (diffuse) light emerging from the LSCs, will not eliminate the problem of light transport. But, interestingly, the LSCs and the light guides do not have to have the same cross-sectional shape as long as the cross-sectional area is preserved. Hence the area constraint is not a major limitation on the design, and it is possible—and often desirable—to have light guides that are as thin and flexible as possible. In this connection we note that very thin LSC sheets are not practical since they would require high dye contents, and that would lead to problems as shown below.

It is very advantageous to produce light internally in a material, because the range of solid angles available to deliver light to the end of the light guide via total internal reflection is much larger than it would be in systems which focus light into light guides. However this is not the main advantage over focusing systems for daylighting via solid light guides. Focusing requires clear skies, because diffuse light cannot be focused, and mechanical tracking of the sun is also needed for a focusing system which adds expense and maintenance costs. And if a cloud shows up on an otherwise clear sky, the light output from the tracking system can switch quickly from very bright to very dark, which is experienced as disconcerting and is hard to compensate with auxiliary lamps.

There are thus two standout attractions for LSC daylighting systems: they produce light under all sky conditions and they do not require any movement. The LSC output falls off only in proportion to the total incident solar flux. Focusing systems, unless filtered, may introduce and transport additional heat from the solar NIR, whereas the fluorescent systems deliver nothing but high efficacy "cool" light without NIR radiation. The absorption strength of the dyes is so good that only 50 to 100 ppm are needed to fully absorb the incident sunlight in a 2-mm-thick sheet. However due to factors we will soon discuss, these systems are different from most solar collectors in that they do not have an output proportional to area or length. This aspect has seldom been appreciated in the past. It is a major limitation, with an impact on optimum designs, and if neglected it can lead to ideas for improvement which will not work in real-size systems. Thus experiments on small samples may provide data which are irrelevant to large-size systems.

Organic dyes for use in LSCs and other solar applications must combine a number of features and have

- Long-term stability under outdoor conditions, with some UV protection if necessary
- High quantum efficiency (QE), which is the ratio between emitted and absorbed photons, and
- Suitable separation between peak emission and peak absorption energies

The aim is to maximize the solar-to-light conversion efficiency, but some loss as heat is unavoidable. Heat loss arises from QE being less than one, implying that some photons in the absorption band do not yield output photons, and from the wavelength shift between peak absorption and peak emission wavelength, known as the Stokes shift.

Figure 5.19 shows absorption and emission spectra of three suitable and commercially available dye molecules. But these emission spectra do not represent the light that emerges at the edge of any extended sheet since the overlap between the spectra in Figure 5.19 leads to attenuation of emitted light in the overlap zone after traveling a moderate distance. Figure 5.20 shows the result of the removal of light in most of the overlap zone from the final spectra [33,34]. This effect reduces the efficiency of the LSC system and shifts the output color [35].

The spectral narrowing and light loss effect apparent in Figure 5.20 depends on average transport length and dye concentration. The loss and color change result from self-absorption and can be modeled accurately [36]. The color rendering properties and lumens provided thus change according to the average distance traveled in the collector, and a point is reached where adding more collector length does not lead to any significant improvement [34]. The optimized dye concentration allows for adequate absorption while limiting this loss. Cutting back somewhat on the amount of dye actually helps and is in part compensated in an LSC stack by emission into lower sheets provided that their colors are stacked in order of decreasing wavelength of emission (i.e., violet on top, then green, then pink) [35].

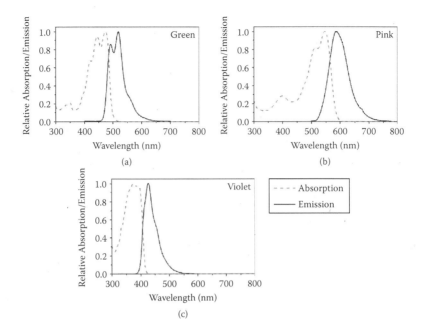

FIGURE 5.19   Normalized absorption and emission spectra of three dyes used in the LSC daylighting system of Figure 5.18.

It is possible to utilize a high proportion of the solar spectrum out to 600 nm, as seen in Figure 5.21, in order to yield good white light output in the $400 < \lambda < 700$ nm range [37]. Some UV absorption is needed to get white light, and hence care is required as this UV can cause dye degradation, mainly in the top sheet. This implies some limitation for the violet output. The main UV problem occurs for $\lambda < 350$ nm, so special filters should be used to block these wavelengths while letting through most radiation at $\lambda > 360$ nm to produce fluorescence.

Hybrid systems combining LSCs with solar cell-powered blue LEDs could lead to interesting lighting devices. The output of blue light would be linearly dependent on the irradiation onto the solar cell, and the fluorescence of green and pink light would have a similar dependence, so the color of the output light would remain constant. Only 50 to 80 lumens of blue light are needed among the 1,300 lumens in full sunlight, that is, 2 to 3 W peak of blue LED power would suffice with current efficacies. This should be compared with the 40 to 60 W of solar power that would be needed for an analogous fully solar-driven LED white light system, for which the LEDs would add a substantial cost to the already expensive solar cells. Hybrid systems with solar cells using light that would otherwise be lost at the edges of the LSC represent another option.

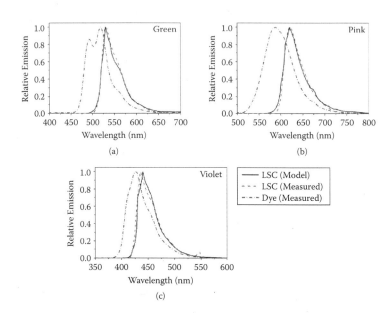

FIGURE 5.20 Measured and calculated normalized emission spectra at the end of a uniformly illuminated 1.2-m-long LSC sheet and as measured for the dye molecules. (After A. A. Earp et al., *Solar Energy Mater. Solar Cells* 84 [2004] 411–426; A. A. Earp et al., *Solar Energy* 76 [2004] 655–667. With permission.)

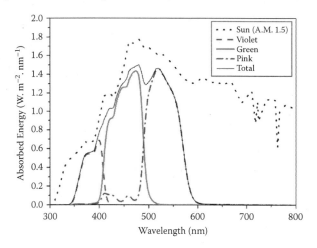

FIGURE 5.21 Solar energy absorbed by each sheet in a LSC stack and their sum. The AM 1.5 spectrum for solar irradiation is shown for comparison. (After A. A. Earp, Ph.D. thesis, University of Technology, Sydney, Australia, 2006. With permission.)

The cost evolution for solar cells, LEDs, and LSCs will tell which systems make sense economically in the future. Presently (2010) the LSC approach is the cheapest. It should also be remembered that a solar cell driven lighting system usually is fixed, so that its light output does not follow the sun's brightness; on the other hand an LSC system mimics direct daylight, which is psychologically advantageous.

Devices combining LSCs, LEDs, and light guides can be used also in a variety of display applications, as discussed in Box 5.1.

5.4.2 Devices with Solar Cells and Mirrors

LSCs look increasingly promising for low-cost concentrating solar power. When LSCs are used with birefringent mirrors, which usually rely on nanostructure (cf. Section 5.3), it may be possible to increase the light output [31], but earlier work was done with small samples and neglected the crucial length dependence which determines the ultimate merits as discussed above. It must be realized that any emitted light that exits the top or bottom surface of the LSC can no longer propagate inside the light guide by total internal reflection, so the gains are similar to those in a mirror light pipe and losses from self-absorption are possible whenever light crosses the guide.

Systems with LSCs and photonic filter structures were studied in recent work [38]. It was found that practical improvements were difficult to accomplish despite favorable predictions from idealized models. Specifically, the mirror reflectance must be at least 87% on average for useful gains with light guides having moderate aspect ratios. If high-quality spectrally selective mirrors are to be used, they must work well at high-incidence angles and it should be realized that the mechanism for concentration changes from TIR trapping to "mirror light pipe" trapping with the mirror being transparent at absorption wavelengths. Cholesteric and other dielectric mirrors can do this as shown in Figure 5.22 with regard to the absorption and emission spectra of one dye [31]. Sheets for driving solar cells do not have to be as long as those for lighting purposes so losses are much less, and in addition one can have arrays with double-sided solar cells located between the edges of adjacent thin sheets. Double-sided cells are designed to convert radiation incident from both sides to electric power and thus are ideal for LSC systems, and overall costs can then be low.

5.4.3 Light Trapping in Light Guides: Getting It All Out

A final piece of important physics is often overlooked in LSC systems but is a key aspect for the achievable energy efficiency. It has to do with what

BOX 5.1   APPLICATIONS IN ENERGY-EFFICIENT AND DAYTIME DISPLAYS

Improved energy efficiency is needed in displays. Fluorescence can contribute by making pixels active rather than wasting most of the backlighting power as is done now in many colored liquid crystal displays (LCDs) which have passive color filters. Other aspects of display, signage, and decoration are outdoor signs and images which do not lose visual clarity under direct solar illumination. Conventional LED and LCD displays and signs can be hard to see because they do not maintain contrast in bright sun. This is particularly serious since many of these devices may be safety-related signage. Pumping in more power may partly improve the contrast, but this will shorten LED lifetimes and waste power. An approach with fluorescent dyes in polymers has been successful and can operate by simply using the sun during the day. Basically letters, images, etc. which are under solar illumination get brighter, and hence give off more light, and thus maintain or even improve their contrast ratio. With LED-based side-lighting for low or no sun conditions, single letters can be lit with high-energy efficiency using light guiding in a rear plate which emits uniformly. Such LED-based approaches are also emerging for backlighting and color in LCD displays.

Recently it has been popular to use strings of LEDs for replacing traditional neon signage and for decoration and highlighting of buildings and features within spaces. These LEDs may also suffer if outdoors during the day, unless extra power is used. Using special light guides with appropriate LEDs at their ends can overcome this problem. Guides, which can be flexible, must transport without backscattering and have controlled side-emission (see next section for details). If doped with appropriate dyes, they can also maintain contrast in the day. The LED output should not be within the dye's absorption band but in its emission band. The attraction of these devices relative to LED strings is the excellent output uniformity with no bright spots (so costly special optics to remove such features is unnecessary). In general, energy efficiency and appearance are much better and costs are lower than for the various alternatives such as neon and LED strings.

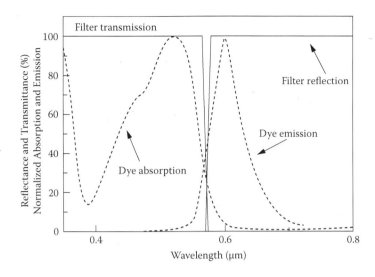

FIGURE 5.22   Spectral absorption and emission for a thin LSC sheet, and spectral reflectance and transmittance for a cholesteric filter. (From W. G. J. H. M. van Sark et al., *Opt. Express* 16 [2008] 21773–21792. With permission.)

happens to light that is generated internally in a solid through which it then travels. The generation can be by photon-to-photon fluorescence, via electron-hole recombination as in a LED, or inside a solid state semiconductor laser. It is a major issue for LEDs and has been one of the main thrusts for their efficacy improvement in recent years (cf. Section 5.6).

Some of the emitted light is trapped because it cannot escape at any of the surfaces or edges if they are smooth. This light is emitted at solid angles that do not lie in any "escape cone" but hit every surface above the critical angle. It is not important for solar cells with LSCs as long as there is intimate contact between the cell and the LSC guide. The only issue is Fresnel reflection at the interface; if there is an air gap or a low-index barrier material between the light guide and cell, then trapping remains a major issue. In the light pipe system of Figure 5.18 this trapped light must be let out in the end at the luminaire without backscattering. In theory a doubling of the net output is possible; in practice a gain of about 60 to 70% is a reasonable goal.

There are various simple ways to ensure that most light gets out; they mainly involve special scattering or geometric effects and will be covered in Section 5.5. The physics of this process, that is, extracting guided light and coupling it into free light waves, is also relevant to the performance of LEDs as covered in Section 5.6. Gains from extracting trapped light will be lower than the ideal factor of two if the basic system has losses (which is always the case) since the trapped light involves modes with more skew rays—which travel much longer on average—than the non-trapped rays for getting to an end. Models of light output must use actual or average paths and their associated attenuation factors in order to be useful.

5.5 LIGHT-DIFFUSING TRANSMITTING MATERIALS

Lighting is normally desired over large areas, but light sources—apart from sunlight—are localized in space. Indoor lighting should usually be moderately uniform since the eye adapts to the brightest spots, which then leads to poor apparent visibility in the darker areas. Glare can be a problem too, and should be avoided.

Sunlight needs to be dispersed because it is very intense and much brighter than what we need, except under a very diffuse sky. It usually enters buildings over limited apertures in a roof or at an angle to the façade and hence it is both localized within a room and intense where it hits directly. Scattering materials can be used in the aperture if it does not have to provide a clear indoors–outdoors contact. In the case of day-lighting via light pipes, as discussed above, there is a need for scattering materials at the light outputs.

Lamplight needs dispersing too, especially for luminaires with multiple LEDs which give very bright spots that can be experienced as distressing or are disliked for aesthetic reasons. Furthermore, LED strips or arrays often look non-uniform. Traditional fluorescent tubes, many compact fluorescent lamps, and all metal halide lamps are very uncomfortable to look at directly as well.

Lamps need to be integrated in fixtures, with or without covering, which can be used to spread the light from the luminaire and avoid bright spots. Unfortunately, the fixtures often reduce the available lumens to values well below what the lamps provide. The light output ratio (LOR) of a luminaire is thus an important parameter for energy-efficient lighting. Many LED fixtures are inefficient and often have LORs which give around one third of the lamp output [39]. Generally speaking, the development of lamp efficiencies seems to progress at a higher pace than for luminaire efficiencies.

LORs can be improved if the interior surfaces of the luminaire strongly reflect any light emitted away from the desired output direction. This reflection can be specular or diffuse. Diffusely reflecting surfaces aids the spreading of the light, while specular surfaces enable output lighting beams to be controlled. In order to diffuse the forward emitted light component, transmitting materials should provide strong forward scattering and weak backscattering. Most standard diffusing pigments such as $BaSO_4$ or $CaCO_3$, however, backscatter strongly as noted in connection with skylights in

Section 5.2. We note, in passing, that forward scattering diffusers may also be important for some types of solar cell arrangements as aids to increased conversion; these issues will be discussed in Section 6.2.

In this section we will look at materials that assist the dispersion of daylight and lamplight. If used with lamps, these materials should also have high transmittance in order to be energy efficient. Combining high transmittance and spreading is nontrivial but achievable with the right materials.

5.5.1 Polymer Diffusers

A number of doped polymers and some co-polymers combine weak backscattering with high transmittance and forward scattering [40,41]. One low-cost approach is to dope with particles whose refractive index n_p is very close to that of the host, denoted n_h, and are large enough to avoid normal Rayleigh scattering. They are labeled TRIMM, which is short for "TranspaRent Index Matched Microparticles." The forward scattering properties have been described in detail for the case of doped PMMA [41,42].

Figure 5.23 uses ray optics to illustrate how a TRIMM system works. Setting $n_h - n_p = \mu_{hp} n_h$, we require that μ_{hp} is small. The particle radius R_p is taken to be large enough that the angular deviation δ covers a narrow range of small values which vary slightly depending on the place where the ray strikes the particle, as given by h. For example, TRIMM particles in PMMA have $\mu_{hp} = 0.011$, so the two indices differ by only 1%. Backscattering is proportional to $(\mu_{hp}/2)^2$, so each time a ray hits a particle it has just 0.003% chance of being turned back.

Figure 5.24 shows some important experimental data for TRIMM-doped PMMA sheets: The high total transmittance drops off very little as the thickness increases from 1 to 4 mm, and most light that does get through is diffuse. This result is great for energy-efficient comfortable lighting. The finding can be extended to much thicker samples, including long light pipes as discussed in the next section. In contrast, normal pigments make backscattering increase strongly so that the transmittance falls off as the thickness or doping level increases, and the color is then shifted to red or orange because the loss of blue light is most pronounced.

An interesting and useful side effect of forward scattering is evident in Figure 5.23(b): Some normally incident rays can be trapped by total internal reflection within a sheet and then travel to a thin edge as imaged in Figure 5.25. The figure shows that some of the incident light exits from a remote edge, while much of the light exits over an area many times larger than the incident spot size.

Nanostructures within the TRIMM particles arise as a result of cross-linking or molecular entanglement and play subtle but key roles

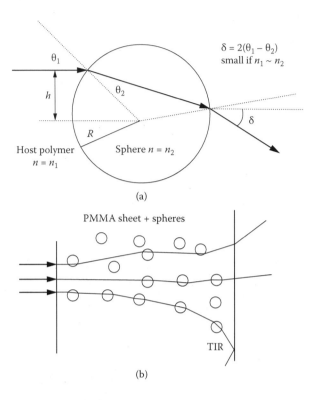

FIGURE 5.23 Ray path through a TRIMM sphere of refractive index n_p in a matrix of index n_h (a), and ray path through a TRIMM-doped clear sheet, including the possibility of some rays being trapped by total internal reflection (TIR) (b). (From G. B. Smith, J. C. Jonsson, J. Franklin, *Appl. Opt.* 42 [2003] 3981–3991. With permission.)

both from a low-cost applications perspective and from an optical physics perspective. The latter is because the nanostructure induces the required slight shift in the refractive index from 1.58 (by 1.1%). With regard to sample preparation, the nanostructure prevents the cross-linked PMMA particles from dissolving in MMA, which is the monomer used for casting. Normal PMMA, on the other hand, dissolves easily in MMA. Cross-linking also allows the particles to be used in injection molding and extrusion, since their melting point is higher than that of normal PMMA molding beads. This is again due to their internally cross-linked nanostructure.

Co-polymers with high diffuse forward transmittance are another option for energy-efficient light spreading. There are several useful available ethylene co-polymers [43], including those made with acid co-monomers such as ethylene acrylic acid (EAA) and ethylene methacrylic acid

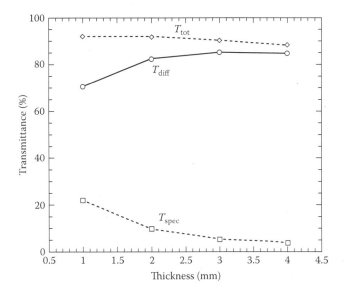

FIGURE 5.24 Total, diffuse, and specular transmittance at $\lambda = 520$ nm as a function of thickness for TRIMM-doped PMMA sheets. (From G. B. Smith et al., *Appl. Opt.* 42 [2003] 3981–3991. With permission.)

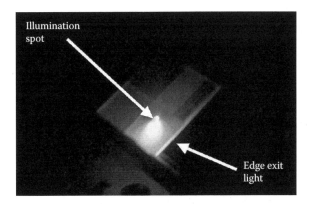

FIGURE 5.25 *A color version of this figure follows page 200.* Light output from a TRIMM-doped PMMA sheet.

(EMAA) as well as various acrylate co-monomers such as ethylene butyl acrylate (EBA), ethyl methacrylate (EMA), and ethylene vinyl acetate (EVA). All of them can have a hemispherical transmittance of 90 to 93% with little spectral variation at visible wavelengths, so they do not induce color shifts in transmitted light. Most scattering-transmitting systems based on doped polymers, however, preferentially reduce the blue component. The diffuse component of the transmittance is difficult to measure in these systems as it is forward peaked near the specular beam direction. The diffuse component is strongest at blue wavelengths but overall transmittance, that is, the sum of diffuse and specular transmittance, does not change spectrally because the diffuse part is forward scattered.

Scattering by resonant nanoparticles used for solar control in laminated windows has to be weak in order to maintain a clear view, as noted in Section 4.5.3. Scattering in polymers usually increases toward the blue, but it is possible to use resonant nanoparticles to create peaks in the scattering also at longer wavelengths. This may mean using surface plasmon resonances as outlined in Sections 3.10 and 3.11 in conducting particles that are larger than 50 nm across. If the resonance peak is in the NIR or red and the particle is large enough to scatter, then scattering will also be strongly peaked near the resonant wavelength. If this peak is in the NIR it may have little impact on visible haze, but it will backscatter solar energy which is preferable to absorption.

Translucent materials, such as those discussed above, are becoming increasingly important in buildings. Box 5.2 discusses a particularly high-profile building, the Sydney Olympics Stadium.

5.5.2 End-Lit Long Continuous Light Sources

We now consider light transport over large distances combined with controlled light emission along the length of the light transport system. What we learned in the last section is now put to use in special ways.

Solid polymer light-guiding systems can be used to deliver light at one end (as in the LSC system in Section 5.4), at select points along their length, or continuously along their whole length as in the experiment shown in Figure 5.26. Light enters at one end from a lamp, and a measuring set-up for monitoring diffuse output as a function of position along the solid light pipe is also shown. Fixed or dynamic colors can easily be arranged using filters or by colored individual LEDs. The shown lighting arrangement is an excellent alternative to neon lighting and has better energy efficiency, flexibility, and no danger of breakage.

Figure 5.27 shows a short end-lit source of this type in use for stair lighting. It has one bright blue LED at each end, and more LEDs can be added if larger brightness wound be required. The system shown has

BOX 5.2   TRANSLUCENT ROOFING: THE SYDNEY OLYMPIC STADIUM

Polymer roofing is commonly made of polycarbonate (PC) in a hollow multi-wall configuration which can be extruded. If the PC is left clear, such structures can cause dramatic and exceptionally uncomfortable bands of glare as reflection from internal walls combine. Standard inorganic pigments can be used for translucency, which leads to glare being eliminated or reduced and also reduces the solar gains, and hence heat load, on people below the roof by backscattering. Solar energy input and lighting can be controlled by varying the pigment concentration.

Large stadiums and other buildings have increasingly used pigmented translucent roofs in recent years. One of the largest of these roofs, at least when it was built in 1997–1998, was for the Sydney Olympic Stadium main stands; it is shown schematically in Figure B5.2.1. Doping levels, and hence the transmittance, were decreased in the PC multi-wall panels from the rear to the front in order to optimally manage glare and heat. Translucency leads to redirection of a lot of light onto the pitch. This has added advantages and makes grass grow more evenly than under standard conditions with an opaque roof, in which case some of the grass is too much in the shade. The redirected light also reduces stark contrasts in TV images covering simultaneously both roof shadow zones and clear sun zones.

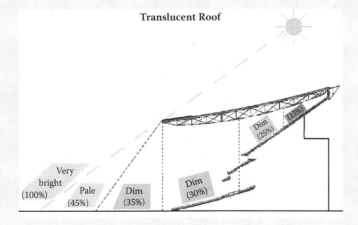

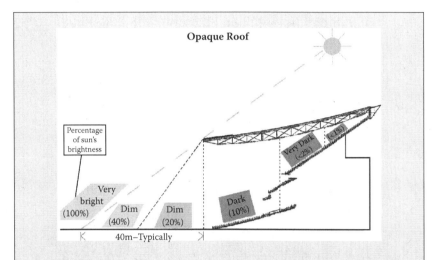

FIGURE B5.2.1   Schematic light levels under the Sydney Olympics Stadium roof, compared to what they would have been with an opaque roof. Note the different gradings of the lux levels on the pitch and in the stands. "Dim" is still relatively bright at 50,000 lux.

FIGURE 5.26   *A color version of this figure follows page 200.* Flexible transparent polymer light guide containing transparent index matched microparticles. An integrating sphere (cf. Section 2.4.2) is used to measure the total light output at different segments along the guide.

FIGURE 5.27 Example of "supersidelighting" with TRIMM-doped polymer light guides end-lit by bright LEDs.

worked well for many years and yields a more uniform output than a string of LED chips.

High energy efficiency is important for any lighting system, and in the present case the entrance optics must be carefully designed to minimize the amount of light from the lamps that does not enter the pipes. Good mirror surfaces and mirror design should bring most light in at angles of incidence on the entrance end within the numerical aperture (NA) of the guide (i.e., within a cone angle $\theta = \sin^{-1}[NA]$). With nothing but air around the guide, the acceptance angle θ_{air} for PMMA and related acrylates is up to 90°. Even a small air gap between the outer jacket and guide can be adequate. If a tight-fitting low-index jacket is used (e.g., Teflon), then $\theta_{air} = 35°$. A fraction of the accepted light will still be reflected at the entrance end by Fresnel refection as it goes from air to PMMA. Unless it is a parallel normally incident beam, this loss can be much higher than 4%.

There is a better coupling approach for the input, as discussed for the luminescent dye system in Section 5.4. If the light never passes into air but remains entirely within solid or polymer, the Fresnel reflection can be reduced or even eliminated in some cases. This possibility is especially worth considering with LEDs as the chips are often covered with a polymer lens anyway to improve the light output control. This provides another advantage for end-lit flexible fibers.

Once light has entered into the pipe, it must emit it from its side in a controlled fashion so it can come out brightly along the whole length. Important issues are

- to ensure that backscattering is weak, since otherwise light will not get far and only the pipe near the entrance will appear bright
- to ensure that the side-scattered emission is adequate, and
- whether light is admitted from both ends or one

Regarding the third item, uniform lighting is easier to accomplish if lamps are at both ends, but if there is only one lamp then it is useful to have a mirror at the opposite end. Double-ended lighting also helps overcome another issue caused by forward scattering: Exit light gets less intense as the exit angle increases from the forward direction, and the light may be weak at exit angles normal to the guide. It is easy to achieve more uniform exit light as a function of exit direction by use of weak scattering in the light pipe jacket material; alternatively, if light is only needed on one side of the pipe, the outward-facing half of the surface of the jacket can be coated with a highly reflecting diffuse white coating. A key advantage of these systems over, say, fluorescent tubes is that light exiting the wrong way can be made to pass back through the tube by reflectance with little loss, which adds to the total energy efficiency.

TRIMM-doping, or suitable co-polymers, can achieve excellent uniform side-illumination, which has been labeled "supersidelight." It has found use in a number of commercial applications and products. The doping concentration can be varied according to the required length of illumination. For a few meters and constant concentration, the intensity will vary but the change may be acceptable for practical purposes. Usually one can obtain a sufficiently uniform light output by putting light sources at both ends or having a mirror at the non-illuminated end.

5.5.3 Light Extraction from Light Pipes and in LCD Displays

Light piping is of growing significance for general lighting, in large area display, and of course in LCD, computer, and TV screens where it is piped inside special flat back panels. High-quality edge-lit PMMA sheet is employed for many applications.

Displays and screens use increasing amounts of lighting energy and need to become more efficient. Polymer diffusers and other polymer structures are important in this context for extracting and spreading light at the end of solid polymer light pipes. Spreading is natural at the end of a pipe if the full acceptance angle is used with external lamps at the input. Getting light out at right angles to the incident beam is needed for LCD systems and also for some fluorescent systems. If light is generated within a solid light guide, such as in the LSCs in Section 5.4, or if there is no air gap between a LED chip and the solid light pipe, then a significant part of the available light can be trapped internally. But if steps are taken to extract this trapped light, a system without air gap will still potentially have much superior lighting efficiency compared to one in which lamp light crosses an air gap because interface reflectance losses are avoided.

Various strategies can be used to extract trapped light from a flat surface or side of an otherwise clear light guide. All of these strategies involve some geometric feature which can be macroscopic, microscopic, or nanostructured. These features are designed to prevent the trapped light from being totally internally reflected and to extract it at the same time. Macroscopic features enable the light to be refracted out. Microscopic features might do the same or might just involve a rough surface which either refracts out or scatters out the trapped light. If the latter takes place, wave optics rather than ray optics is in play, for example, diffraction.

Another way of thinking of light extraction is that the evanescent fields locked in at the edge of a waveguide or optical fiber can be coupled into external propagating waves by perturbing the waveguide surface, thus allowing energy to escape from the guide. A simple approach which works well is to coat the back side of the guide with a highly reflective, highly diffuse white paint wherever the light should emerge. This exploits the high transparency of the guides, and the paint-covered points or strips effectively become "lamps." Large back-lit display panels, which are edge-lit with tubes or LED strings and use TIR light piping, can exploit the same principles by using multiple spots of glass frit or paint, which usually is distributed less densely near the lamps to achieve even illumination. LCD displays use macroscopic structures, sometimes coupled with surface microstructures. It is important again for energy efficiency that scattering in the wrong direction is either avoided or compensated for by having a good back reflector.

5.6 ADVANCED ELECTRONIC LIGHTING CONCEPTS

Lamp technology is currently (2010) undergoing a revolution. Driving forces include

- the need to achieve very large decreases in CO_2 emissions by use of energy-efficient lamps
- rapid recent advances in a variety of novel lamp technologies
- large savings in energy that will occur if all lamps are intrinsically "smart," that is, easily dimmable and switchable via electronic control, and

- simplicity, longer lifetime, and increased efficiency for solar powered lighting, provided that associated lamps are electronic and do not require high ac voltages

By use of "smart" lamps it is possible to maximize the benefits from daylight and from advances in switchable glazing, and also to adjust to occupancy patterns. Dimming is not possible with many current lighting systems and often, when dimming can be accomplished, the energy savings are not large because of the physical nature and ancillary electronics used to control the light output. But in electronic lighting the light output is generally directly linked to the current flow and hence the input power.

Semiconductor-based and organic solid-state lighting is in focus for current progress and will be discussed extensively below. But other electron-based lamp technologies are coming to the market, too. This latter group has some features in common with traditional filament bulbs and either use novel materials as energy-efficient filaments or a new type of fluorescence which does not rely on mercury vapor-based UV light emitting discharges as is the case for traditional fluorescent tubes and compact fluorescent lights (CFLs). Electron-stimulated luminescence (ESL) lamps are at the forefront of these LED competitors now, but electroluminescence and cathodoluminescence in nanostructures—along with microplasma systems for high intensity lamps—are also under serious commercial consideration. Cost advantages are likely for some of these systems and, for example, ESL lamps at market launch seem to be price competitive with CFLs on a lumens and lifetime basis. We will mainly consider bright LEDs in this section, but it is important to keep in mind that many interesting competing technologies are on the horizon. Traditional tungsten filament lamps will be found only in museums within a few decades, and it would not be surprised if they were followed soon after by compact fluorescent lamps and fluorescent tubes.

5.6.1 Semiconductor Light-Emitting Diodes

Figure 5.3 introduced the concept of "intelligent solid state lighting" (ISSL, or intelligent SSL), with light-emitting diodes as a technology for the immediate future. A brief introduction to how LEDs and OLEDs work, in this and the following sections, is therefore in order and will lay a foundation for understanding why nanostructures are becoming increasingly important for lifting their efficacy and brilliance, and for lowering the costs of these emerging lighting technologies. LEDs relying on nanostructures for their high performance are called "nanophotonic" LEDs and are predicted to become the system of choice for general

FIGURE 5.28 *A color version of this figure follows page 200.* Two applications of colored LEDs. (From K. Dowling, LED Essentials, Department of Energy, Webinar October 10 [2007]; http://apps1.eere.energy.gov/buildings/publications/pdfs/ssl/webinar_2007-10-11.pdf.)

LED-based lighting. Barely out of the starting blocks in 2009 with commercial products, the nanophotonic LED market is expected to grow at a compound rate of 91% over the next 4 years to a US$ 2.7 billion value by 2014 [44].

Standard LEDs are already dominating various specialty markets for signaling, display, medical and dental, and decorative lighting, and Figure 5.28 illustrates how colored LEDs have revolutionized traffic lighting and building décor [5]. Other applications were shown in Figures 5.26 and 5.27. In fact colored lighting has entered a new age, and cumbersome neon lights and energy-inefficient color filters on normal lamps are headed for redundancy. LEDs may soon dominate much small-scale solar-powered lighting, especially in developing nations. As noted in the introduction to this chapter, SSL devices may be the most expensive lighting to buy, but the total cost over their lifetime—counted per million lumen hours—is rapidly becoming competitive to fluorescent tubes and CFL lamps for well-run and properly mounted systems. Progress in luminaire design, and in the construction of the LED chips and their enclosures, has contributed to this development.

The improvement in average lamp and luminaire efficacy has been remarkable in the last 5 years, and Figure 5.29 shows that the improvement has been 75% on average [39]. However, it is also noticeable that

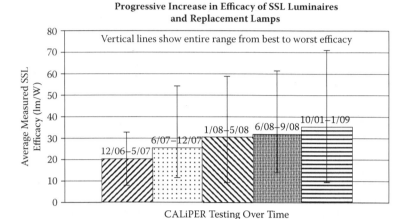

FIGURE 5.29 Average and spread of measured lamp and luminaire effi-cacies for solid state lighting luminaires and replacement lamps over the last 5 years. (From DOE Solid-State Lighting CaliPer Program. Summary of Results: Round 7 of Product Testing, January 2009; http://apps1.eere. energy.gov/buildings/publications/pdfs/ssl/caliper_round_7_summary_ final.pdf.)

the range of efficacies in products has widened and currently ranges from 10 to 70 lumens/watt; the lower efficacies have more to do with cur-rent market emphases than technology. The range in the performance is expected to narrow as LED technology matures and begins to impact general lighting. In a worst-case scenario, the efficacies will stabilize around the upper levels of today's technology, but probably the aver-age efficacies will become higher because there is still much scope for improvement given the further progress in relevant semiconductor mate-rials along with commercial adoption of emerging nanoconcepts.

An LED is a semiconductor *p-n* junction device in which electron and hole carriers, injected by an external voltage into the junction region, recombine and eventually emit light. Before recombination, the electron–hole pair may be bound together as an exciton which can decay directly across the band gap or via some defect. The color, or energy, of the emitted light depends on the band gap of the semiconductor and on defect energy levels if they are involved. The fundamental structure is seen in Figure 5.30 [5].

Materials that are widely used in LEDs include *p*- and *n*-type GaN, AlGaN, InGaN, AlGaInN, GaAs, GaAsP, and various heterostructures of such compounds. The big breakthrough for blue and UV emission, and for using phosphors as in Figure 5.30 to produce bright white light, came with the ability to make good electrical quality GaN and InGaN

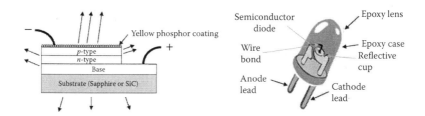

FIGURE 5.30   Basic *p-n* junction-based LED with some of its emission exciting a yellow phosphor to produce white light. (From K. Dowling, LED Essentials, Department of Energy, Webinar October 10 [2007]; http://apps1.eere.energy.gov/buildings/publications/pdfs/ssl/webinar_2007-10-11.pdf.)

which have wide enough defect-free band gaps. Just as in fluorescent lamps, these phosphors need to be activated by high energy (i.e., blue or UV) photons to emit. It should be noted that these devices currently need expensive crystalline substrates, such as sapphire and *c*-SiC, to ensure that GaN grows epitaxially. Large cost savings and higher production rates would be possible if that epitaxy was achievable with low-cost substrates. Epitaxy on glass can be promoted by pre-coating it with one or two thin intermediate layers, with ZnO being one possibility under investigation. ZnO is also of interest as a low-cost UV emitting diode, though it may only find uses in heterostructures unless a method is found to make high-quality *p*-type ZnO. Another major technical issue with LEDs is that, in simple layer structures, only a small fraction of the light escapes while most is trapped within one of the layers. However, various schemes have been devised to extract much more light, with nanostructures providing the best solutions, as discussed below.

5.6.2 Organic Light-Emitting Diodes

OLEDs and their white-light versions (WOLEDs) are also junction devices and, apart from the electrodes, are all-organic and require conductors similar to those used in organic solar cells [45]. Light emission involves electron hole recombination, as with LEDs, but the device construction is more complex. In OLEDs, one or more organic layers act as either electron transport and injection material or hole transport and injection material. Electron injection fills a high-energy molecular orbital to form a HOMO level, and this electron forms an exciton with a hole in a lower-energy unoccupied molecular orbital, called a LUMO (or vice versa for hole injection). Light emission is then possible when the excitons give up their energy to dye or phosphor molecules, which can occur with high quantum efficiency; the processes are known as "electrolumines-cence." White or mixed color light is possible if several dyes are used. An intermediate organic conductor between the external electrode and the light-emitting layer creates a barrier of a few volts, as in a semiconductor junction LED, which enhances the efficiency compared to what would be the case for injection directly via metal or oxide electrodes.

As with LEDs, light extraction is a key issue and attention to this and other materials aspects has led to OLEDs with reported efficacies of 90 LW^{-1}, which for the first time surpasses that in the best fluorescent tubes [46]. The design and materials of these latest devices are such that almost all charge carriers injected by the power supply contribute to the production of light or individual photons; in scientific terms, we say that the quantum efficiency is almost 100%. Further gains will require even better optical extraction efficiency and better initial current injec-tion efficiency. It appears that 100 LW^{-1} is over the horizon, so that if long service lifetimes are achievable these OLEDs could become very important for general lighting purposes. The color of their light is excel-lent, and they can produce light uniformly in large area panels, which is distinctly different in appearance from the almost point-like emission from inorganic LED chips.

5.6.3 Nanostructures for Improved LED Performance

A variety of nanostructures are under investigation for improved LED performance. They can be used

- to increase the percentage of light extracted or out-coupled from LED chips
- to enhance the luminescent output via larger surface areas of emitters
- through surface fluorescence in quantum dots
- for more efficient use of electrons injected into the device via nanostructured electrodes, and
- for novel nano-phosphors

Thin films and epitaxial film growth have important roles to play in these developments.

A lot of the light generated within the active junction region of basic LED planar chips does not make it to the outside but is ultimately lost by absorption and scattering. The loss is amplified many times over because internal light trapping and wave-guiding occurs in much the same way that light generated internally in fluorescent dye-doped polymers for polymer light guiding is trapped internally (cf. Sections 5.4 and 5.5).

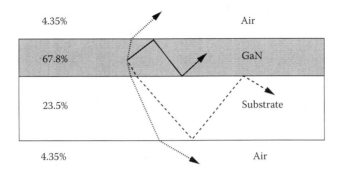

FIGURE 5.31 Sketch of a basic GaN LED. The percentage of light in each layer is shown. (Redrawn from M. Zoorob, G. Flinn, *LEDs Mag.* [August 2006] 21–24.)

Light trapping is aggravated in semiconductors because of their large refractive indices, as seen schematically in Figure 5.31 which shows that as much as 67.8% is trapped in the smooth GaN layer and that 23.5% is trapped in the sapphire substrate [47].

To extract more light, one or more of the interfaces can be modified or the structure and shape of the active layer itself can be altered. One basic approach to limit wave guiding and light trapping is to have a graded refractive index at each interface, which could be achieved with a nano- or microstructure, or to have additional thin layers. The general optical principles and geometric solutions to accomplish this have been applied to LSC light pipes [48] and apply equally well to semiconductor systems. The aim is to limit TIR and Fresnel losses at exit surfaces and maximize extraction in desired directions. Apart from graded refractive indices, it is possible to use special geometries tailored to ray optics, such as micro-lenses and macroscopic roughness, as well as nanoscale wave effects. The purpose of the latter is to couple the waveguide modes shown in Figure 5.31 into external traveling waves. Major gains in efficiency have come recently from efforts in this area, but more is possible by exploiting nanostructures. Options include photonic quasi-crystals at the exit surface [49], nanostructured electrodes, and the use of surface plasmon resonance at the exit surface [50].

Ordered or photonic lattice surface nanostructures can be used not only to extract more light but also to send it into predetermined far-field beam patterns. Some photonic crystal patterns considered for this purpose are shown in Figure 5.32 [49].

The photonic crystal can also be part of the active light-emitting system, for example the *p*-doped GaN layer can be a photonic crystal on top of a planar *n*-doped GaN layer, and extraction of 73% of the generated light has recently been reported for such structures [51]. The

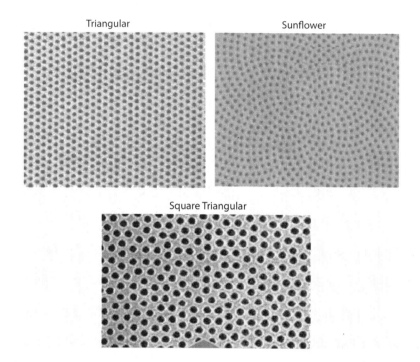

FIGURE 5.32   Photonic quasicrystal structures, and their designations, for use as light extractors in LEDs. (From M. D. B. Charlton et al., *Proc. SPIE* 6486 [2007] 64860R 1–10. With permission.)

number of allowed modes in any waveguide drops as the planar layer gets thinner. Generated light can only travel in these modes, and the 73% extraction efficiency in a 700-nm-thick GaN layer was achievable because the photonic crystal only had to pick up a few modes so that the crystal design could be simplified. A bottom reflector of silver was also used in this study [51]. LED encapsulant materials normally let out more light than bare chips do, but they also spread the light more. However, these efficient nanostructures can bypass the need for encapsulants, thereby allowing a tighter exit beam profile.

Apart from photonic crystals with well-defined lattice structures, other types of columnar, nanowire, or surface-textured systems can be used to enhance the fluorescent output. Their large surface area is the main source of extra yield, but they may also enable more excitons to build up and emit light before being lost as heat. InGaN columns can act as one-dimensional quantum confinement structures as can various other nanowires including organic ones [1], and organic nanowires incorporating two dyes can have a combined output which is white. Figure 5.1, at the beginning of this chapter, showed a "live" fluorescent

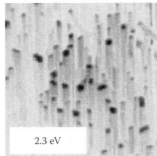

FIGURE 5.33 Cathodoluminescence images of light emitted at two energies from ZnO nanocolumns. The side of the images is 6.7 μm, and the width of the hexagonal columns varies from 100 to 600 nm. (From M. Foley et al., *Appl. Phys. Lett.* 93 [2008] 243104 1–3. With permission.)

array of the latter type. The crystalline organic materials used in these devices may also be candidates for electroluminescence.

Another interesting aspect of semiconductor nanocolumns is that they usually grow with preferred orientation, especially on oriented substrates such as sapphire. The tips and edges of these columns can then have different fluorescent intensities, because they represent different crystal facets; this is seen in Figure 5.33 for ZnO, a material of large current interest for UV-emitting LEDs [52]. The excitation in Figure 5.33 was with a beam of electrons, but light is detected. This process is called cathodoluminescence. Green emission at 2.3 eV is strong from the side-walls and weak from the tips, and UV emission is also stronger from the column side-walls than from the tips.

Enhanced light extraction efficiency from LEDs has been convincingly demonstrated using conducting nanoparticles deposited onto the surface of the LEDs. Light output enhancement factors of up to 12 have been found for thin silicon-on-insulator LEDs [50,53] as can be seen in the comparison of the light output from nanoparticle-coated and uncoated sections of a LED in Figure 5.34. Optimized extraction of light is the reverse of increased absorption in solar cells, though with the physics reciprocal, and both work well. A thin dielectric spacer layer between the nanoparticles and the active silicon is needed in devices, but such a layer is there anyway for quenching surface recombination.

How can metal particles, which normally absorb light, enhance light output? The origin of the enhancement shown in Figure 5.34 lies in resonant scattering, which peaks at the effective surface plasmon resonant frequency of the nanoparticle–substrate combination as discussed in Chapter 3. For silver on silicon, the largest enhancement occurs at $\lambda \approx 770$ nm, and for gold it is at $\lambda \approx 840$ nm. These emission maxima are

FIGURE 5.34 Light output from a LED that has been partly covered with silver nanoparticles. (From S. Pillai et al., *Appl. Phys. Lett.* 88 [2006] 161102 1–3. With permission.)

significantly red-shifted compared to the surface plasmon resonances for Ag and Au nanoparticles in air which take place at $\lambda \approx 350$ and $\lambda \approx 480$ nm, respectively. The plasmon resonances for particles on waveguides can change as a result of coupling to the waveguide modes and because of the high refractive index of the medium under the nanoparticles.

New developments in nanoscience may lead to high-efficiency white-light emitting phosphors for use in some LEDs and other new lighting systems. Silicon microparticles and bulk silicon are not efficient light emitters because Si is an indirect semiconductor, but nanocrystalline Si is different and may be able to produce efficient bright emission. The energy levels and wave vectors in sufficiently small nanocrystals are not those of normal band electrons but are governed by quantum confinement of the electrons. Such Si nanoparticles can be applied from solution.

5.6.4 Emerging Lamp Technologies

To conclude the discussion on electronic processes for efficient lamps, it is important to have a look at systems that do not rely on electron and hole recombination in device junctions. Various other electro-optic and thermo-optic mechanisms are under consideration or on the verge of commercialization. All of them must be energy efficient and have long service lives to warrant serious consideration. They should also have some other attractions, such as better color rendering than compact fluorescent bulbs, no flicker, no mercury, and easy full-range dimming.

The electron-stimulated luminescence lamp is the most advanced of these alternatives. It is based on cathodoluminescence that takes place when a beam of electrons is accelerated toward a luminescent coating on the inside surface of a bulb. The localized electron-hole excitations produced by the incident electrons decay via recombination and produce white light. The efficacy is now (2010) around 30 LW^{-1} and is expected to reach 40 LW^{-1}. An efficacy of 30 LW^{-1} is about three times that of incandescent bulbs of the same lumens and matches CFLs in this regard, but ESL lamps are better than CFLs in many other ways, including in visual appeal. The ESLs are at present much cheaper than high-quality LED lamps.

Other approaches could lead to the resurrection of light bulbs similar to traditional filamental types but with novel light-generating mechanisms that make them much more energy efficient. One approach is to use doped silicon oxide as a thin film on chips arrayed along the "filament" and cause it to produce red, green, and blue light which in combination gives white light.

Nanotechnology and new materials perhaps will mean that the traditional bulb's days are not over yet. It has after all considerable aesthetic and visual appeal over CFLs. The use of new thermal-based radiant emission systems cannot entirely be ruled out as a low-cost option for moderately energy-efficient lighting. Thus some refractory thin films applied to hot tungsten filaments can lead to a much larger emission of visible radiation than bare tungsten for the same power input. For example SiC, discussed in Chapters 3 and 7 in the contexts of optimized radiation control and sky cooling, could be used as a simple coating to enhance the output relative to that of bare tungsten. In addition, bulb lifetimes can be extended by ceramic coatings as the normal erosion mechanism for the filaments will be significantly reduced.

Other approaches include having nanostructured surfaces on the filaments to increase the emittance of thermal radiation in the visible wavelength range [54]. Surface nanostructures will only slightly affect the IR thermal emittance via increased conduction electron scattering, but they can have a large impact on the absorptance and hence emittance at visible wavelengths. By one or more of the above technologies, lighting with "light bulbs" might again become competitive in terms of cost, lifetime, and energy-use criteria. New approaches to high-intensity lamps, which are usually based on large electrode cathodic arc discharges, are also emerging. Such lamps are usually for specialty outdoor lighting and produce from 20,000 to 200,000 lumens.

5.6.5 Concluding Remarks

Nanostructures and novel thin films and coatings implemented in LEDs, OLEDs, and non-traditional light bulbs will lead to major improvements in lighting quality and performance, with very large savings in energy and CO_2 emissions. They will enable a variety of new control and operational philosophies as well as easier integration with daylight and solar power. The brief coverage in this chapter is only an introductory glimpse into a vast area of current research and commercialization activity and is not all-inclusive.

The rate of change during the first decade of the 2000s, in what has been a very conservative area of technology for half a century, is quite astounding. It is testimony to what scientists, engineers, and business people can achieve together when provided with the necessary science, plus incentives and resources. To some extent these recent developments have piggy-backed on the wealth of knowledge and resources previously built up for the electronics, optics, information technology, and communications industries. The rapid growth in nanoscience has also contributed very significantly. If similar efforts and expertise are applied to other areas of renewable energy and energy savings technology, there is every reason to believe that analogous rapid and revolutionizing progress will occur.

REFERENCES

1. Y. S. Zhao, H. Fu, F. Hu, A. Peng, W. Yang, J. Yao, Tunable emission from binary organic one-dimensional nanomaterials: An alternative approach to white-light emission, *Adv. Mater.* 20 (2008) 79–83.
2. R. Fouquet, P. J. G. Pearson, Seven centuries of energy services: The price and use of light in the United Kingdom (1300–2000), *Energy J.* 27 (1) (2006) 139–177.
3. R. Fouquet, *Heat, Power and Light: Revolutions in Energy Services*, Edward Elgar, Cheltenham, U.K., 2008.
4. www.colorkinetics.com/support/whitepapers/CostofLight.pdf.
5. K. Dowling, LED Essentials, Department of Energy, Webinar October 10 (2007); http://apps1.eere.energy.gov/buildings/publications/pdfs/ssl/webinar_2007-10-11.pdf.
6. J. D. Balcomb, The coming revolution in building design, in *Proc. PLEA: Passive and Low-Energy Architecture*, Lisbon, Portugal, June 1, 1998; pp. 33–37.
7. J. A. Leech, W. C. Nelson, R. T. Burnett, A. Aaron, M. E. Raizenne, It's about time: A comparison of Canadian and American time-activity patterns, *J. Exposure Anal. Environm. Epidem.* 12 (2002) 427–432.
8. M. Fontoynont, Perceived performance of daylighting systems: Lighting efficacy and agreeableness, *Solar Energy* 73 (2002) 83–94.
9. J. McHugh, R. Dee, M. Saxena, Visible light transmittance of skylights, California Energy Commission, PIER, 2004; Contract Number 400-99-013.

10. L. Heschong, J. McHugh, Skylights: Calculating illumination levels and energy impacts, *J. Illum. Engr. Soc.* 29 (winter) (2000) 90–100.

11. M. S. Rea, editor-in-chief, *The IESNA Lighting Handbook: Reference and Application*, 9th ed., Illum. Engr. Soc. North America, New York, 2000.

12. P. D. Swift, G. B. Smith, Cylindrical mirror light pipes, *Solar Energy Mater. Solar Cells* 36 (1995) 159–168.

13. P. D. Swift, R. Lawlor, G. B. Smith, A. Gentle, Rectangular-section mirror light pipes, *Solar Energy Mater. Solar Cells* 92 (2008) 969–975.

14. I. R. Edmonds, G. I. Moore, G. B. Smith, P. D. Swift, Daylight enhancement with light pipes coupled to laser-cut light-deflecting panels, *Lighting Res. Technol.* 27(1) (1995) 27–35.

15. www.skydome.com.au.

16. P. D. Swift, G. B. Smith, J. Franklin, Hotspots in cylindrical mirror light pipes: Description and removal, *Lighting Res. Technol.* 38 (2006) 19–31.

17. http://www.bendinglight.co.uk/assets/pdf_downloads/How_Serraglaze_Works.pdf.

18. J. Reppel, I. R. Edmonds, Angle selective glazing for radiant heat control in buildings: Theory, *Solar Energy* 62 (1998) 245–253.

19. I. R. Edmonds, J. Reppel, P. Jardine, Extractors and emitters for light distribution from hollow light guides, *Lighting Res. Technol.* 29(1) (1997) 23–32.

20. C. G. Granqvist, Materials for good day-lighting and clean air: New vistas in electrochromism and photocatalysis, in *50th Annual Technical Conference Proceedings*, Society of Vacuum Coaters, Albuquerque, NM, 2007; pp. 561–567.

21. C. G. Granqvist, Indoor light balancing, World Intellectual Property Organization WO/2008/048181 (2008).

22. B. Bläsi, C. Bühler, A. Georg, A. Gombert, W. Hoßfeld, J. Mick, P. Nitz, G. Walze, V. Wittwer, Microstructured surfaces in architectural glazings, in *Proceedings ISES Solar World Congress*, Gothenburg, Sweden, June 14–19, 2003.

23. V. Boerner, S. Abbott, B. Bläsi, A. Gombert, Nanostructured holographic antireflection films, in *Proc. Displays and Vacuum Electronics*, Garmisch-Partenkirchen, Germany, May 3–4, 2004, VDE-Verlag, Berlin, Germany; ITG-Fachbericht Vol. 183, pp. 211–214.

24. A. Gombert, B. Bläsi, C. Bühler, P. Nitz, J. Mick, W. Hoßfeld, M. Niggemann, Some application cases and related manufacturing techniques for optically functional microstructures on large areas, *Opt. Engr.* 43 (2004) 2525–2533.

25. G. Walze, A. Gombert, P. Nitz, B. Bläsi, Rigorous validation of the lateral Goos-Hänchen shift in microstructured sun shading systems, *Proc. SPIE* 6197 (2006) 61970Z 1–9.

26. A. Gombert, C. Buhler, W. Hossfeld, J. Mick J., B. Blasi, G. Walze and P. Nitz, A rigorous study of diffraction effects on the transmission of linear dielectric micro-reflector arrays, *J. Optics A: Pure Appl. Opt.* 6 (2004) 952–960.

27. M. F. Weber, C. A. Stover, L. R. Gilbert, T. J. Nevitt, A. J. Ouderkirk, Giant birefringent optics in multilayer polymer mirrors, *Science* 287 (2000) 2451–2456.

28. http://ab-initio.mit.edu/photons/tutorial/L1-bloch.pdf; http://ab-initio.mit.edu/photon/tutuorial/spie-course-new.pdf.

29. D. Krüerke, N. Gough, G. Heppke, S. T. Lagerwall, Electrically tuneable cholesteric mirror, *Mol. Cryst. Liquid. Cryst.* 351 (2000) 69–78.

30. A. Goetzberger, W. Greubel, Solar energy conversion with fluorescent collectors, *Appl. Phys.* 14 (1977) 123–139.

31. W. G. J. H. M. van Sark, K. W. J. Barnham, L. H. Slooff, A. J. Chatten, A. Büchtemann, A. Meyer, S. J. McCormack, R. Koole, D. J. Farrell, R. Bose, E. E. Bende, A. R. Burgers, T. Budel, J. Quilitz, M. Kennedy, T. Meyer, C. De Mello Donegá, A. Meijerink, D. Vanmaekelbergh, Luminescent solar concentrators: A review of recent results, *Opt. Express* 16 (2008) 21773–21792.

32. A. Goetzberger, Fluorescent solar energy concentrators: Principle and present state of development, in *High-Efficient Low-Cost Photovoltaics*, edited by V. Petrova-Koch, R. Hezel, A. Goetzberger, Springer Series in Optical Sciences, Springer, Berlin, Germany, 2009; Vol. 140, pp. 159–176.

33. A. A. Earp, G. B. Smith, J. Franklin, P. Swift, Optimization of a three-colour luminescent solar concentrator daylighting system, *Solar Energy Mater. Solar Cells* 84 (2004) 411–426.

34. A. A. Earp, G. B. Smith, P. D. Swift, J. Franklin, Maximising the light output of a luminescent solar concentrator sheet, *Solar Energy* 76 (2004) 655–667.

35. P. D. Swift, G. B. Smith, Color considerations in fluorescent solar concentrator stacks, *Appl. Opt.* 42 (2003) 5112–5117.

36. P. D. Swift, G. B. Smith, J. B. Franklin, Light to light efficiencies in luminescent solar concentrators, *Proc. Soc. Photo-Opt. Instrum. Engr.* 3789 (1999) 21–28.

37. A. A. Earp, Shedding Natural Light without Windows or Skylights, Ph.D. thesis, University of Technology, Sydney, Australia, 2006.

38. M. Peters, J. C. Goldschmidt, P. Löper, B. Bläsi, A. Gombert, The effect of photonic structures on the light guiding efficiency of fluorescent concentrators, *J. Appl. Phys.* 105 (2009) 014909 1–10.

39. DOE Solid-State Lighting CaliPer Program. Summary of Results: Round 7 of Product Testing, January 2009; http://apps1.eere.energy. gov/buildings/publications/pdfs/ssl/caliper_round_7_summary_ final.pdf.

40. A. Tagaya, Y. Koike, Highly scattering optical transmission polymers for bright display, *Macromol. Symp.* 154 (2000) 73–82.

41. G. B. Smith, J. C. Jonsson, J. Franklin, Spectral and global diffuse properties of high-performance translucent polymer sheets for energy efficient lighting and skylights, *Appl. Opt.* 42 (2003) 3981–3991.

42. J. C. Jonsson, G. B. Smith, C. Deller, A. Roos, Directional and angle-resolved optical scattering of high-performance translucent polymer sheets for energy-efficient lighting and skylights, *Appl. Opt.* 44 (2005) 2745–2753.

43. G. M. Wallner, W. Platzer, G. M. Lang, Structure–property correlations in polymeric films for transparent insulation wall applications. Part I: Solar optical properties, *Solar Energy* 79 (2005) 583–592.

44. C. B. Les, Nanophotonics market: Upward bound, *Photonics Spectra* (10) (2009) 32.

45. C. W. Tang, S. A. VanSlyke, Organic electroluminescent diodes, *Appl. Phys. Lett.* 51 (1997) 913–915.

46. S. Reineke, F. Lindner, G. Schwartz, N. Seidler, K. Walzer, B. Lüssom, K. Leo, White organic light-emitting diodes with fluorescent tube efficiency, *Nature* 459 (2009) 234–238.

47. M. Zoorob, G. Flinn, Photonic quasicrystals boost LED emission characteristics, *LEDs Mag.* (August) (2006) 21–24.

48. G. B. Smith, J. B. Franklin, Sunlight collecting and transmitting system, U.S. Patent 6059438 (2000).

49. M. D. B. Charlton, M. E. Zoorob, T. Lee, Photonic quasi-crystal LEDs: Design, modelling, and optimization, *Proc. SPIE* 6486 (2007) 64860R 1–10.

50. S. Pillai, K. R. Catchpole, T. Trupke, G. Zhang, J. Zhao, M. A. Green, Enhanced emission from Si-based light-emitting diodes using surface plasmons, *Appl. Phys. Lett.* 88 (2006) 161102 1–3.

51. J. J. Wierer, Jr., A. David, M. M. Megens, III-nitride photonic-crystal light-emitting diodes with high extraction efficiency, *Nat. Photonics* 3 (2009) 163–169.

52. M. Foley, C. Ton-That, M. R. Phillips, Cathodoluminescence in ZnO nanorods, *Appl. Phys. Lett.* 93 (2008) 243104 1–3.

53. S. Pillai, K. R. Catchpole, T. Trupke, M. A. Green, Surface plasmon enhanced silicon solar cells, *J. Appl. Phys.* 101 (2007) 093105 1–8.

54. A. Y. Vorobyev, V. S. Makin, C. Guo, Brighter light sources from black metal: Significant increase in emission efficiency of incandescent light sources, *Phys. Rev. Lett.* 102 (2009) 234301 1–4.

CHAPTER 6

Heat and Electricity
Solar Collectors and Solar Cells

This chapter deals with devices for solar absorption that produce heat (photothermal conversion in solar collectors) and electricity (photoelectric conversion in solar cells). These devices can be building-integrated on roofs, walls and even windows, and they can also be free standing. Figure 6.1 shows some illustrative examples, specifically of a solar collector installation (a), a wall essentially covered with solar cells (b), and solar cells integrated in a wall glazing (c). The latter example shows solar cells that are very well protected by the glazing, but obviously these windows do not permit a good visual indoors–outdoors contact. Both photothermal and photoelectric devices lead to heating of the air around them, which then contributes to the urban heat island effect in big cities (cf. Section 7.2). The focus of this chapter is on nanostructural features of essential materials for applications to heat and electricity.

6.1 SOLAR THERMAL MATERIALS AND DEVICES

Solar collectors for hot water production have been used at least since the 1950s. Roof mounting is commonplace, and the collectors are oriented to maximize the inflow of solar energy. Figure 6.2 shows the principles of the flat plate collector comprising an absorber plate positioned in a box which is thermally well insulated on the back and sides and has a transparent cover—normally of glass—toward the sun. Obviously, the glass should not absorb solar energy excessively, so it is preferable that it has low iron content (cf. Figure 4.2) and is antireflection coated. The absorber plate is thermally linked to a heat transfer medium that is normally water; it goes through "riser" tubes and emerges heated at the

FIGURE 6.1 *A color version of this figure follows page 200.* Photos illustrating building-integrated devices for photothermal and photoelectric conversion of solar energy. Panel (a) shows a roof-mounted solar collector arrangement from Sweden, used to heat a number of private houses; panel (b) is a solar-cell-covered façade at Solar-Fabrik in Freiburg, Germany; and panel (c) is a solar cell installation forming an architectural element on a building belonging to Fraunhofer Institute of Solar Energy Systems in Freiburg, Germany.

upper end. Construction aspects and thermodynamics of solar collectors have been described in detail in the literature [1].

The flat plate collector is only one of many different types. Another, which is used on a large scale, particularly in China, has the absorber placed in a vacuum tube [2]. This design cuts down on the thermal losses but adds complexity. There is also a large variety of concentrating solar collectors, with low concentration and possible for building integration as well as with high concentration and suitable for large-scale power generation. High concentration is achieved by tracking reflectors that follow the movement of the sun, either using linear trough collectors with focusing onto tubular absorbers or using reflector fields with three-dimensional tracking and concentration onto centrally positioned tower-mounted absorbers. Direct absorption in fluids, with or without blackening by added particles, has received some attention, as has

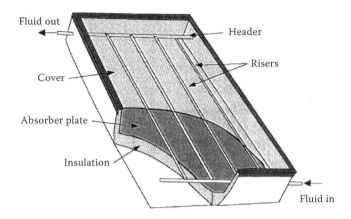

FIGURE 6.2   Principle design of a flat-plate solar collector.

absorption in gases containing absorbing particles [3]. The discussion below is mainly geared toward buildings-related applications.

6.1.1 Spectral Selectivity and Its Importance

What are the desirable optical properties of the surface of the absorber plate (i.e., of the surface that should absorb the impinging solar energy and convert it into useful energy in a heat transfer medium)? This issue was discussed at length in Chapter 2 but is repeated briefly here for convenience. The primary requirement is that solar energy, at $0.3 < \lambda < 3$ µm, should be absorbed (i.e., $A_{sol} = 1$); second, the absorbed energy should not be reradiated as useless heat to the ambience so that thermal emission, at $3 < \lambda < 50$ µm, should be avoided (i.e., $E_{therm} = 0$). Figure 6.3 emphasizes again that the overlap between solar and thermal properties is almost nil so that optimization in the two spectral ranges can be achieved separately. The absorber is nontransparent, and hence the desired spectral reflectance is

$$R(\lambda) = 0 \text{ for } 0.3 < \lambda < 3 \text{ µm} \qquad (6.1)$$

$$R(\lambda) = 1 \text{ for } 3 < \lambda < 50 \text{ µm} \qquad (6.2)$$

These properties are shown in the lower part of Figure 6.3. A surface with a reflectance profile approximating this is referred to as a "selective surface" or, more specifically, a "selectively solar absorbing surface." The magnitude of A_{sol} and its deviation from unity, and the value of E_{therm} and its deviation from zero, constitute quality measures of the surface as further discussed below. A good selective surface should have,

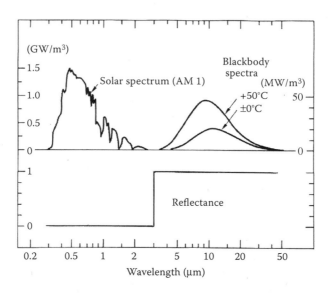

FIGURE 6.3 The upper part shows a solar spectrum for one air mass (AM1) and blackbody spectra for two temperatures. The lower part shows reflectance for an idealized surface designed for absorbing solar energy with minimum thermal loss. This figure is a companion to Figure 4.5.

say, $A_{sol} > 95\%$ for normal incidence of the solar rays and $E_{therm} < 10\%$ for hemispherical thermal emission. The change of the optical properties takes place at a "critical" wavelength λ_c that was put, somewhat arbitrarily, at 3 μm in Figure 6.3. This value of λ_c is good for applications in which the heat transfer medium is not too hot (in practice < 100°C), whereas the high temperatures needed in a thermal power station may require $\lambda_c \sim 2$ μm or even shorter.

It is illustrative to show the effectiveness of the selective absorber over the blackbody absorber via a calculation of the ratio of the energy absorbed and retained by the two surfaces. This is done in Figure 6.4 for a concentration equal to unity (appropriate for a flat plate collector) and for concentration of the solar energy by 10 times [4]. In the absence of concentration and at room temperature, the blackbody absorber only gives some 60% of the energy that is produced and retained by the ideal selective absorber, and the blackbody absorber becomes practically useless at higher temperatures. For a solar concentration of 10 times, the difference between the two types of surfaces is not at all as large at low to moderate temperatures, and the difference is even less significant at higher concentrations. Clearly, spectral selectivity is a key concept for the regular flat plate solar collector, which is the design of most interest for building integration. It may be less important in a highly concentrating

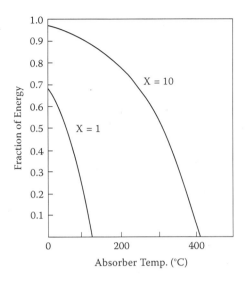

FIGURE 6.4 Fraction of energy absorbed and retained for a blackbody as a function of temperature, as compared with the case of an ideal spectrally selective absorber with $\lambda_c = 2 \, \mu m$, for two values of the solar concentration X. (From B. O. Seraphin, in *Solar Energy Conversion: Solid-State Physics Aspects*, Springer, Berlin, 1979; *Topics in Applied Physics* Vol. 31, chap. 2, pp. 5–55. With permission.)

set-up for power generation, but spectral selectivity nevertheless is not insignificant because the absorber constitutes only one part in a more complicated device, and improving it even to a minor degree may make sense economically.

Spectrally selective transmittance of the cover glass of the solar collector in principle is an alternative to the spectrally selective absorber. For this latter case one would like to have, ideally, a glass with $T_{sol} = 1$ and $R_{therm} = 1$ so that thermally emitted radiation from the solar absorber is reflected back to the same surface. The infrared reflecting coating on the downward facing glass surface could be made of doped SnO_2, In_2O_3, or ZnO, as discussed in Section 4.4. However, in order to compete with a good spectrally selective absorber, the IR reflector must be antireflected, not only for perpendicularly incident radiation, as in Figure 4.18, but over a wide angular range.

6.1.2 Principles for Spectral Selectivity

Spectrally selective surfaces can be designed according to a number of different principles [5,6], as indicated in Figure 6.5. The most straightforward

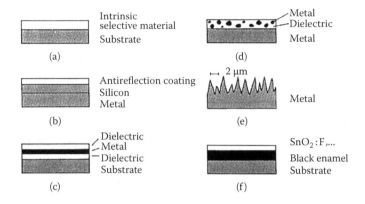

FIGURE 6.5 Schematic designs of six different approaches to spectrally selective absorption of solar energy. (From G. A. Niklasson, C. G. Granqvist, in *Materials Science for Solar Energy Conversion Systems*, edited by C. G. Granqvist, Pergamon, Oxford, 1991; pp. 70–105. With permission.)

of these is to have a material whose intrinsic optical properties have the desired spectral selectivity. This approach has not been fruitful, though, and no material with sufficiently good properties has yet been identified. Another possible way to create spectral selectivity is to coat a low-reflecting metal with a semiconductor having a band gap corresponding to λ_c. Silicon, with a band gap of ~1.1 eV, is a candidate material; another possibility is PbS with a band gap of ~0.4 eV [7]. A problem with silicon is that its refractive index is so high that reflective losses become excessive. Antireflection treatment is possible, but the weak absorption in Si, which demands a large thickness, makes the approach unattractive with regard to thermal applications. Instead, the desired properties have been obtained by use of thin films with a suitable nanostructure in the composition and/or in the surface roughness as discussed next.

Multilayer absorbers backed by metal can be tailored to have the desired optical properties. Basically, the design principles are the same as those underlying the metal-based transparent heat reflectors discussed in Section 4.3.4, although the coating should have maximum solar absorptance (rather than maximum T_{lum} or T_{sol}) and maximum thermal reflectance. Three-layer coatings of the type $Al_2O_3/Mo/Al_2O_3$ have been rather successful and can be manufactured with large area technologies [8].

Nanostructures with a mixture of metal particles in a dielectric host represent another approach to spectral selectivity. The materials are often referred to as "ceramic-metallic mixtures" or "cermets." Their principal optical properties can be illustrated from an effective medium calculation according to Equation 3.19, wherein for simplicity we consider an

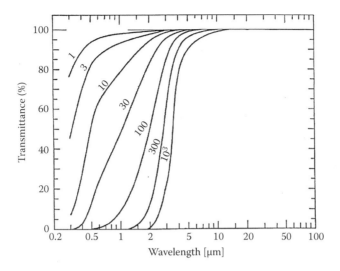

FIGURE 6.6 Computed spectral transmittance for a layer comprised to 1% of Cr particles in air. Data are given (in μm) for the film thicknesses shown. (From C. G. Granqvist, *Phys. Scripta* 16 [1977] 163–164. With permission.)

essentially nonreflecting medium with 1% of Cr particles surrounded by air [9]. As seen from Figure 6.6, the transmittance is low at short wavelengths and high at long wavelengths and the change in transmittance depends on the thickness of the layer. For practically useful surfaces—such as those discussed below—the coating is

- Backed by a reflecting metal
- The particles are embedded in a dielectric material (normally an oxide)
- The filling factor of the metallic component is comparable to that of the dielectric
- The nanostructural entities are nonspherical

The latter three items make it possible to decrease the coating thickness to only a fraction of a micrometer in a coating for spectrally selective solar absorption.

Surface roughness leads to other possibilities to create spectral selectivity. If the surface has metallic protrusions separated by distances of the order of the wavelengths for solar irradiation, this radiation will penetrate into the structure and undergo multiple reflections and thereby become absorbed, whereas radiation with longer wavelengths will not "see" the surface as rough but as smooth and hence reflecting. These

features are clearly illustrated in the recent calculations reported on in
Figure 6.7 [10]. They regard a tapered subwavelength grating comprised
of a square lattice array of tungsten pyramids with a 250-nm-period in
both transverse dimensions and a height of 500 nm. The spectral reflectance was computed (by a technique we do not discuss here) as a function
of light incidence represented by polar and azimuthal angles θ and ϕ,
respectively, defined in the same way as in Sections 2.3 and 4.7. Panels (b)

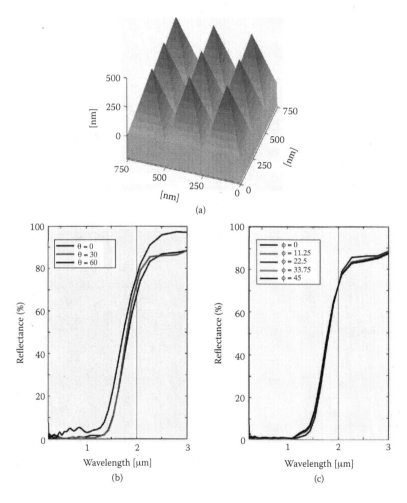

FIGURE 6.7 Computed spectral reflectance for a surface consisting of
tungsten nanopyramids according to panel (a). Data are shown in (b)
and (c) as a function of polar angle θ and azimuthal angle ϕ, respectively.
(After E. Rephaeli, S. Fan, *Appl. Phys. Lett.* 92 [2008] 211107 1–3.
With permission.)

and (c) in Figure 6.7 demonstrate a very low reflectance—irrespective of the direction of the light incidence—for solar energy and a sharp onset of reflectance for $\lambda > \lambda_c$. The approximate independence of light incidence is, of course, a very important asset for a stationary solar collector such as the one depicted in Figure 6.1a. If the height of the pyramids was put to 250 nm, the value of A_{sol} was significantly degraded, and the same was the case if the periodicity was significantly larger than 250 nm [10].

The final approach to spectral selectivity is different and is based on a *blackbody-type surface coated with a film with selective transmittance*, specifically transmitting at $\lambda < \lambda_c$ and reflecting at $\lambda > \lambda_c$. Such a surface can consist of black enamel. The heavily doped wide band gap semiconductors discussed in Section 4.4 are natural candidates for the top layer, and $SnO_2{:}F$ stands out as particularly interesting on account of its ruggedness and possibility for large-scale manufacturing by low-cost spray pyrolysis.

Particularly detailed studies of selectively solar absorbing surfaces have been made on Co-Al_2O_3 nanocomposite films prepared by co-evaporation [11]. The results of that study are summarized in Box 6.1.

6.1.3 Selectively Solar-Absorbing Coatings Based on Nanoparticles: Some Practical Examples

The case study on well-characterized Co-Al_2O_3 films discussed in Box 6.1 has set the scene for appreciating spectrally selective surfaces used in real life, or at least developed with that in mind. The choice of Co and Al_2O_3 is less restrictive than one perhaps would believe, and the data for Co are representative of transition metals in general (including Cr and Ni); Al_2O_3 is a typical oxide with good thermal stability and hence well suited for solar-related applications.

In fact, there are numerous selectively solar absorbing coatings in practical use. They are made by a variety of coating technologies such as electroplating, anodization and other chemical conversion techniques, evaporation, sputtering, etc. [6]. All of these coatings acquire their optical properties by a combination of metal-dielectric nanostructures and surface roughness integrated with multilayer features (i.e., by a concoction of at least three of the mechanisms illustrated in Figure 6.5). Rather than trying to cover the different coatings in detail, we will next present a series of Ni-based coatings prepared by different technologies. By doing this we will also be able to give a view on the development of the field during the past couple of decades and of the refinements that have been accomplished.

Electrochemical methods were used in the earliest work to produce selectively solar absorbing surfaces. Thus electroplating was employed to

BOX 6.1   CASE STUDY FOR Co-Al$_2$O$_3$ NANOCOMPOSITE FILMS

Here we present some data from a very detailed study of Co-Al$_2$O$_3$ films in order to elucidate critical parameters for the use of metal-dielectric-based films as spectrally selective absorbers of solar energy [11]. The films were made by *e*-beam co-evaporation of Co and Al$_2$O$_3$ under well-controlled conditions onto unheated substrates. Electron microscopy showed that a nanostructure with hcp Co nanoparticles embedded in Al$_2$O$_3$ prevailed as long as the Co volume fraction f_{Co} remained below ~0.3. At higher f_{Co}s, the two components formed a two-phase mixture of a more random nature. The average size of the Co nanograins was ~1 nm for small f_{Co}s and rose approximately linearly with the Co content to ~2.5 nm for f_{Co} ≈ 0.6. The effective dielectric function was evaluated from measurements of $T(\lambda)$ and $R(\lambda)$ on samples with $0.1 < f_{Co} < 0.6$. These data could be reconciled with theories for optical homogenization (cf. Section 3.9) at least for $f_{Co} < 0.3$ and assuming that a minor part of the Co was atomically dispersed in the Al$_2$O$_3$ matrix.

Empirical data on the complex dielectric function were then used to compute solar absorptance for normally incident solar radiation as well as hemispherical thermal emittance at 100°C. Figure B6.1.1 shows these quantities as a function of thickness for uniform Co-Al$_2$O$_3$ films backed by a substrate characterized by the dielectric function for Al and Ni. Requiring a high value of A_{sol} and a low value of E_{therm} leads to Co-Al$_2$O$_3$ films with high f_{Co}s, and it is found that the Al substrate yields a particularly low value of E_{therm}. There is an optimum film thickness for giving maximum A_{sol}; its value is around 70 nm, and the beneficial effect on the absorption is due to antireflection of the underlying metal.

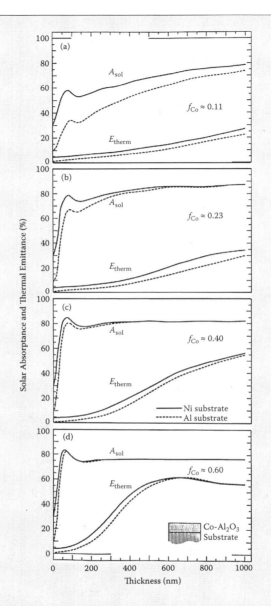

FIGURE B6.1.1 Computed normal solar absorptance and hemispherical thermal emittance at 100°C for Co-Al$_2$O$_3$ as a function of film thickness. The Co contents and substrate metals shown here were studied. (From G. A. Niklasson, C. G. Granqvist, *J. Appl. Phys.* 55 [1984] 3382–3410. With permission.)

The computational study in Figure B6.1.1 clearly leads to a rather well defined design of a spectrally selective surface. This design was then produced experimentally and is further investigated in Figure B6.1.2, showing data on spectral reflectance for a Ni surface coated with a 70-nm-thick Co-Al$_2$O$_3$ film having $f_{Co} \approx$ 0.6. Spectral selectivity with $\lambda_c \approx 2$ μm is manifest. The reflectance shows a peak at around 25% at a wavelength of ~0.5 μm which obviously is undesired and tends to deteriorate A_{sol} significantly. However, applying an antireflection coating consisting of 70 nm of Al$_2$O$_3$ leads to a double minimum in the reflectance and a much improved solar absorptance. A still lower value of A_{sol} could have been achieved by having a graded composition of the Co-Al$_2$O$_3$ film, but it should be remembered that the exercise at hand aims at showing principles rather than optimized products—which we deal with shortly.

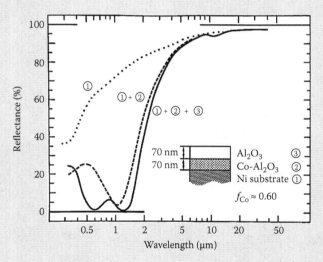

FIGURE B6.1.2   Measured spectral reflectance of a Ni surface (dotted curve), after deposition of ~70 nm of Co-Al$_2$O$_3$ with $f_{Co} \approx 0.6$ (dashed curve), and after overcoating with 70 nm of Al$_2$O$_3$ (solid curve). (After G. A. Niklasson, C. G. Granqvist, *J. Appl. Phys.* 55 [1984] 3382–3410. With permission.)

We finally widen the perspective by considering angular dependent properties, specifically $A_{sol}(\theta)$ and $E_{therm}(\theta)$ where θ is the angle with regard to the surface normal. These data are shown in Figure B6.1.3, where we also demonstrate the internal consistency of the evaluations by plotting closely agreeing data as computed from the complex dielectric function and as derived from angular dependent spectral reflectance for s- and p-polarized light. It is seen that the optimized design leads to $A_{sol} \approx 95\%$ at normal incidence, and that this value does not fall off drastically until high angles are reached. The slow drop at increasing angles obviously is favorably for stationary solar collectors. E_{therm}, on the other hand, is about 5% at small angles and rises gradually for increasing angles and has a peak value of ~15% at $\theta \approx 85°$. This latter feature makes the hemispherical value of E_{therm} increase to ~7%, i.e., a by fraction that is noticeable though not very prominent. Similar results for A_{sol} and E_{therm} have been obtained for Ni-based coatings such as those discussed next [6,12].

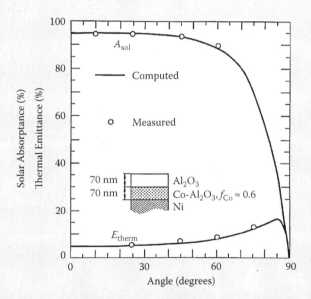

FIGURE B6.1.3 Angular-dependent solar absorptance and thermal emittance at 100°C as evaluated from reflectance data (circles) and as computed from dielectric functions (solid curves) for the optimized two-layer coating sketched in the inset. (After G. A. Niklasson, C. G. Granqvist, *J. Appl. Phys.* 55 [1984] 3382–3410. With permission.)

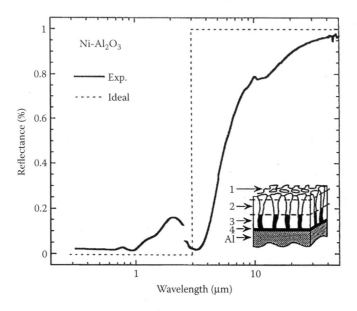

FIGURE 6.8   Spectral reflectance of Ni-pigmented alumina made by anodization followed by ac electrolysis. The data combine two sets of measurements in different wavelength ranges. Dotted lines indicate the ideal performance. Inset shows the nanostructure with a compact barrier layer of Al_2O_3 (4), a Ni-Al_2O_3 composite layer (3), and a porous Al_2O_3 layer (2) whose pores widen at the top surface (1). (After Å. Andersson et al., *J. Appl. Phys.* 51 [1980] 754–764. With permission.)

make "black nickel" and "black chrome," both consisting of a layer with metallic nanoparticles in an oxide-based matrix of unspecific nature. A later generation of Ni-based coatings used a two-step procedure with initial anodization of aluminum in dilute phosphoric acid so that the surface layer of the metal was transformed into ~0.7 μm thick porous Al_2O_3 with channels perpendicular to the surface and extending through the pore layer [13]. Nanostructures of this kind are described briefly in Appendix 1. In a second step, metal was precipitated inside the pores by ac electrolysis in a bath containing nickel sulfate. Metal particles were then formed as nanorods with diameters of 30 to 50 nm and a length of ~300 nm [14]. With suitable anodization parameters, the metal fraction was ~30%. Essentially, the coating comprises Ni-pigmented Al_2O_3 under a porous Al_2O_3 antireflection layer, and the bottom of the coating has a layer of compact Al_2O_3 serving as diffusion barrier. The inset of Figure 6.8 illustrates this structure, whereas the main part of the figure shows spectral reflectance. The reflectance resembles the desired profile indicated by the dotted lines. Commercial production of such coatings

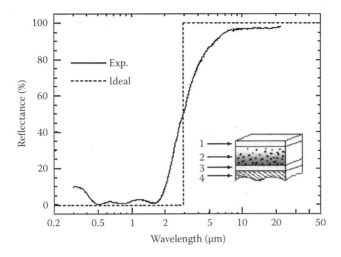

FIGURE 6.9 Spectral reflectance of a Ni-based coating made by sputter deposition and roll-to-roll manufacturing. Dotted lines indicate the ideal performance. Inset shows the nanostructure with Al substrate (4), barrier layer (3), graded Ni-NiO composite layer (2), and antireflecting Al_2O_3 layer (1).

typically yielded $A_{sol} \approx 96\%$ and $E_{therm} \approx 15\%$. The manufacturing involves treatment of large amounts of hazardous chemicals and generally has been superseded by alternatives with "greener" attributes.

The durability and degradation mechanism of the Ni-pigmented anodic Al_2O_3 coatings have been investigated in considerable detail [5,6]. Prolonged treatment at high temperatures led to a gradual decrease of the solar absorptance, which could be reconciled with uniform oxidation (rather than a progressing oxidation front) of the Ni nanoparticles. Detailed work has led to lifetime predictions, which have been found to agree well with real ageing data [6,15,16].

Sputter deposition was developed into a manufacturing technology for spectrally selective coatings during the 1990s [17]. The inset of Figure 6.9 illustrates the composition of a fully developed coating with a graded Ni-NiO film backed by Al with a highly reflecting barrier layer and having an 85-nm-thick antireflecting Al_2O_3 layer at the top. The graded layer was made by the special sputter technology shown in Figure A1.8. Deposition used roll-to-roll coating onto 30-cm-wide ribbons comprised of two Al sheaths with a flattened Cu tube in the center. After sputter deposition, this ribbon was cut into suitable lengths, and the Cu tube was opened by pressurized air to turn it into "riser" tubes (cf. Figure 6.2). The spectral reflectance of the coated Al-based ribbon is closer to the ideal performance than the electrochemically deposited

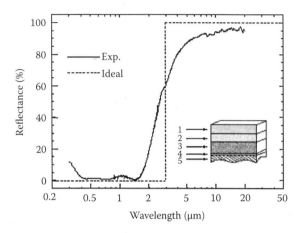

FIGURE 6.10 Spectral reflectance of a Ni-based coating made by sol-gel deposition. Dotted lines indicate the ideal performance. Inset shows the nanostructure with Al substrate (5), barrier layer (4), Ni-Al$_2$O$_3$ composite layers with two compositions (3 and 2), and antireflecting SiO$_2$ layer (1).

coating in Figure 6.8 and corresponds to A_{sol} = 97% and hemispherical E_{therm} = 5% for 100°C [18].

During the 2000s, another Ni-based coating, based on *sol-gel deposition*, has emerged [19]. Its cross section was illustrated as an example of a sol-gel coating in Figure A1.2, which showed a 85-nm-thick bottom layer of 80% Ni + 20% Al$_2$O$_3$ and an upper 60-nm-thick layer of 40% Ni + 60% Al$_2$O$_3$ under a 65-nm-thick antireflecting SiO$_2$ layer. The Ni-Al$_2$O$_3$ layers were heat treated at 550°C and the SiO$_2$ layer at 400°C. The coating was deposited onto Al having a barrier layer. The optical performance is reported in Figure 6.10. Just as for the previous film, the spectral reflectance corresponds to A_{sol} = 97% and hemispherical E_{therm} = 5% for 100°C.

Evaporation serves as an alternative to sputter deposition, and an interesting type of selectively solar absorbing coatings has been developed by reactive evaporation to make TiO$_x$N$_y$ (known as "TiNOX") coatings [20]. A_{sol} ≈ 95%, E_{therm} ≈ 5%, and stability up to 400°C, have been reported.

The coatings discussed thus far are examples of selective absorbers suitable for buildings-related applications. As mentioned above, there are also many large-scale applications for solar-driven power generation, and then the absorber should work at high temperatures so that λ_c ≈ 2 μm may be appropriate. Figure 6.11 shows an example of such a coating deposited onto stainless steel and designed to work on tubular collectors

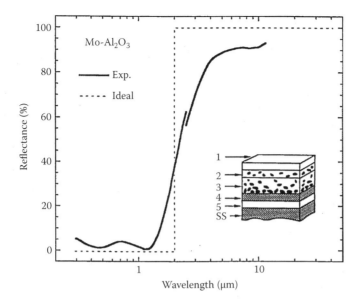

FIGURE 6.11 Spectral reflectance of a Mo-Al$_2$O$_3$-based coating made by sputter deposition. The data combine two sets of measurements in different wavelength ranges. Dotted lines indicate the ideal performance. Inset shows a five-layer structure with a stainless steel (SS) substrate, Al$_2$O$_3$ (5), Mo (4), graded Mo-Al$_2$O$_3$ (3), graded Mo-SiO$_2$ (2), and a top layer of SiO$_2$ (1). (After M. Lanxner, Z. Elgat, *Proc. Soc. Photo-Opt. Instrum. Engr.* 1272 [1990] 240–249. With permission.)

at temperatures exceeding 300°C [21]. The nanostructure embodies two Mo-dielectric composite layers backed by an IR reflecting, and hence low-emittance, Mo layer. There is also an antireflecting SiO$_2$ layer at the top and a diffusion barrier of Al$_2$O$_3$ at the bottom. The coating shown has $A_{sol} = 97 \pm 1\%$, $E_{therm} \approx 10\%$ at 50°C and $E_{therm} = 17 \pm 1\%$ at 350°C.

6.1.4 Colored Absorbers and Paints: Novel Developments

It was demonstrated above that spectrally selective solar absorbers with close to ideal properties can be produced successfully not only in the laboratory but as commercial products. So what remains to be done? Current development progresses in two different directions: one of them—at first sight perhaps paradoxically—addresses the fact that the earlier discussed spectrally selective surfaces are too good! The latter statement can be interpreted as saying that the surfaces, being almost ideal solar absorbers, also look very black to the eye which often is not

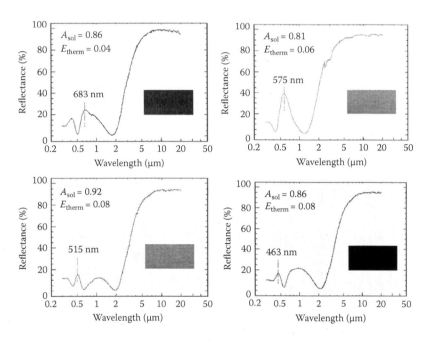

FIGURE 6.12 *A color version of this figure follows page 200.* Spectral reflectance for TiAlO$_x$N$_y$ films made by sputter deposition onto Al. Different colors (shown) were obtained by selecting film thicknesses leading to reflectance maxima at the wavelengths shown in the luminous wavelength range. Corresponding values of A_{sol} and E_{therm} are stated. (From D. Zhu, S. Zhao, *Solar Energy Mater. Solar Cells* [2010] in press. With permission.)

desired for aesthetic reasons. Hence, there is a need to develop colored selective absorbers for which the coloration is achieved without an excessive loss of solar absorptance. The other development line is related to cost issues and deals with spectrally selective paints.

We first consider colored coatings produced by sputtering and show data for TiAlO$_x$N$_y$ films backed by Al in Figure 6.12 [22]. Single homogeneous films lead to two interference minima when their thickness is ~200 nm. The precise location of the intervening reflectance maximum is governed by the thickness, which can be tuned so that the perceived color is red, yellow, or dark blue. A green color is somewhat more difficult to obtain and requires a double layer coating with TiAlO$_x$N$_y$ films having different compositions and hence refractive indices. Corresponding values of A_{sol} lie between 81 and 92%, whereas the emittance values, based on near-normal reflectance measurements, lie between 4 and 8%. These results are not as good as those for optimized spectrally selective

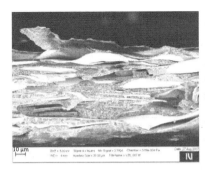

FIGURE 6.13   Cross section (left) and top view (right) of a thickness-insensitive spectrally selective paint with Al flakes. (From B. Orel, private communication [2009].)

solar absorbers, but the colors allow applications in situations for which black surfaces may not be acceptable.

Selectively solar absorbing paints represent a low-cost alternative to the coatings applied by other techniques. The paint formulations typically contain a strongly absorbing pigment such as $FeMnCuO_x$ or $CuCr_2O_4$ as well as highly reflecting metal flakes, typically of Al in a polyurethane binder [23]. Figure 6.13 depicts a cross section and a top view of such a coating. The flakes align more or less in parallel and serve as "artificial low-emittance substrates" so that the optical properties are thickness insensitive and independent of the substrate. In the absence of metal flakes, the coating would have a thermal emittance typical for the organic binder (i.e., $E_{therm} \approx 90\%$). A proper paint formulation, however, leads to coatings with $80 < A_{sol} < 90\%$ and $30 < E_{therm} < 40\%$. Figure 6.14 shows typical reflectance spectra for three paint layers, and demonstrates that spectral selectivity is at hand, though far from the optimum. The samples are visibly colored and appear green, red, and blue. Very recently, paint layers with self-cleaning properties have been developed [24].

6.2 PHOTOVOLTAIC MATERIALS AND DEVICES

Solar cells, also called photovoltaic (PV) cells, convert solar energy into electricity and are widely touted today (2009) as a technology for large-scale electricity production. The efficiency of the solar cells is less than that for converting solar energy into heat, but electrical energy, of course, can be used for many more purposes than thermal energy. What is not converted to electricity in the solar cell turns into heat, and solar cells therefore become heated during operation. This heating tends to erode

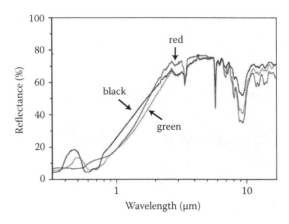

FIGURE 6.14 Spectral reflectance for visibly green, red, and blue layers of thickness-insensitive spectrally selective paint. Corresponding values of A_{sol}/E_{therm} are 84/36, 83/40, and 81/32%, respectively. (From B. Orel, private communication [2009].)

the photoelectric efficiency as well as cause degradation and diminish the operating life time of the cells. Systems combining electricity production with heat production—using a heat transfer fluid for cooling the solar cells—have been discussed and implemented many times, and may be technically elegant, but such systems have not gained much popularity owing to their technical complexity.

The heating that would ensue from widespread deployment of building integrated photovoltaics, known as BIPV, has not received much attention so far but clearly is an effect that contributes to urban heat islands (cf. Section 7.2) and hence stands in opposition to the global cooling one might accomplish via increasing urban albedos (surface "whiteness") [26]. On the other hand, if BIPV power displaces coal power it provides net environmental benefits quickly despite its additional heat output [27]. Net benefits require the accumulated CO_2 reductions, which arise from PV cells replacing coal power, to surpass the impact of associated heat emissions. But also necessary to consider are alternative mitigation strategies, such as global cooling by raising roof albedos. Consider, for example, a 1 kW_p solar cell installation (where the subscript indicates peak output). It might occupy about 10% of a home's roof area; what if instead the whole roof were specially painted to raise its solar albedo by 30 to 40%? It would take 50 to 80 years for the accumulated CO_2 reductions from these solar cells to match the initial albedo offsets from the complete white roof [27]! Doubling the cell area would cut these times into half, but the cost differential—which is very large—would also double. In addition, the production of silicon for solar cells generates a lot of heat and CO_2 emissions, and this issue is so important that it is specifically addressed for some emerging routes for solar cell production. One possibility is to use concentrated solar radiation in the melting and refinement process.

Below we first give a brief overview of solar cell types and then discuss how nano-features can boost the efficiency of silicon-based solar cells and how nanotechnology leads to new types of solar cells, such as dye-sensitized and organic cells.

6.2.1 Technology Overview

In principle, all that is required for an electricity-generating material is a capacity for having excited states into which electrons can be brought from a ground state via photon absorption, and means to extract electrons from these states. There is a plethora of options but—despite intense research and development over many years—attention has been focused only on a few of these. A common feature of solar cells is that the side exposed to the sun must have some current collecting arrangement, either a metallic pattern, a transparent conducting film (cf. Section 4.4) or a combination of these. We do not go into details about solar cell technology here but refer the reader to excellent texts on the subject [28–33].

Figure 6.15 gives a bird's eye view on the development of a number of solar cell technologies that has followed the first solar cell created in 1954 [34]. The highest efficiencies are for advanced, and expensive, designs with superimposed two-junction and three-junction cells (see below), which can have a conversion efficiency η exceeding 40%. Other technologies, including today's dominating Si cells and various thin-film technologies, have efficiencies with maximum values in the 20 to 25% range. It is the latter that are of interest for BIPV in the foreseeable future.

Silicon is an excellent material for solar cells and it is by far the most used one today [32]. There are different types of Si-based solar cells, as apparent from Figure 6.15, with single crystalline ones reaching $\eta \approx$ 25%. Silicon is nontoxic, very abundant, and has a thorough technological base as a result of its ubiquitous uses in microelectronics. In essence a slab or film of Si is n-doped and p-doped in its two surface regions, and this structure is put between a transparent front electrode and an opaque metallic back electrode. The device is supported by glass or by a foil of a metal or polymer. The Si can be of various types:

- Crystalline and cut out from a single crystal ingot
- Polycrystalline and prepared from a multi-crystalline ingot or via "ribbon growth"

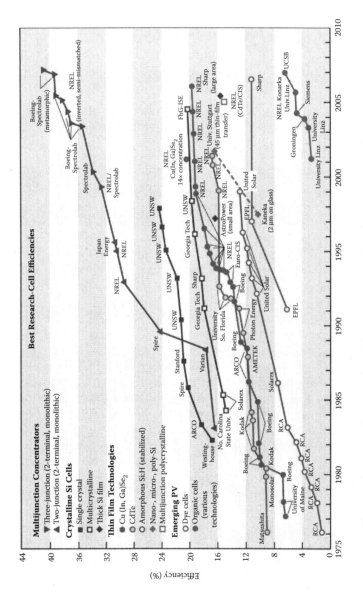

FIGURE 6.15 Best efficiency for research-type solar cells of a variety of kinds versus year when the results were presented. The symbols have designations mentioning the laboratories responsible for the data; the symbols are joined by straight lines to guide the eye. Acronyms used are NREL (National Renewable Energy Laboratory, United States), UNSW (University of New South Wales, Australia), FhG-ISE (Fraunhofer Institut für Solare Energiesysteme, Germany), EPFL (École Polytechnique Fédérale de Lausanne, Switzerland), and UCSB (University of California–Santa Barbara). (After J. Berry, private communication [2009].)

- Amorphous and hydrogenated, typically being a film made by glow discharge deposition from silane gas

High values of η demand low reflection losses, which can be accomplished by an antireflecting film or—for single crystalline cells—by using anisotropic etching to make pyramid-type texture. Single crystalline and polycrystalline cells normally have metallic "fingers" as current collectors, which partially shade the cell. Higher efficiency (but more expensive) cells use "buried" vertical electrodes formed inside laser-etched grooves. These shade less and collect more current. SnO_2-based transparent conductors are often used as electrodes in thin-film devices. A thin TiO_2 layer can serve as protection against harmful exposure to hydrogen radicals in some techniques for making amorphous Si solar cells.

Compound semiconductors of the III-V type—based on GaAs, (Al,Ga)As, InP, or (In,Ga)P—can display high conversion efficiencies and good resistance to ionizing radiation. However, they are costly and mainly of interest for space applications; no information for this type of cells in shown in Figure 6.15.

II-VI compounds represent another option, and solar cells based on CdTe and Cd(S,Te) are noteworthy for their robust and cheap manufacturability and are of much interest for terrestrial applications [35]. Potential environmental problems regarding large-scale deployment of a technology involving Cd-containing materials must not be overlooked, though. Figure 6.15 shows that $\eta \approx 17\%$ has been achieved.

Ternary I-III-IV$_2$ compounds attract much interest for PV applications [35,36]. In particular, $CuInSe_2$, $Cu(In,Ga)Se_2$, and $Cu(In,Ga)S_2$ are notable for their ability to yield high efficiency in thin-film solar cells with values up to ~20% shown in Figure 6.15. The sulfur-containing films seem to allow cheaper manufacturing techniques than the selenide-containing ones. However, the fabrication of ternary oxides with strict demands on composition is always demanding with regard to large-scale manufacturing.

It is customary to speak of first, second, and third generation solar cells, as illustrated in Figure 6.16 [31]. Here "generation I" refers to the well-established crystalline Si wafer technology and "generation II" refers to today's thin-film technologies. "Generation III" draws on a number of advanced concepts that in the future may allow solar cells to get closer to the thermodynamic limit of ~93% while the cost is kept low. "Generation III" solar cell technologies include multi-junction devices, arrangements leading to the creation of multiple electron-hole pairs, cells utilizing hot carrier effects, thermo-photovoltaics, as well as different thermoionic and thermoelectric approaches; these concepts are generally considered to be outside the scope of the present book, and the reader is referred to the literature [31] for in-depth information.

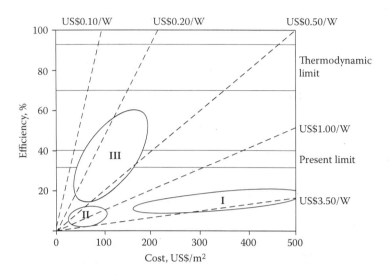

FIGURE 6.16 Efficiency versus cost (in 2003 US$) for solar cell technologies belonging to generations I–III. (From M. A. Green, *Third Generation Photovoltaics: Advanced Solar Energy Conversion*, Springer, Berlin, Germany, 2003. With permission.)

Some emerging PV technologies, represented in the lower right-hand corner of Figure 6.15, may also be considered "generation III"; they are considered below.

6.2.2 Nanofeatures for Boosting the Efficiency of Silicon-Based Solar Cells

Thinner wafers would lead to a reduction in the production cost for Si solar cells and a lowering of their energy payback time. An even more radical approach to saving on the amount of Si is to use devices or wires in which the *p-n* junction is radial rather than planar [37]. Thickness reduction is a nontrivial issue, though. Surface texturing at the micro scale allows thinner *c*-Si, but if the thickness is too small there will nevertheless be an excessive number of photons that pass through the cell without being absorbed, so η becomes undesirably small.

Nanostructures can improve the situation and allow thinner *c*-Si cells. One approach uses micrometer-thick *c*-Si under glass arranged so that the light coupling into the Si is enhanced either by having the glass surface nanostructured or by having the transparent conducting oxide electrodes between the smooth glass and the Si film deposited so as to be nanostructured. A problem with *c*-Si layers having thicknesses of a few micrometers is that they require a very long high-temperature anneal—as much as about 24 h—in order to achieve the desired electrical quality, so energy, emissions, and urban heat island impacts may be as high as for Si wafers, even if total cost is down. Lower production energy would be needed for this type of cell if the c-Si layer and the nanostructured TCO layer were deposited pyrolytically on glass sheet in conjunction with its production in a float line. This idea is under serious consideration by some glass manufacturers. Another attraction of such a production route is that it opens the way toward commodity-scale manufacturing with associated low costs for solar cells. As pointed out before in Section 4.1.2, float glass is made on the scale of billions of square meters per annum, and new plants are easily built. The electrical quality of the c-Si made in this way will be the main challenge.

Nanostructured TCOs for solar cells are already being deposited pyrolytically onto glass sheet during its production on a float line by use of a modification of the processes for making low-emittance windows, as described in Figure A1.1. The physics behind enhanced light coupling to Si due to nano-TCOs depends on the scale of the nanostructure. A combination of scattering and surface plasmon resonance coupling can be involved. Plasmon resonances can lead to coupling into the Si layer both by scattering and more directly if the Si layer's light transport modes and the nanoparticle's SPR modes form a coupled resonant system. These types of effects were discussed in Section 3.11. The effects are most obvious when metal nanoparticles are deposited onto a very thin Si-oxide layer on the Si. This idea was first discussed for Cu, Ag, and Au particles [38], and subsequently for Ag nanoparticles on the LED structure shown in Figure 6.17 [39]; such a LED structure can be easily adapted for use as a solar cell. Enhanced photocurrents arise at wavelengths that depend on the particle shape and size, as well as on the oxide and waveguide thicknesses in structures such as that in Figure 6.18. Overcoating the Ag particles with a thin insulator, as shown for ZnS, red-shifts and enhances the coupling and protects the silver. Details of the relevant SPR mechanism are elaborated in Box 6.2.

Surface texturing of the Si itself with pyramids is used to enhance light coupling, but such texturing can degrade the electrical properties. In the plasmonic approach, however, the nanoparticles are not in contact with the Si, so electrical degradation does not occur.

The structures discussed above involve Si layers with thicknesses of at least a few micrometers. It may also be possible to enhance the coupling into much thinner Si layers by use of SPR nanoparticles, but the coupling mechanism is different from that with thicker layers since nanothin Si cannot act as a waveguide for light. In this latter case, the extra coupling is via the increased electric field strength close to the resonant

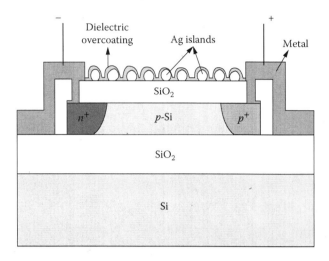

FIGURE 6.17 Schematic of a simple LED or thin crystalline Si solar cell structure. (From S. Pillai et al., *Appl. Phys. Lett.* 88 [2006] 161102 1–3. With permission.)

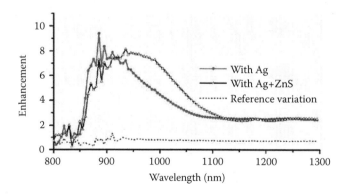

FIGURE 6.18 Enhancement of light coupling in the case of bare and ZnS coated Ag nanoparticles for the structure in Figure 6.17. The actual data refer to electroluminescence. The dotted curve shows data taken in the absence of nanoparticles. (From S. Pillai et al., *Appl. Phys. Lett.* 88 [2006] 161102 1–3. With permission.)

**BOX 6.2   HOW CAN RESONANT METAL
NANOPARTICLES, WHICH NORMALLY
CONVERT LIGHT INTO HEAT, ENHANCE THE
CONVERSION OF LIGHT TO ELECTRICAL
CURRENT WHEN PLACED ON A LAYER OF Si?**

The trick is to have the surface plasmon resonant energy in the metal nanoparticles convert into waves of traveling light inside the Si layer before it can dissipate as heat within the nanoparticle. It is also necessary to produce electric current before any reverse coupling of this light back into other nanoparticles occurs inside the Si. There are two distinct situations.

In the first of these the Si layer is thick enough to act as a waveguide, which means that certain wavelengths or modes can travel sideways through the layer long enough to be absorbed by the Si as in Figure B6.2.1. The illuminated resonant metal nanoparticles behave like little dipoles or antennas and radiate in all directions, in effect scattering a lot of light into the Si at an angle so that light travels much further on average inside the Si than if the nanoparticles had not been there. The effective dipole sits a few nanometers above the oxide layer. Models for this effect have been established [41] and show that a point dipole just above an Si layer can couple very strongly (with nearly 100% efficiency). Finite-size nanoparticles are not as efficient but still couple well. Both shape and size of the nanoparticles influence the coupling efficiency [42].

A nanocylinder and a nanohemisphere couple more strongly, at about 90% efficiency, than a nanosphere if they all have the same radius. The latter's optical coupling efficiency varies with particle size, and 150-nm-diameter and 100-nm-diameter spheres are about 50 and 70% efficient, respectively. The Si layer is rather thick, around 1 μm, but this is much less than a typical Si-wafer thickness. It is interesting that the resonant properties are rather independent of the choice of metal. While Cu, Ag, and Au nanoparticles on glass have their resonances at widely different wavelengths, the enhancement effect is peaked at almost the same wavelength in the NIR when the nanoparticle is located immediately above an Si waveguide layer [38]. The waveguide coupling is responsible for this effect as it couples to the tails of the SPRs.

The second situation regards a less-than-60-nm-thick semiconductor layer under surface plasmon resonant nanoparticles. This was the case for the data presented in Figure 6.19. Very strong evanescent or localized electric fields occur just outside the resonant particles. These fields extend only a short distance, but if the particle is close enough to a semiconductor layer whose thickness matches the extent of the local field, then enhanced absorption becomes possible in the semiconductor at wavelengths where transmittance would normally take place. As noted above, charge transport is a concern in such layers.

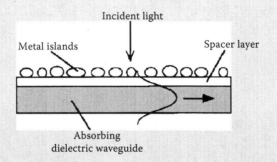

FIGURE B6.2.1   Metal nanoparticles on a thin spacer layer on top of a layer which acts as a waveguide. (From H. R. Stuart, D. G. Hall, *Appl. Phys. Lett.* 69 [1996] 2327–2329. With permission.)

nanoparticles. It has been estimated that Si solar cells as thin as 15 to 50 nm may be made up to 40% optically efficient by using an overlayer of plasmonic nanoparticles, as evident from the data in Figure 6.19 [40]. The challenge lies in achieving good electrical properties in such layers.

6.2.3 Dye-Sensitized Solar Cells

Dye-sensitized solar cells (DSSCs) are based on electrochemical conversion of solar energy. The pivotal work was by Grätzel in 1991 [43], and the cells are often referred to as "Grätzel cells," although, as always, there was a large body of earlier work to build on. The highest value of η so far is 11.1%, which was measured under standard AM1.5 solar irradiation onto an aperture that was ~0.2 cm^2 in size [44]. Apertures of these magnitudes are typical for many efficiency assessments, and much lower efficiencies are typically observed for module sizes of more practical interest. Figure 6.15 shows that $\eta \approx 10\%$ was reached rapidly, but the progress manifested since the mid-1990s seems to have been rather slow despite huge research efforts worldwide [45]. Low manufacturing cost is often put forward as a specific asset for these cells, and was done so already in the first publication on DSSCs; however, since no

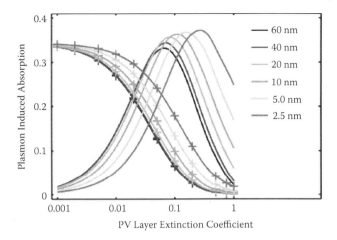

FIGURE 6.19 *A color version of this figure follows page 200.* Absorption induced in various thicknesses of very thin Si by neighboring Ag nanoparticles (peaked curves to right). The x-axis is the absorption coefficient of the Si material. The particles themselves absorb according to the plots with crosses. (From C. Hägglund, B. Kasemo, *Opt. Express* 17 [2009] 11944–11957. With permission.)

reliable technology for mass fabrication seems to be at hand (in 2010), such statements must be regarded as tentative, at best.

Figure 6.20 is a schematic representation of the functional principles of a DSSC [46]. A nanoporous and nanocrystalline oxide layer is at the heart of the device. It is normally made of connected TiO_2 anatase nanoparticles, but other oxides, such as ZnO and SnO_2, may be used, too. The surfaces of the nanoparticles have an attached monolayer of a charge transfer dye ("sensitizer"). When the latter is exposed to solar irradiation, it is photo-excited to a state denoted S^* in Figure 6.20, and an electron is injected into the conduction band of the oxide. The original state of the dye is then restored by electron donation from an electrolyte, which normally is a liquid organic solvent containing the iodide ion/triiodide ion (I^-/I_3^-) system. The regeneration of the sensitizer by iodide intercepts the recapture of the conduction band electron by the oxidized dye. The iodide, in its turn, is regenerated at the counter electrode (cathode), which is usually Pt-coated glass, and the circuit is completed via the external load. The voltage generated under illumination corresponds to the difference between the Fermi level of the oxide and the redox potential of the electrolyte, and electric power is generated without any permanent chemical transformation.

The use of a liquid electrolyte is problematic from a device perspective, but it is possible to replace it—albeit then having a decrease in

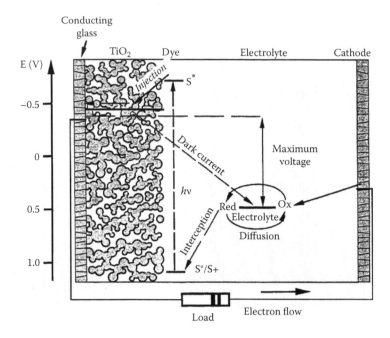

FIGURE 6.20   Functional principles for a DSSC, showing the excitation of the dye by photons with the energy $h\nu$, the redox reaction in the electrolyte, and the ensuing development of a current with a certain maximum voltage. An energy scale is shown to the left. (After A. Hagfeldt, M. Grätzel, *Acc. Chem. Res.* 33 [2000] 269–277. With permission.)

efficiency—with a wide band gap inorganic p-type semiconductor or a hole-transmitting solid. The nature of the TiO_2 layer is critical for the proper functioning of the DSSC—in particular, for its sustaining good electrical transport through the nanoporous structure; details of this transport still remain poorly understood. Figure 6.21 illustrates typical nanostructures of TiO_2 films used in DSSCs. Part (a) refers to the surface of a layer made by a compression technique [47], and part (b) shows a cross section through an as-deposited TiO_2 films made by sputter deposition at oblique incidence toward a rotating unheated substrate. The nanostructure formed by the latter technique has been termed parallel penniform [48]. Comparable efficiencies were observed on cells based on these two types of TiO_2 films. Nanowires may be used as an alternative to the porous oxide films, but the efficiencies obtained so far have been low. Plastic substrates may allow for cheap manufacturing, and DSSCs of this type have reached $\eta = 7.4\%$ [49].

The TiO_2 in the DSSC absorbs predominantly in the UV and hence can convert only a tiny fraction of the solar energy into electricity. Virtually

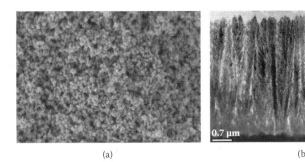

(a)   (b)

FIGURE 6.21   Scanning electron micrograph of a TiO$_2$ layer made from P25 nanoparticles using a standard procedure (a), and cross-sectional transmission electron micrograph for a 4.2-μm-thick sputter-deposited TiO$_2$ film. (Panel a from H. Lindström et al., *Solar Energy Mater. Solar Cells* 73 [2002] 91–101. With permission; panel b from M. Gómez et al., *Solar Energy Mater. Solar Cells* 62 [2000] 259–263. With permission.)

all of the solar absorption is in the dye, and very large research efforts have gone into the development of dyes covering the solar spectrum, at least for λ < 0.9 μm. Furthermore, the dye should be firmly grafted to the TiO$_2$ surface and inject electrons into the conduction band with a quantum efficiency near unity and simultaneously be stable against degradation for long enough times. Dye development requires advanced organic chemistry, and only some indications can be made here.

Polypyridyl complexes of ruthenium are particularly successful sensitizers, with *cis*-RuL$_2$-(NCS)$_2$—where L stands for 2,2'-bipyridyl-4,4'-dicarboxylic acid—is the premier sensitizer, known as the N3 dye. It has been challenged by "black dye" more recently; this is tri(cyanato)-2,2',2"-terpyridyl-4,4',4"-tri(carboxylate)Ru(II). Figure 6.22 shows the spectral response of the photocurrent for DSSCs sensitized by N3 and "black dye," and it is apparent that the latter gives superior coverage of the NIR and hence higher efficiency.

Plasmonic effects can increase the absorption in the dye and in principle lead to larger efficiencies [50]. Nanoparticles of Ag or Au have the desired plasmonic properties, but a problem with regard to DSSC applications is that the I$^-$/I$_3^-$ system is extremely corrosive, even for Au. A conceivable solution is to cover the nanoparticles with pinhole-free TiO$_2$ [51].

Another way to increase the efficiency is to increase the amount of light trapping in the TiO$_2$, which can be made by adding scattering submicron particles [44]. Clearly, this is the same strategy as the one used to improve conventional thin-film solar cells via "milky" ZnO:Al transparent conductors (cf. Section 4.4.2). Figure 6.23 shows that when the haze

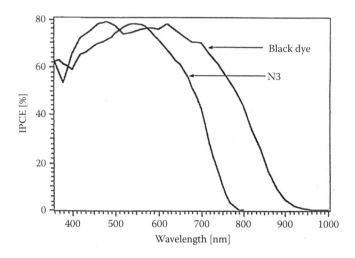

FIGURE 6.22 Spectral incident photon to current conversion efficiency (IPCE) for DSSCs sensitized by two different dyes. (From A. Hagfeldt, M. Grätzel, *Acc. Chem. Res.* 33 [2000] 269–277. With permission.)

(ratio of diffuse to total optical transmittance) was enhanced by addition of particles with a diameter of 0.4 μm, the photo-response is widened in a manner analogous to the one shown in Figure 6.22. It is noteworthy that even as little as 10% haze gives a substantial improvement.

The majority of the work on DSSCs is related to TiO_2, and this material turns up in many green nanotechnologies. Widening the absorption to include the luminous part of the spectrum is of great interest, and this issue is discussed in Box 6.3.

6.2.4 Organic Solar Cells

Organic solar cells (OSCs) attract much interest today as power sources, primarily for non demanding applications. Their development has largely paralleled that of OLEDs (cf. Section 5.6), which use similar chemicals and structures. But it remains to be seen whether OSC development will follow that of OLEDs, which are already commercial and growing in importance. Possibilities to achieve very low-cost solar cells through solution processing, as well as flexibility and light weight, are strong technology drivers.

OSCs can be designed according to several different principles and involve many types of materials, but so-called bulk heterojunctions comprised of a conjugated polymer blended with a fullerene derivative have been widely studied recently, especially with regard to the system poly(3-

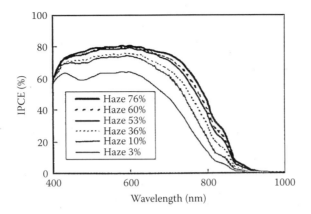

FIGURE 6.23 Spectral incident photon to current conversion efficiency (IPCE) for DSSCs containing light scattering particles. (From Y. Chiba et al., *Jpn. J. Appl. Phys* 45 [2006] L638–L640. With permission.)

hexylthiophene) (P3HT) as conjugated polymer and [6,6]-phenyl-C_{61}-butyric acid methyl ester (PCBM) as fullerene derivative [54–56]. The P3HT has a band gap of ~1.9 eV and enables efficient light absorption for $\lambda < 0.65$ μm at film thicknesses as small as ~200 nm. OSCs of this type have yielded $\eta \approx 5\%$ [57], and even larger efficiencies, up to 6.5%, have been reported recently for tandem cells incorporating alternative conjugated polymers and fullerene derivatives separated by a Ti oxide layer [58]. Figure 6.24 illustrates a typical device design with an active composite between a TCO-coated substrate and an Al back electrode.

The mechanism for photoelectric conversion in the OSCs is complicated but can be described in terms of four basic steps:

- Absorption of light in the conjugated polymer to generate excitons (i.e., strongly bound electron–hole pairs, which are prevalent as a consequence of the low dielectric constant)
- Diffusion of these excitons
- Dissociation of the excitons with charge generation
- Charge transport in the polymer and in the fullerene derivative

These steps are illustrated in Figure 6.24. The exciton dissociation takes place almost exclusively at the interfaces between the two materials, with P3HT serving as electron donor and PCBM as electron acceptor. This model is oversimplified, though, and does not explain why it is energetically favorable for the exciton to dissociate and charge separate. A detailed discussion of several different theoretical issues is given in the literature [56].

BOX 6.3   MODIFIED TiO₂ WHICH ABSORBS AT VISIBLE WAVELENGTHS: A "HOLY GRAIL" IN SOLAR-ENERGY-RELATED MATERIALS SCIENCE

The "holy grail" in the tales of King Arthur was sought for its divine and miraculous powers, but it also came to symbolize much wasted effort and tragedy, with little or no reward. The words are sometimes used to describe an ultimate and problematic goal. Titanium dioxide is an important core material related to solar energy and energy efficiency and crops up as such in a range of applications throughout this book. These many-fold uses stem from its being a wide band gap semiconductor with very high refractive index and UV photocatalytic capability. The latter includes the ability to generate hydrogen from water directly under solar irradiation, though not efficiently enough to be commercially viable. TiO_2 is transparent at visible wavelengths but absorbs solar light in the UV. If it could be modified so that its electrical and photocatalytic properties were retained, but it absorbed at visible solar wavelengths, then many exciting new opportunities would arise. The use of dyes on the surface of TiO_2 nanoparticles in the DSSC solar cells is the only real useful advance to date in this field, but this approach has severe limitations. Doing away with the dye would make life much easier, but is it possible?

This prospect of visibly absorbing TiO_2 has entranced and enmeshed in unforeseen difficulties many scientists for nearly four decades, with a long trail of claims of relevant breakthroughs that either have never come to useful fruition or have proven to be experimental artifacts. One wonders how many research dollars (many millions probably) have been raised in the quest for this technological grail. Many dopants have been tried, including Te, Cr, N, C, Nb, and Eu, along with many doping techniques from chemical to ion implantation. Another approach is to plasma treat with hydrogen to create oxygen defects. Recent work has shown [52,53] that nitrogen defects can introduce tails into the band gap that extend into visible wavelengths. However, such tails can be reminiscent of amorphous semiconductors, and associated electrical characteristics are likely to be degraded. Nanostructures may be worth some effort. For example core-shell systems with interfaces between TiO_2 and another nanothin visibly absorbing material might deserve a look, and slight lattice distortions at such interfaces can be of interest. Oxygen defects on some crystalline planes of nanorods might be visibly active, as is found in ZnO nano-rods. Empirical work has not so far provided a solution, so advanced theoretical band structure and interface studies are urgently needed to see if any real progress is possible.

Visibly absorbing TiO_2 has clearly proven to be a holy grail of solar-energy-related materials science, with much to offer but little to show for decades of effort. Will a solution eventually emerge? Or will we be forever tantalized by an illusory prospect? Will good theoretical science show the way forward, or draw an end to this saga? Only the future, and possibly some inspired nanotechnology, will tell!

We return to visibly absorbing TiO_2 in the context of photocatalysis and air cleaning in Section 8.2.

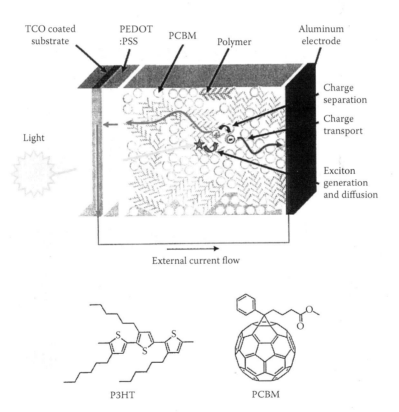

FIGURE 6.24 Schematic rendition of a polymer-fullerene solar cell (upper part) and of its molecular components (lower part). (From C. J. Brabec, J. R. Durrant, *MRS Bull.* 33 [2008] 670–675. With permission.)

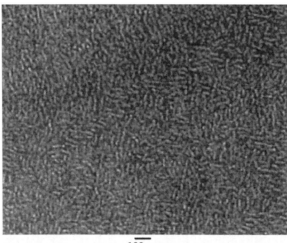

100 nm

FIGURE 6.25 Transmission electron micrograph depicting the nanostructure of a 1:1 blend of P3HT and PCBM after annealing at 150°C
for ½ h. (From K. Sivula et al., *J. Am. Chem. Soc.* 128 [2006] 13988–
13989. With permission.)

The nanostructure of the active layer in the OSC is of the greatest
importance for achieving a high efficiency. Ideally, the nanostructure
should comprise a two-component mixture with both donor and acceptor forming continuous structures with maximum interfacial area and
a mean domain size commensurate with the exciton diffusion length,
which is 5 to 10 nm. In practice, judicious choices of external parameters
such as solvent type and annealing temperature are needed to reach the
desired nanoscale interpenetrating network for phase segregated P3HT
and PCBM. Figure 6.25 shows an example of the nanostructure that can
be accomplished experimentally [59].

The efficiency of the OSC hence depends critically on its manufacturing details, and structural control at the nanoscale is necessary.
Figure 6.26 is a direct proof of this and shows IPCE data on a 1:1
blend of P3HT and PCBM in its pristine state and after heat treatment.
Clearly, the heating has improved the properties significantly and $\eta =$
2.7% was recorded for this particular OSC [60]. Plasmonic effects can
be used to improve the light absorption on OSCs, as in other types of
solar cells [61,62].

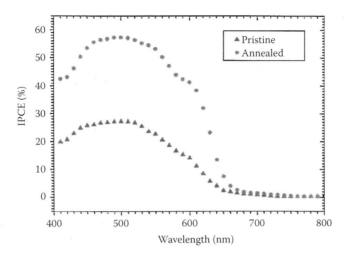

FIGURE 6.26 Spectral incident photon to current conversion efficiency (IPCE) for an OSC in pristine state and after annealing at 120°C for 1 h. (From X. Yang et al., *Nano Lett.* 5 [2005] 579–583. With permission.)

REFERENCES

1. G. M. Morrison, Solar collectors, and solar water heating, in *Solar Energy: The State of the Art*, edited by J. Gordon, James & James Science Publishers, London, 2001, chap. 4 and 5, pp. 145–289.
2. Z.-Q. Yin, Development of solar thermal systems in China, *Solar Energy Mater. Solar Cells* 86 (2005) 427–442.
3. T. P. Otanicar, P. E. Phelan, J. S. Golden, Optical properties of liquids for direct absorption solar energy systems, *Solar Energy* 83 (2009) 969–977.
4. B. O. Seraphin, Spectrally selective surfaces and their impact on photothermal solar energy conversion, in *Solar Energy Conversion: Solid-State Physics Aspects*, Springer, Berlin, 1979; *Topics in Applied Physics* Vol. 31, chap. 2, pp. 5–55.
5. G. A. Niklasson, C. G. Granqvist, Spectrally solar-absorbing surface coatings: Optical properties and degradation, in *Materials Science for Solar Energy Conversion Systems*, edited by C. G. Granqvist, Pergamon, Oxford, U.K., 1991; pp. 70–105.
6. E. Wäckelgård, G. A. Niklasson, C. G. Granqvist, Selectively solar-absorbing coatings, in *Solar Energy: The State of the Art*, edited by J. Gordon, James & James Science Publishers, London, 2001, chap. 3, pp. 109–144.

7. N. Etherden, T. Tesfamichael, G. A. Niklasson, E. Wäckelgård, A theoretical feasibility study of pigments for thickness-sensitive spectrally selective paints, *J. Phys. D: Appl. Phys.* 37 (2004) 1115–1122.

8. J. A. Thornton, J. L. Lamb, Thermal stability studies of sputter-deposited multilayer selective absorber coatings, *Thin Solid Films* 96 (1982) 175–183.

9. C. G. Granqvist, Coatings of ultrafine chromium particles: Efficient selective absorbers of solar energy, *Phys. Scripta* 16 (1977) 163–164.

10. E. Rephaeli, S. Fan S., Tungsten black absorber for solar light with wide angular operation range, *Appl. Phys. Lett.* 92 (2008) 211107 1–3.

11. G. A. Niklasson, C. G. Granqvist, Optical properties and solar selectivity of coevaporated Co-Al$_2$O$_3$ composite films, *J. Appl. Phys.* 55 (1984) 3382–3410.

12. T. Tesfamichael, E. Wäckelgård, Angular solar absorptance of absorbers used in solar thermal collectors, *Appl. Opt.* 38 (1999) 4189–4197.

13. Å. Andersson, O. Hunderi, C. G. Granqvist, Nickel pigmented anodic aluminum oxide for selective absorption of solar energy, *J. Appl. Phys.* 51 (1980) 754–764.

14. E. Wäckelgård, A study of the optical properties of nickel-pigmented anodic alumina in the infrared region, *J. Phys: Cond. Matter* 8 (1996) 5125–5138.

15. B. Carlsson, K. Möller, M. Köhl, U. Frei, S. Brunold, Qualification test procedure for solar absorber surface durability, *Solar Energy Mater. Solar Cells* 61 (2000) 255–275.

16. B. Carlsson, K. Möller, M. Köhl, M. Heck, S. Brunold, U. Frei, J.-C. Marechal, G. Jorgenson, The applicability of accelerated life testing for assessment of service life of solar thermal components, *Solar Energy Mater. Solar Cells* 84 (2004) 255–274.

17. E. Wäckelgård, G. Hultmark, Industrially sputtered solar absorber surface, *Solar Energy Mater. Solar Cells* 54 (1998) 165–170.

18. S. Zhao, E. Wäckelgård, Optimization of solar absorbing three-layer coatings, *Solar Energy Mater. Solar Cells* 90 (2006) 243–261.

19. T. Boström, G. Westin, E. Wäckelgård, Optimization of a solution-chemically derived solar absorbing spectrally selective surface, *Solar Energy Mater. Solar Cells* 91 (2007) 38–43.

20. M. Lazarov, R. Sizmann, U. Frei, Optimization of SiO$_2$-TiO$_x$N$_y$-Cu interference absorbers: Numerical and experimental results, *Proc. Soc. Photo-Opt. Instrum. Engr.* 2017 (1993) 345–356.

21. M. Lanxner, Z. Elgat, Solar selective absorber coatings for high service temperatures, produced by plasma sputtering, *Proc. Soc. Photo-Opt. Instrum. Engr.* 1272 (1990) 240–249.

22. D. Zhu, S. Zhao, Chromaticity and optical properties of colored and black solar-thermal absorbing coatings, *Solar Energy Mater. Solar Cells* (2010) in press.

23. R. Kunič, M. Koželj, A. Šurca Vuk, A. Vilčnik, L. Slemenik Perše, D. Merlini, S. Brunold, Adhesion and thermal stability of thickness insensitive spectrally selective (TISS) polyurethane-based paint coatings on copper substrates, *Solar Energy Mater. Solar Cells* 93 (2009) 630–640.

24. I. Jerman, M. Koželj, B. Orel, The effect of polyhedral oligomeric silsesquioxane dispersant and low surface energy additives on spectrally selective paint coatings with self-cleaning properties, *Solar Energy Mater. Solar Cells* 94 (2010) 232–245.

25. B. Orel, private communication (2009).

26. H. Akbari, S. Menon, A. Rosenfeld, Global cooling: Increasing world-wide urban albedos to offset CO_2, *Climatic Change* 94 (2009) 275–286.

27. I. Edmonds, G. Smith, Conversion efficiency and ground albedo dependence of technologies for mitigating global warming, to be published (2010).

28. A. L. Fahrenbruch, R. H. Bube, *Fundamentals of Solar Cells: Photovoltaic Solar Energy Conversion*, Academic, New York, 1983.

29. R. H. Bube, *Photovoltaic Materials*, Imperial College, London, 1998.

30. M. A. Green, *Solar Cells: Operating Principles, Technology, and System Applications*, 2nd edition, Bridge Printery, Sydney, Australia, 2002.

31. M. A. Green, *Third Generation Photovoltaics: Advanced Solar Energy Conversion*, Springer, Berlin, Germany, 2003.

32. A. Goetzberger, C. Hebling, H.-W. Schock, Photovoltaic materials: History, status and outlook, *Mater. Sci. Engr.* R 40 (2003) 1–46.

33. K. L. Chopra, P. D. Paulson, V. Dutta, (2004), Thin-film solar cells: An overview, *Progr. Photovoltaics: Res. Appl.* 12 (2004) 69–92.

34. J. Berry, private communication (2009).

35. A. Romeo, M. Terheggen, D. Abou-Ras, D. L. Bätzner, F.-J. Haug, M. Kälin, D. Rudmann, A. N. Tiwari, Development of thin-film $Cu(In,Ga)Se_2$ and CdSe solar cells, *Progr. Photovoltaics: Res. Appl.* 12 (2004) 93–111.

36. B. J. Stanbery, Copper indium selenides and related materials for photovoltaic devices, *Crit. Rev. Solid State Mater. Sci.* 27 (2002) 73–117.

37. B. Tian, X. Zheng, T. J. Kempa, Y. Fang, N. Yu, G. Yu, J. Huang, C. M. Lieber, Coaxial silicon nanowires as solar cells and nanoelectronic power sources, *Nature* 449 (2007) 885–889.

38. H. R. Stuart, D. G. Hall, Absorption enhancement in silicon-on-insulator waveguides using metal islands, *Appl. Phys. Lett.* 69 (1996) 2327–2329.

39. S. Pillai, K. R. Catchpole, T. Trupke, G. Zhang, J. Zhao, M. A. Green, Enhanced emission from Si-based light emitting diodes using surface plasmons, *Appl. Phys. Lett.* 88 (2006) 161102 1–3.

40. C. Hägglund, B. Kasemo, Nanoparticle plasmonics for 2D-photovoltaics: Mechanisms, optimization, and limits, *Opt. Express* 17 (2009) 11944–11957.

41. K. R. Catchpole, S. Pillai, Absorption enhancement due to scattering by dipoles into silicon waveguides, *J. Appl. Phys.* 100 (2006) 044504 1–8.

42. K. R. Catchpole, A. Polman, Design principles for plasmon enhanced solar cells, *Appl. Phys. Lett* 93 (2008) 191113 1–3.

43. B. O'Regan, M. Grätzel, A low-cost, high-efficiency solar cell based on dye-sensitized colloidal TiO_2 films, *Nature* 353 (1991) 737–740.

44. Y. Chiba, A. Islam, Y. Watanabe, R. Komiya, N. Koide, L. Han, Dye-sensitized solar cells with conversion efficiency of 11.1%, *Jpn. J. Appl. Phys* 45 (2006) L638–L640.

45. H. Tributsch, Dye sensitization solar cells: A critical assessment of the learning curve, *Coord. Chem. Rev.* 248 (2004) 1511–1530.

46. A. Hagfeldt, M. Grätzel, Molecular photovoltaics, *Acc. Chem. Res.* 33 (2000) 269–277.

47. H. Lindström, E. Magnusson, A. Holmberg, S. Södergren, S.-E. Lindquist, A. Hagfeldt, A new method for manufacturing nanostructured electrodes on glass substrates, *Solar Energy Mater. Solar Cells* 73 (2002) 91–101.

48. M. Gómez, E. Magnusson, E. Olsson, A. Hagfeldt, S.-E. Lindquist, C. G. Granqvist, Nanocrystalline Ti-oxide-based solar cells made by sputter deposition and dye sensitization: Efficiency versus film thickness, *Solar Energy Mater. Solar Cells* 62 (2000) 259–263.

49. T. Yamaguchi, N. Tobe, D. Matsumoto, H. Arakawa, Highly efficient plastic substrate dye-sensitized solar cells using a compression method for preparation of TiO_2 photoelectrodes, *Chem. Commun.* (2007) 4767–4769.

50. C. Hägglund, M. Zäch, B. Kasemo, Enhanced charge carrier generation in dye sensitized solar cells by nanoparticle plasmons, *Appl. Phys. Lett.* 92 (2008) 013113 1–3.

51. S. D. Standridge, G. C. Schatz, J. T. Hupp, Toward plasmonic solar cells: Protection of silver nanoparticles via atomic layer deposition of TiO_2, *Langmuir* 25 (2009) 2596–2600.

52. T. Lindgren, J. M. Mwabora, E. Avendaño, J. Jonsson, A. Hoel, C. G. Granqvist, S.-E. Lindquist, Photoelectrochemical and optical properties of nitrogen-doped titanium dioxide films prepared by reactive DC magnetron sputtering, *J. Phys. Chem. B* 107 (2003) 5709–5716.

53. Y. Nosaka, M. Matsushita, J. Nishino, A. Y. Nosaka, Nitrogen-doped titanium dioxide photocatalysts for visible response prepared by using organic compounds, *Sci. Technol. Adv. Mater.* 6 (2005) 143–148.

54. C. J. Brabec, J. R. Durrant, Solution-processed organic solar cells, *MRS Bull.* 33 (2008) 670–675.

55. B. R. Saunders, M. L. Turner, Nanoparticle–polymer photovoltaic cells, *Adv. Colloid Interface Sci.* 138 (2008) 1–231.

56. B. C. Thompson, J. M. J. Fréchet, Polymer-fullerene composite solar cells, *Angew. Chem. Int. Ed.* 47 (2008) 58–77.

57. W. Ma, C. Yang, X. Gong, K. Lee, A. J. Heeger, Thermally stable, efficient polymer solar cells with nanoscale control of the interpenetrating network morphology, *Adv. Funct. Mater.* 15 (2005) 1617–1622.

58. J. Y. Kim, K. Lee, N. E. Coates, D. Moses, T.-Q. Nguyen, M. Dante, A. J. Heeger, Efficient tandem polymer solar cells fabricated by all-solution processing, *Science* 317 (2007) 222–225.

59. K. Sivula, C. K. Luscombe, B. C. Thompson, J. M. J. Fréchet, Enhancing the thermal stability of polythiophene: Fullerene solar cells by decreasing effective polymer regioregularity, *J. Am. Chem. Soc.* 128 (2006) 13988–13989.

60. X. Yang, J. Loos, S. C. Veenstra, W. J. H. Verhees, M. M. Wienk, J. M. Kroon, M. A. J. Michels, R. A. Janssen, Nanoscale morphology of high-performance polymer solar cells, *Nano Lett.* 5 (2005) 579–583.

61. S.-S. Kim, S.-I. Na, J. Jo, D.-Y. Kim, Y.-C. Nah, Plasmon enhanced performance of organic solar cells using electrodeposited Ag nanoparticles, *Appl. Phys. Lett.* 93 (2008) 073307 1–3.

62. D. Duche, P. Torchio, L. Escoubas, F. Monestier, J.-J. Simon, F. Flory, G. Mathian, Improving light absorption in organic solar cells by plasmonic contribution, *Solar Energy Mater. Solar Cells* 93 (2009) 1377–1382.

CHAPTER 7

Coolness
High-Albedo Surfaces and Sky Cooling Devices

Cooling is as important as heating—at least. Cooling is needed to make buildings comfortable, and it is necessary for the preservation of food and for innumerable other purposes. Yet cooling is seldom talked about as much as it deserves to be, and one problem in that respect is with language. We provide warmth to heat a room, but what do we provide if we want to make it cool? Is it "coolth"? This word is not entirely unknown in English but is archaic and arcane at best. We have chosen to speak of "coolness" in this chapter. Coolness can be obtained by many techniques, and two are of particular interest for this book: the coolness connected with prevention of excessive temperatures through reflection of incoming solar radiation by surfaces having high albedo ("whiteness"), and the coolness that can be captured under a clear sky—that is, what we referred to as "sky cooling" in earlier chapters.

The energy needed for cooling increases in the world, and it does so steeply [1,2]; it is becoming increasingly important, especially for the peak demand of electricity. Solid numbers are not always easy to come by, but it is known that the increase in the energy for air conditioning and refrigeration has risen by an average of 17% per annum in the European Union between 1995 and 2003 [3]. Such an increase, of course, is unsustainable and must be curtailed. Another indication of the same growth trend, from a different part of the world, is that more than half of the electricity demand in some major Chinese cities is for air conditioning, refrigeration, and water cooling [4].

Our emphasis in the past chapters has been on using and controlling natural energy flows during the daytime when solar radiation is present. We now also consider the natural bounty that the clear sky has to offer, especially at night. To most people, including many engineers and scientists, what we will reveal in this chapter at first seems too good to be true.

303

FIGURE 7.1 "Dew-rain" harvest system outside Ajaccio, Corsica, France. The white foil is designed so as to collect water efficiently. (From www.opur.fr.)

But large cooling capability without external power is almost effortless once you know which materials and structures to use. This cooling is almost "unreal" in both senses of the word—the literal "not believable" and the more modern meaning "amazing." Cooling of this kind can lead to many new applications; one of them is illustrated in Figure 7.1, which is a photo of a foil-type device capable of spontaneously reaching subambient temperatures and going below the dew point so that water is collected [5]. This "dew rain" can then be used for human consumption, irrigation, etc. The energy for cooling the device is not visible; it consists of flows of heat energy in the form of thermal radiation that is always going out from the earth and coming in from the atmosphere, with more going out than in during the night.

We cannot see the flows of infrared energy with our eyes, but infrared cameras and simple hand-held noncontact thermometers can detect them. Point such a thermometer at the ground and then up into the clear night sky! The measured temperatures are very different and can be lower for the sky reading by tens of degrees Celsius. The thermometer can measure the sky to be at $-40°C$ when the ground is at around $+20°C$. So the clear sky is very much colder than the air! But why are they different? If there is a cloud about and you point the thermometer at it, the cloud will appear much warmer than the clear sky.

These simple observations, whose origins were outlined in Chapter 2, hint at the possibility of cooling to temperatures well below those of the ambient without any need for external power, and they also emphasize the great advantages of a clear sky over a cloudy sky. To understand these ideas in a simple way, imagine for a moment that your eyes could "see" in the infrared at night. Emerging from the earth would then be all of the "IR colors" given off by most bodies at ambient temperature. The radiation falling onto the earth from above, however, would look distinctly different because one band of "color" that is very strong in the outgoing radiation is weak and almost missing. This means that the atmosphere is more or less unable to radiate some "colors" so a cooler sky results. By the way, starlight would stand out with a distinct "color," as it is only for wavelengths where the incoming atmospheric radiation is weak that IR "starlight" can easily get through to the earth's surface.

Chapters 2 and 3 showed that it is the trace gases in the atmosphere that cause these radiative properties; in particular the amounts of water, CO_2, and ozone were found to be critical for getting the net outflow of heat from the earth by thermal radiation just right for most forms of life. We saw that the atmosphere is delicately tuned with regard to the amount of radiant heat that gets through into space, and that much of this radiation gets trapped and reradiated back to earth. The materials and structures discussed in this chapter are aimed at making maximum use of this "sky window"—which is more "open" in dry geographic locations than humid ones—while minimizing the impact of incoming radiation, which includes both thermal radiation and daytime solar energy.

7.1 TWO COOLING STRATEGIES

Cooling can be achieved by use of two classes of surfaces that are distinctly different with regard to function and relevant spectral range. Thus, cooling can be accomplished by

- Reflection of excessive solar energy
- Radiation of energy directly into outer space, where it is permanently lost

This chapter will discuss both types of surfaces.

7.1.1 High-Albedo Surfaces

The first class of surfaces reduces solar heat gains by increasing the albedo (i.e., by reflecting an increased amount of solar energy). This works for an individual building and vehicle as well as for a whole urban environment. In particular, there are very great benefits if the surfaces within a global urban environment become more solar reflective. Roofs occupy 20 to 25% of the urban areas while road surfaces occupy 30 to 44%, and in total they make up 55 to 65% of the urban area exposed to the sun. The surfaces are huge in absolute numbers and amount to ~3.8 ×10^{11} and ~5.3 × 10^{11} m^2 for roofs and paved areas, respectively, according to a recent study [6]. Typical roof materials absorb 75 to 90% of the incident solar energy, and typical road surfaces absorb 70 to 90%, whereas fresh bitumen and asphalt has a solar absorptance of ~95%. The potential to reduce the solar heat gain is great for both elements, and reasonable practical goals are that the albedos are increased by 0.25 and 0.15 for roofs and roads, respectively [6]. In combination, the average solar reflectance over all urban areas then rises by 0.1. The resulting gain in the earth's solar reflectance is 0.03% if this takes place worldwide in villages, towns, and cities.

The reduction in heat gain, and hence in temperature rise due to solar absorption, can be equated to an amount of additional incoming heat from a rise in atmospheric greenhouse gases so that the net impact on earth's climate is zero. It has been estimated that a rise in urban albedo across the globe by an average of 0.1 would offset the addition of 44 Gt (gigatonnes) of CO_2 to the atmosphere [6]. At current growth rates in emissions, the change in albedo would offset about 11 years of emissions increase, which would be highly beneficial by allowing new energy technologies some breathing space to take hold. If carbon offsets are worth US$ 25 per ton, then the effort is worth 1.1 trillion U.S. dollars, which would pay for much of what is needed to create this impact. Much of the albedo change could be achieved by normal activities, with just a slight change in materials used. Some excellent coatings that are well suited to increase the albedo will be analyzed in Section 7.3.

The mentioned 44 Gt savings in CO_2 emissions is in essence a one-off reduction in heat gain, but the use of high-albedo building materials will also have a continuous effect and reduce the need for cooling of the buildings by 20% or more. Some examples from roof albedo changes on actual buildings (to be discussed shortly), in fact, show even larger effects. The cooling energy benefit also extends to the whole city or town when its average local air temperature falls. Urban "greening" from parks and from gardens on rooftops, walls, balconies, and public and private grounds can also save in both heating and cooling needs, in addition to giving albedo gains. The connection between the urban microclimate, energy use, and health benefits, it seems, is not widely recognized.

Like the adverse effects of the fluorescent lamp on daylight use from the 1950s, which were noted in Chapter 5, air conditioning technologies have sadly desensitized many of us to natural ways of thermal management and, as with daylight, we then also lost many psychological and health benefits.

7.1.2 Sky Cooling

The second class of surface of main interest in this chapter pumps heat away by radiative cooling to the atmosphere and dumps the heat into space. The energy is removed directly by transmission through the sky window and indirectly via absorption in the atmosphere and re-radiation (about a third of which goes outward, as seen in Section 2.1). High levels of cooling can be achieved with surfaces of this type as long as there is no incident solar energy. But it is also possible to achieve useful sky cooling in the daytime, though some special arrangements must then be made as will be discussed later. A combination of high solar reflection and efficient sky cooling in one surface can lead to daytime cooling and is thus of great relevance for this chapter; solar reflection keeps the building cool, while sky cooling contributes to make its radiative output surpass the solar heat gain so that subambient cooling starts earlier in the afternoon than would be the case without sky cooling.

One can even go further with some surfaces designed for sky cooling and combine heating and cooling in a single surface or single stacked system having suitable spectral properties. This can be done sequentially with daytime heating and night-time cooling but, most fascinating, is it is even possible to *simultaneously* supply coolness and heat in a specially designed single-stacked system. This idea can be varied to provide— instead of the heat–cold combination—a joint supply of daylight and cooling, or electric power and cooling.

To evaluate the benefits and relative costs of albedo increases, sky cooling, and other novel cooling concepts, it is necessary to compare savings per dollar invested to that for advanced conventional systems and other options for natural cooling. This is discussed further in Box 7.1.

7.2 CITY HEATING, GLOBAL COOLING, AND SUMMER BLACKOUTS

The world's population does not stop increasing. At the moment (2010), it amounts to about 6.8 billon people, and it is not expected to stabilize until after 2200 when the population is just above 10 billion [9]. More than half of the population now lives in cities; 670 million inhabit the 100 largest cities, and the mean area per person is 560 m² [6]. Naturally, this agglomeration of people and activities influences the cities' climates.

7.2.1 Urban Heat Islands

Decreasing the temperatures in the cities is particularly important since they trap heat and hence get hotter than their surrounding nearby

BOX 7.1   ON THE EFFICIENCY OF ALTERNATIVE COOLING TECHNIQUES

There are many ways to obtain cooling, and electrically powered compressors, absorption cycle chillers, evaporative cooling systems, and thermoelectric cooling can all be adapted for use with solar power or solar heat; it is also possible to have hybrid systems combining sky cooling with more conventional methods. Many of these options can cool without drawing electric power from the grid. Some need no electric power at all apart, perhaps, for fans or pumps, and some just increase the performance efficiency.

Electrically powered compressors for cooling can be driven by solar cells, but this is very expensive. Solar heat can be used for absorption cycle refrigeration or air conditioning, but these systems, which rely on heat-induced chemical changes, have much lower efficiencies than electrically powered systems. A key concept here is the coefficient of performance (COP), which measures how much supplied energy E_{in} is needed to remove a given amount of heat Q_H. Clearly, COP = Q_H/E_{in} should be as high as possible. One should note that all of these systems dump an amount of heat Q_{out} equal to $Q_H + E_{in}$ into the local environment that is larger than the heat Q_H removed from a room or fridge. If the COP has a value one, as is usually the case for absorption cycle cooling, the thermal impact locally is severe and is more than twice the heat that is pumped away. And if the exhaust heat is extracted by water evaporation, then there is a high demand on water resources and an elevation of local humidity, both of which are undesirable.

One of the authors (GBS) was involved in solar thermal absorption cycle air conditioning of a large medical center in a Sydney suburb, plus a university-based facility, in the late 1970s [7]. The system worked well but was far too expensive to run and maintain, and it also used a lot of water. Other solar-based systems were evaluated around that time and had typical COPs in the range 0.5 to 0.6 [8]. The COP of absorption chillers has not advanced very much in the past 30 years. Modern two-stage absorption coolers with >200 kW cooling capacity can have COP ≈ 1.1, but single-stage coolers still only achieve COP ≈ 0.7.

Sky cooling, in contrast to the cooling discussed above, does not have any adverse impact on the local environment and actually helps ameliorate the urban heat island effect discussed next, whereas these other options—even if solar driven—will exacerbate it.

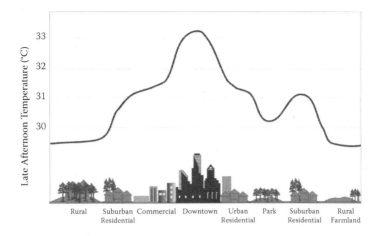

FIGURE 7.2 Schematic of the urban heat island effect, showing temperature rise for different types of urban landscape. (From http://www. urbanheatislands.com. With permission.)

countryside. This is shown schematically in Figure 7.2, which indicates that the temperature difference can be as large as ~4°C between a downtown area with high-rise buildings and its rural setting. Exceptionally, the energy increase can be even larger, and specific data are available for a number of big cities [10]. This "urban heat island" (UHI) effect has several reasons, the major ones being

- Low albedos of roofs and roads
- High thermal mass
- Commercial and industrial activity
- Pumping of heat from inside buildings to just outside them

For the specific case of Greater London, the air conditioning load was estimated to be 25% above that of the surrounding rural region due to the UHI effect [11].

As discussed above, increasing the albedos of roofs and roads will lower the solar heat gain, and thereby the UHI. The reduced need for air conditioning will lead to

- Reduction in local heating outside buildings and cars
- Power savings
- Reduced leakage of greenhouse refrigerant gases
- Less water needed for cooling towers

Cooler urban environments also dramatically reduce urban smog [12,13] and thus have health benefits.

7.2.2 Global Cooling by Increased Albedo

The discussion in Section 7.1.1 showed that albedo changes can be very important for offsetting CO_2 emissions. Here, we analyze this effect further, mainly by way of two illustrative examples presented in Box 7.2. The first example deals with air cooling and leads to the main conclusion that a 20% reduction globally of the power for air conditioning would give the same effect on CO_2 emissions as an increase of the urban albedo effect discussed above by as little as 0.1%, or in absolute numbers, an increase of the urban albedo by 0.0001. We will show below that there are techniques that can raise the albedo much further. The second example is related to power stations and demonstrates again the overwhelming effect of even minor changes in the urban albedo.

The very large effect of albedo changes may come as a surprise to most people and, to get some perspective on them, let us consider the other ways by which we heat up the atmosphere. All waste heat, which is a by-product from power plants, industry, car and truck exhausts, and air conditioning, can be reduced if each of these sources becomes more efficient or, in some cases, if there is a switch to nonheat-generating processes. No doubt, efficiency gains will reduce the projected heat burden in coming years. It is worth looking at this likely impact on global cooling, relative to the albedo effect, to emphasize the latter's magnitude because this raises a key issue in the way we humans respond, and the need for new ways of thinking as we seek to combat global warming. We sense the heat coming from air conditioner compressors, car exhausts, steel plants, and foundries, and we see its effects in the steam coming off the cooling towers of power plants. These heat outlets are localized. In contrast, the impacts of raised solar reflectance are subtle and disperse. It may thus be surprising that the albedo effect on the UHI and on global cooling by far outweighs what one might expect to gain from improved efficiencies in these more obviously "hot" systems. Of course, both the one-off carbon offsets and ongoing CO_2 reductions from each of the "hot" sources are still very beneficial.

These analyses irrevocably point at the major potential impact of paints and surface coatings on global warming. Their joint direct and indirect effects will, if implemented, outweigh those from wind, solar cells, and solar thermal energy by an enormous margin for many years to come. Add in the relative very low cost per unit of CO_2 emissions, both offset and cumulatively avoided, and surface coatings should become an immediate high-priority focus area. Thus, Section 7.3 on paints should be seen as a core section in this book.

BOX 7.2 QUANTITATIVE ALBEDO EFFECTS: TWO EXAMPLES

The first example is in regard to air cooling. Typical cooling units add additional heat to the neighborhood given by input energy E_{in} = [(heat removed from a space)/COP] = Q_H/COP, where COP is the coefficient of performance. World air conditioning in 2003 needed 395 TWh per year, and by 2030 that is projected to rise to 1,269 TWh [14]. The additional external heat load globally due to air conditioning is thus now around 45 GW at any one time and, assuming an average COP equal to three (i.e., high performance chillers), 180 GW is continuously heating up outside urban air. This heat load depends on the season and is peaked in the middle of the day, so instantaneous values per urban square meter can be very much higher than the 24-hour average.

A reduction in the power for air conditioning by 20%, say, would give a direct global cooling that is equivalent to a global CO_2 offset by 40 Mt (megatonnes), which is 0.1% of the offset from the albedo effect of 44 Gt discussed in Section 7.1.1. This 0.1% offset may seem small, but it is still very worthwhile, and there are two main benefits: locally, there is a positive effect on the UHI effect and associated smog reduction, and globally there is a net build-up of CO_2 reductions relative to business-as-usual (BAU) due to the associated reduction of cooling power needs of ~80 TWh each year. Here, BAU corresponds to unchanged urban albedos. This benefit is 80 Mt per annum, assuming coal fired power, so it takes just 6 months to match and then surpass the environmental benefit from the decrease in direct cooling. After 5 years, the accumulated contribution to atmospheric cooling relative to BAU (a 440 Mt reduction) is still only 1% of the benefit given by an increase of the albedo by 0.1. It should be remembered that if a 20% heat output reduction today (2010) decreases the global and UHI heat load by 79 TWh, such a percentage decrease in 2030 will diminish the global heat load by 250 TWh, with a direct emission offset of 126 Mt of CO_2 and an ongoing accumulating reduction relative to BAU of 253 Mt of CO_2 per year.

To emphasize the albedo effect once more, we now look at a second example: power stations. The total atmospheric heat load just from power stations—given typical efficiencies—is now probably around 35,000 TWh each year (~24,000 TWh from coal and ~11,000 TWh from oil and gas), or at any one time around 4 TW of heat. This massive total is less than the direct heating reductions that follow from raising the urban albedo by 0.1. A 10% reduction in power station heat output worldwide—either from efficiency gains, more gas power, replacement by renewables, or some combination of these—would offset the warming of nearly one extra Gt of CO_2 added to the atmosphere. To achieve an equivalent reduction from raised albedo would require lifting its average by a very modest 0.0025! Note that, as in the example above of air conditioning reductions, CO_2 reductions from this combination of more efficient power stations and renewables would continue to compound year after year, and would surpass the 10% direct heating offsets of ~1 Gt in 8 to 11 months. However, they would take over 10 years to surpass the benefits of raising average urban albedos by 0.1! What is even more interesting is the relative financial benefits, or return on investment, of the albedo approach versus the renewables approach, both in terms of investment capital per ton of CO_2 reduction and relative accumulated returns over 10 years, assuming a fixed price of CO_2 per ton avoided or offset. These financial considerations enhance the value of the albedo approach even more.

Similar comparative analyses could be pursued in other examples applied to reductions of waste heat outputs from industrial processes, especially from metal and ceramic processing, and from vehicle exhaust emissions. The less heat pumped into the air, the more CO_2 can be added without a rise in temperature. The near-to-medium term scope is thus probably up to 50 Gt of CO_2 offsets, with any further upside to come from the effect of an albedo rise being too conservative.

7.2.3 Avoiding Summer Blackouts

The UHI effect is a growing problem. It not only detracts from the pleasure of being outside but actually compounds the demands on cooling systems as hot air infiltration adds to cooling loads and also degrades chiller efficiencies. The more heat derived from cooling one can pump into outer space the better, which is why it is so important to make more use of solar reflectance and sky cooling.

The air conditioning demand reaches a peak when the weather is at its hottest, near midday or early afternoon. Then, wholesale electric power gets very expensive, inefficient generators are fired up, and blackouts ensue when demand exceeds supply. Power supply from solar cells or solar thermal power stations is advantageous for this situation, as their output peaks when the load maximizes and traditional power costs are high. However, among all current options, it is even more energy efficient and financially attractive to spread the load over time and use conventional power stations at night for cooling. Chillers work more efficiently at night, and one can further reduce demand by using supplementary night sky cooling. A number of state-of-the-art air conditioning systems cool glycol-water mixtures to near ice temperature overnight and then use the stored cold fluid the next day in various ways such as to cool overhead beams. The cooled thermal mass of the building can involve using interior walls or floors, columns or beams filled with chilled water-glycol, or suitably located phase change materials. The latter can be micro-encapsulated waxes. Greater use of night cooling with conventional compressors plus storage is thus to be encouraged. This also opens the way for hybrid operation with advanced night sky cooling systems. These do not seem to have been considered before, but have much to offer. They will reduce compressor power needs and also pump much of the exhaust heat into space instead of into nearby air. Such hybrid systems are well suited to both large buildings and to homes. The sky cooling devices introduced in Section 7.4 may also be applicable in homes for collecting and storing cold fluid overnight to fully supply cooling needs the next day. The implication is that if cooling loads in homes are not too high—which presumes good design—these systems could completely eliminate the need for electrically powered cooling, apart from a small amount of power for fans or pumps.

Other sources of natural cooling also need to be considered in buildings, including natural ventilation, cool underground soil, and evaporative cooling. In some parts of the world, snow and ice can be accumulated during the winter and used for cooling during the summer. But ultimately these sources still pump heat or water vapor into the local environment, just as electrically powered cooling does. Night sky cooling avoids this.

7.3 HIGH-ALBEDO PAINTS FOR COOL BUILDINGS

The thermal performance of façades and roofs in buildings is a key to determine the need for heating or cooling energy and for the thermal comfort of the occupants. The improvement potential is huge, and it has been estimated that retrofitting of façades and roofs to make them more thermally efficient can reduce today's (2010) demand for heating and cooling by 50 to 60% [14]. The need for action is urgent, particularly in warm countries undergoing rapid development, since improved living standards tend to rapidly accelerate the use of electrically powered air conditioning [4]. Increased use of air conditioning in cars and other forms of transport also emphasizes the need for better control of solar heat gains. Both coatings and insulation materials are of importance for improving the energy efficiency.

Chapters 4 and 5 demonstrated that windows and luminaires are significant elements for the façade's impact on the internal energy demand. These are important for the building's apertures but, of course, the opaque parts of the building envelope matter very much, too, and, as we will discuss here, new types of paints have much to offer and can be as significant as good thermal insulation. Furthermore, as explained earlier in this chapter, improved roof coatings can contribute substantially to raising the urban albedo and hence offset large amounts of added CO_2, as well as contribute to demand savings and create nicer microclimates.

For a given color of the paint, different combinations of selective coatings and insulation may yield the same overall improvement of the thermal performance of a wall or roof of a building in a given geographic location. Therefore, the final decision of what paint to actually use will come down to relative costs, availability of the different options, aesthetic appeal, ease of application and installation, and whether the building is under construction or being retrofitted. Routine maintenance of painted surfaces may provide an opportunity for easy improvement in performance once the technologies and approaches we address here become more widely available or lower in cost. The number of white and colored paint products being marketed as "solar" or "sun reflecting" has increased greatly in recent times as people are becoming more interested in green technology and aware that solar and thermal control can yield large energy savings.

7.3.1 How Cool Can a Solar Exposed Roof Get?

A surface having a neutral color can range from very bright white with $A_{sol} \sim 10\%$, through dull white and light gray with A_{sol} between 40 and 60%, to near black with $A_{sol} \sim 90\%$. So, whitish-looking surfaces are not necessarily very good solar reflectors and may absorb as much as half of the incident solar energy in some cases. Also, dull white solar irradiated surfaces with $A_{sol} \approx 50\%$ can get hot and yield significant internal heat transfer to whatever is thermally connected to such a surface.

It is not uncommon that roofs reach 60°C to 75°C under clear skies, so it is easily realized that roofs should have high solar reflectance combined with high thermal emittance, at least for warm and hot climates. The high emittance not only helps keep down daytime temperatures on roofs and walls but allows the roof, and often the interior and building mass, to cool to a temperature a few degrees below that of the ambient at night. With highly solar reflecting roofs and walls it can take some time after sunrise to overcome the stored coolness. This type of spectral selectivity minimizes solar heat gain and maximizes emitted radiation and hence is the exact reverse of what is needed for the efficient solar thermal collectors discussed in Section 6.1; the ideal spectrum for the surfaces of present interest was introduced in Section 2.9.

Normal metal roofing is neither strongly solar reflecting nor strongly thermally emitting, so it has far from ideal properties. A well-known treatment for roofs involves coatings that are heavily pigmented with TiO_2 microparticles. This pigment has a high refractive index so the paints backscatter strongly and hence reflect the solar heat, and suitable additional constituents can render them efficient thermal radiators. A variety of such high-performance basic white paints are available. Further inorganic and other additives can improve them optically and thermally and sometimes also enhance their long-term durability; such improvements can be especially important for roofs, given their high solar exposures and large temperature changes.

Coatings with high solar reflectance and high thermal emittance, and related pigmented polymer foils, have been discussed over the years for both daytime use and for night sky cooling [15–18]. The best "cool" roof paints combine a hemispherical solar reflectance of ~90% with a blackbody thermal emittance of ~95%, and are able to maintain these characteristics for many years with only slight changes; these paints have a diffuse appearance, which is important since they do not produce glare. The same coating will be mentioned as an excellent candidate for day–night sky cooling in Section 7.4.

Glare can be disturbing, and there has been confusion on this issue, which may have prevented the uptake of coatings with low solar absorptance. In general, a highly diffuse but strongly reflective coating is much less of a problem for glare than a partly smooth but much darker coating (or even a plane glass sheet). In other words, glare from reflected solar light is a large problem for surfaces with a moderate specular component if they are viewed from a point to which the mirror component of the reflected beam is directed. And it is difficult not to look at that bright spot! A highly diffuse white surface, on the other hand, spreads the reflected energy into the whole outgoing hemisphere, so viewing from any one direction can be maintained with relative ease. Therefore, bright white paint rarely causes glare unless it has a semisheen or glossy surface, and a glossy black surface may cause more glare than a diffuse white one.

The practical energy savings of a coating with high solar reflectance and thermal emittance can be very large. Box 7.3 reports specific data for three different commercial buildings, demonstrating that the power for air conditioning can be decreased to a fraction of what is demanded with standard roofing, and that the average roof temperature can be decreased by 15°C to 20°C or even more.

BOX 7.3   PRACTICAL EXPERIENCE FOR
BUILDINGS WITH "COOL" PAINTED ROOFS

Figure B7.3.1 shows power for air conditioning, measured during 1.5 years for two nearly identical supermarkets located in the same region and exposed to the same warm-to-hot climate [19]. When the standard finish was used, the power reached 40 to 45 MWh per month during several months. A "cool" paint had a dramatic effect and limited the power to between 15 and 22 MWh per month.

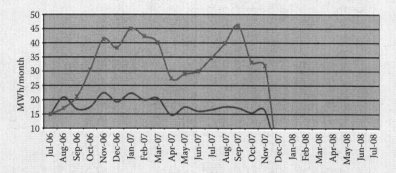

FIGURE B7.3.1   Air conditioning power usage in two almost identical nearby supermarkets near Brisbane, Australia, during a 1.5-year-long measurement campaign. Upper and lower curves refer to buildings having standard external finish and coated with "cool" paint, respectively. (From J. Bell et al., *Proceedings CD and Summary Book of the CIB International Conference on Smart and Sustainable Built Environment*, Brisbane, Australia, [November 19–21, 2003] [ISBN: 1-74107-040-6]; Paper T606.)

The huge energy savings observed in Figure B7.3.1 cannot all be attributed to direct reductions in heat gain. There are other benefits, too, which are related to two main influences, namely

- Significantly improved microclimate, especially just at and above the roof
- The "cool" coating's ability to reach temperatures below those of the ambient overnight

The cooler microclimate can increase the efficiency of roof-mounted cooling units and/or provide cooler air-to-air exchange systems. The quantitative levels of power savings are dependent on the installation of the cooling unit and how it is operated, the way internal air exchange occurs, and the roof and ceiling insulation. The conclusion is that the very large savings evident from Figure B7.3.1 may not always occur, but at least half of the shown savings can be expected in most cases.

Figure B7.3.2 reports additional data demonstrating the large impact of a "cool" roof finish, here applied to a newspaper office/production facility [20]. When a standard roof material was used, the temperature frequently reached 50°C to 70°C. Replacing this material with a "cool" white paint led to a very significant decrease in the roof temperature: the temperature difference was ~17°C on average and was exceptionally as large as ~25°C.

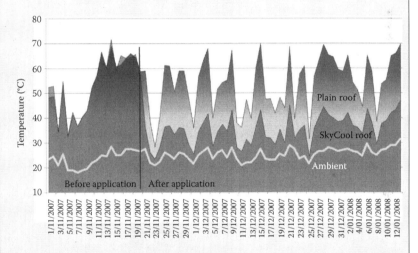

FIGURE B7.3.2 Maximum roof temperatures on a newspaper office/production facility in Sydney, Australia, during a 2.5-month-long measurement campaign. Data are shown for a standard plain roof, for a roof coated with "cool" white paint, and for the ambient. (From SkyCool Pty Ltd., Sydney, Australia [2007], private communication.)

A third example of the great benefits of "cool" paints is a 35,000 m² roof on the Melbourne Airport. This project also involved additional cooling at night by thermal storage in building elements and contents, whose thermal mass then reduced the warming on the following day. The project led to reductions in CO_2 emission exceeding 4,000 tons during 18 months.

The "cool" paints used on the supermarket, office/production facility, and airport discussed in Box 7.3 raised the roof albedo by some 0.4
to 0.5. The magnitude of the change makes it interesting to return to the
discussion in Section 7.1 about the possibility of using increased urban
albedos to offset CO_2 emissions on a global scale. As argued there, an
overall change of the roof albedo by 0.25, together with a change of the
road albedo of 0.15, would lead to an offset of these emissions by an
amount corresponding to 11 years of expected increases. The conclusion
then is that the "cool" paints can have a significant impact on the global
CO_2 emissions if they are implemented in the urbanscape on a large
enough scale.

The "cool" paints discussed above are visibly white. This is fine on
shopping malls, supermarkets, warehouses, factories, airport terminals,
some offices, and the like, but there are many roofs and walls that must
be colored for aesthetic diversity and appeal. The same applies to cars.
Is it still feasible to increase the albedo in this case? The answer is "yes"
and, as discussed below, it is possible to have visibly identical black paints
with a solar reflectance of 5 and 55%, that is, an albedo gain of 0.5 can
be achieved. All "cool" paints should have maximum NIR reflection,
and this property should be maintained while the visible reflectance
is altered as desired for color. Most standard paints do not have high
NIR reflectance, though, so we need to examine ways of achieving that.
Various nanostructures are one option, as discussed below.

7.3.2 Colored Paints with High Solar Reflectance

It is important to understand how the hemispherical solar reflectance
$R_{H,sol}$ can be divided into components for ultraviolet, luminous, and near-
infrared reflectance. Using the data in Chapter 2, it can be found that

$$R_{H,sol} = 0.05R_{H,UV} + 0.43R_{H,lum} + 0.52R_{H,NIR} \qquad (7.1)$$

that is, 52% of the reflectance is in the infrared. This shows that high
solar reflectance can only be obtained if the NIR reflectance is very large.
And if a distinct color is required, not just a pale one, then the luminous
reflectance should be confined to a narrow band. The latter feature can
be achieved with certain dyes, select pigments which may incorporate
nanofeatures, and some nanoparticles.

The ideal colored "cool" paint for warm or hot climates would thus
have, neglecting the UV, a maximum solar reflectance of $0.43R_{H,lum} +$
0.52. Thus, a black paint with $R_{H,lum} \sim 0.05$ could still have $R_{H,sol}$ as
high as 0.55. But a conventional carbon black coating can achieve $R_{H,sol}$
~ 0.06 at best. It then follows that a black roof, wall, or car body that

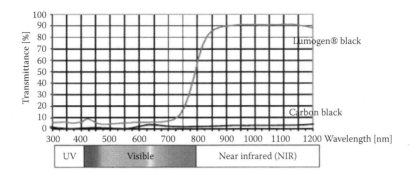

FIGURE 7.3   Spectral transmittance for a "cool" black and a conventional carbon black. (From B. Schuler, *Proceedings of Convation06*, Sydney, Australia, 2006. With permission.)

reflects most of the NIR reflects almost ten times more solar energy than a conventional black paint! This qualitative difference can lead to large reductions in roof temperatures, from ~65°C to around 35°C, with ensuing lower internal heat gains.

Figure 7.3 shows transmittance spectra of a "cool" black paint and of a normal black paint based on carbon [21]. The "cool" paint is transparent in the NIR and, if it is applied to a reflecting backing, the reflectance spectrum corresponds to the transmittance spectrum of a coating at double thickness. The conventional black, on the other hand, has a low reflectance throughout the spectral range shown.

The data in Figure 7.3 do not extend beyond 1.2 μm in wavelength. Most of the NIR solar energy lies at $0.7 < \lambda < 1.2$ μm, so it is most important that the reflectance is high in this range. Nevertheless, the optical properties are of some interest also at longer wavelengths, since many pigments and binders may have absorption features there. Figure 7.4 shows specific examples of "cool light brown" and "cool green" paints [22]. Both paints have optical properties in good agreement with those for their standard counterparts within the luminous range. Clearly, the brown paint reflects moderately well in the entire NIR spectrum, whereas the green paint displays unwanted low reflectance for $1.2 < \lambda < 1.8$ μm.

7.3.3 Mechanisms and Nanostructures for Colored "Cool" Paints

This section will discuss a number of strategies for combining color and high NIR reflectance of the kind illustrated in Figures 7.3 and 7.4. Nanostructures will be seen to play important roles for these strategies. The focus will be on roof and wall paints, but it should be noted that

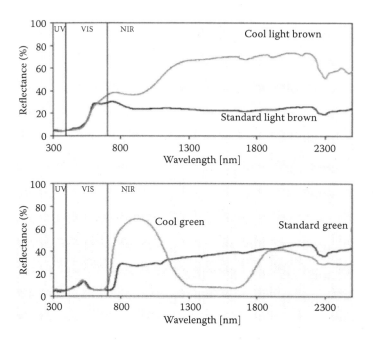

FIGURE 7.4　Spectral luminous and NIR reflectance for two "cool" colored paints and for their standard counterparts. (From A. Synnefa et al., *Solar Energy* 81 [2007] 488–497. With permission.)

white and colored paints with opposite infrared properties are of interest for surfaces inside rooms, and they can reduce inward heat flows from hot walls and roofs; these paints require a low emittance. Thus, ideal paints for indoor and outdoor applications have radically different thermal properties.

Oxide-coated metal flakes can be used for "cool" paints. Aluminum flakes are added to a transparent binder to create a paint coating with a cross section as illustrated in Figure 7.5a. Such paints would be neutral in color and highly solar reflecting. A variety of colors can be accomplished by having each flake coated with an oxide of precisely controlled thickness in the nano range, as seen in Figure 7.5b. The visual impression is created by optical interference, and different oxide thicknesses in the range between 20 and 200 nm lead to a variety of colors. The thin-oxide layers are largely transparent in the NIR, so reflection in that range occurs from the Al flake. In practice, the coatings are produced in a hot fluidized bed reactor using an iron-containing organometallic compound so that each flake is coated with iron oxide [23,24].

A single-layer coating on the Al flakes gives colors which look similar from most directions. With double layers, however, the perceived color

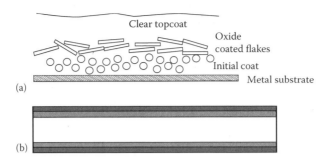

FIGURE 7.5   Cross-sectional views of a colored "cool" paint layer containing oxide-coated metal flakes (a) and of an individual flake with a double-layer coating (b).

changes as the view direction is altered, while the high NIR reflection is preserved. This latter application uses Al flakes that have been pre-coated with nano-thin SiO_2 layers via sol-gel coating before the iron oxide layer is applied. This two-layer coating is notable for its use of two of the earth's most abundant oxides. A clear top overcoat, shown in Figure 7.5a, imparts a high emittance—as is normally desired—but a different and thinner top coat would make it possible to achieve a low emittance.

Coated metal flakes in paint binders have other applications as well, and most surfaces, including those on wood and fabrics, can nowadays be coated with a commercial paint which provides a low thermal emittance of about 0.2 to 0.3. Metal flakes can also be incorporated directly into or onto the polymer or on natural fibers used to weave fabrics. The aim of the flakes, apart from decorative innovation and shiny fabrics, can be to enhance the thermal insulation. For daytime cooling, such a coating can be used on inner surfaces to reduce incoming heat and on outer surfaces to reflect solar heat. At night, both types of surfaces reduce heat loss. The metal flakes also reduce the wearer's thermal image at night, and military applications of this kind have probably been the main interest to date.

Paints incorporating metal flakes can find additional applications for solar thermal systems, as discussed in Section 6.1.4. They may also be of interest as low-cost coatings combining high solar reflection with high-performance angular selective sky cooling. The desired structure looks like that in Figure 7.5, but the overall thickness should be up to 1.5 μm (i.e., much larger than for a "cool" paint for which 0.2 μm suffices).

Inorganic pigments which transmit most of the NIR give another option for colored "cool" paints. Here, one uses a strongly NIR-reflecting substrate directly under a pigmented colored layer that is at least moderately NIR transparent. This transparency is the main feature, and the pigments are normally in nanoparticle form. A variety of special inorganic pigments are available, and their visual appearance is close to that of many standard paints that have much higher solar absorptance [22]. Paints of the type indicated here clearly avoid the need for metal flakes and can be used on any type of base surface including wood shingles, ceramic tiles, polymer sheet, and metal sheet, so this paint can be of great practical value considering the vast array of existing buildings that need to be made more energy efficient easily and at low cost. The under-layer or substrate for the paint can either be an undercoat or the metal onto which the coating is applied.

Plasmonic nanoparticles that impart color but transmit most of the NIR is the third and final option for colored "cool" paints to be considered here. This option has received less consideration than the other two but has interesting features, including ease of incorporation into fabrics. Select nanoparticles which display surface plasmon resonances can have narrow and strong absorption bands in the visible. If they also transmit at NIR wavelengths, then they can be used for "cool" paints. Obviously, there are some analogies to the laminated windows discussed in Section 4.5.3. In an application, the nanoparticles are added into an otherwise transparent thin overlayer, just as for the inorganic pigments. The nanoparticles of practical interest must be very durable, which largely excludes the noble metals and aluminum. Plasmonic nitrides and some borides are robust, though, and thus have interesting possibilities. ZrN, TiN, and HfN nanoparticles can impart color.

The best possibilities may be for core-shell nanoparticle systems in which the shell or core is conducting, provided they can be mass-produced at low enough cost. Again, the shell material at least must be robust. The beauty of the core-shell nanoparticles, as explained earlier in Section 3.9, is that they can be color tuned just by varying the core-shell size ratio. One such system has been considered for nanoinks [25]. Figure 7.6 shows spectral absorption and extinction for two systems: (a) Al cores and Si shells, and (b) Ag cores and Si shells. Absorption dominates but there is some scattering. A narrow absorption peak stands out clearly; its location can be varied within the visible spectrum when the thickness of the metal shell is changed. Scattering also peaks near the resonance.

7.4 SKY COOLING TO SUBAMBIENT TEMPERATURES

This aspect of green nanotechnology has received attention only spasmodically over a number of years [26], and the applications have emphasized passive cooling of buildings and water collection. This is the *demand* or efficiency side of the energy equation. But sky cooling also has potential as a widespread *supply* side or active cooling technology—not just for passive removal of heat though that will be always be one aspect.

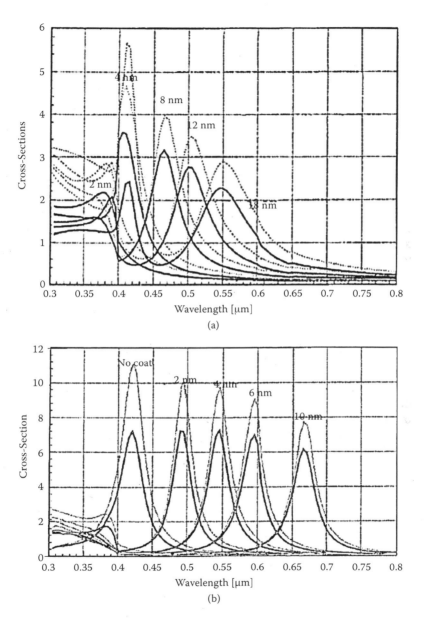

FIGURE 7.6 Spectral absorption (solid curves) and extinction (dashed curves) for shells of Si on (a) Al and (b) Ag cores of 44 nm diameter. The Si shells have the thicknesses shown. (From M. R. Kuehnle, H. Statz, Encapsulated nanoparticles for the absorption of electromagnetic energy, U.S. Patent Publication US 2005/0074611 A1 [2005].)

"Active" means cooling and storing cold fluids, or hybrid integration into active systems such as directly in the condenser circuits of normal chillers or indirectly to improve external heat removal. An additional potential that is not often considered is for electric power supply side, as discussed later in Section 7.6. Cooling is needed in all electric power systems, as outlined in Box 2.2, and this adds directly to global warming; sky cooling can have major implications for which renewable sources to use and where to locate them. Water collection from the atmosphere, and as an aid to water condensation in distillation, is still another area where sky cooling will have an impact [27]; this is discussed in Section 7.5.

But before one embarks on active sky cooling to maintain comfortable interiors in buildings, the base cooling load should be minimized by having good thermal insulation, appropriate roof and façade technologies (such as high-albedo paints, as discussed above), and advanced fenestration as discussed in Chapter 4. The overall design is also essential for new or renovated buildings. Indeed, substantial reductions in cooling loads can be obtained by diminishing direct and indirect gains from solar irradiation, and if this is done well, the desired cooling can be accomplished by use of moderate areas for emitting radiation to the sky.

Why has the great potential of sky cooling not been successfully exploited to date, despite our ready access to it? First, it is not a widely understood or appreciated field, and few scientists are active in it. But more important, there has been little effort to develop products based on sky cooling, possibly apart from arrangements for water collection. The diverse technological scope beyond these applications is not well understood, and this "knowledge gap" has yet to be bridged.

Clearly, the field of sky cooling has so far fallen short of its potential by a wide margin, but it has too much to offer to be neglected. Thus, practical cooling at a low cost down to 15°C below the coldest ambient temperature of the night has been demonstrated, and we will discuss how even lower temperatures are achievable.

7.4.1 Sky Radiance

Chapter 2 introduced the transparency of the atmosphere as a function of wavelength λ and zenith angle θ_z and emphasized the existence of a sky window capable of yielding subambient cooling. The sky window was found to gradually close as θ_z approached 90°. We now go into more detail on the spectral and directional properties of the sky radiance. This information is needed in order to devise optimum materials and device designs for sky coolers, and for modeling cooling performance for different atmospheric conditions. The sky radiance emerges from all directions and is measured in units of $Wm^{-2}Sr^{-1}$.

Outside the sky window, the atmosphere acts essentially as a black-body radiator with regard to both wavelength and angle. Inside the sky window, the radiance is weak for small zenith angles, which is a consequence of the very low temperature of outer space. Nevertheless, there is a small atmospheric absorptance within the sky window, which adds somewhat to the zenith radiance at small zenith angles but adds a lot at large zenith angles. This absorptance originates from a sharp absorption feature due to ozone as well as weak residual absorption from CO_2 and water vapor, and it must be included in the analysis. The angular dependence occurs because, effectively, the atmospheric layer thickness d_a contributing to the radiance is $d_a(\theta_z) = d_a(0)/\cos\theta_z$. Hence, the atmospheric transmittance in the sky window, denoted $T_{a,sw}(\theta_z,\lambda)$, drops for increasing values of θ_z while there is an associated rise in the emittance $E_{a,sw}$ in the sky window according to

$$E_{a,sw}(\theta_z,\lambda) = 1 - T_{a,sw}(\theta_z,\lambda) = 1 - [T_{a,sw}(0,\lambda)]^{1/\cos\theta_z} \qquad (7.2)$$

This equation agrees with Equation 2.22. Further discussion on this matter can be found elsewhere [28].

But how, in detail, does the intensity of the sky radiance depend on atmospheric conditions? Figure 7.7 reports data obtained from computations based on an elaborate representation of the atmospheric constituents and their height profiles, and illustrates a number of important features: Part (a) reports the angular dependence of the sky radiance for the case of a moderately dry atmosphere characteristic of a clear mid-latitude summer climate, and shows results for $0 < \theta_z < 90°$ [29]. The sky window is seen to be gradually closed as θ_z goes up, and the variation is consistent with Equation 7.2.

The role of atmospheric humidity, which is implicit in Figure 7.7a, is elaborated in Figure 7.7b, which shows zenith radiance for three amounts of precipitable water [30,31]. This parameter is the length a water column would have if all of the water along a straight path perpendicular from the earth's surface to the top of the atmosphere would condense. Figure 7.7b shows that the sky window closes as the amount of precipitable water goes up. Hence, sky cooling is increasingly efficient as the humidity drops. The lowest amount of precipitable water in the figure is 1.5 cm, but it can be as small as 0.3 to 0.4 cm under particularly favorable conditions. The sharp spectral feature at 9.7 μm, due to ozone, is of particular interest as it gives the strongest contribution to the radiance in the sky window under clear and dry conditions. This feature has implications for the selection of materials, and a material that absorbs entirely within the sky window range, but not near 9.7 μm, can reach much lower temperatures than one whose absorption band straddles the ozone peak.

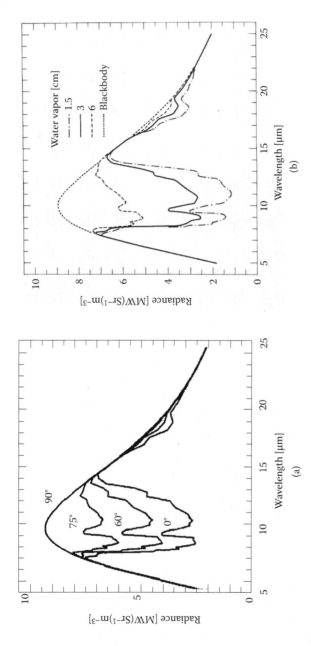

FIGURE 7.7   Spectral radiance for a clear mid-latitude summer sky at different zenith angles including the vertical and the horizontal (part a) and spectral zenith radiance from clear skies containing different amounts of precipitable water (part b). (Part a from Berdahl, R. Fromberg, *Solar Energy* 29 [1982] 299–314. With permission. Part b from P. Berdahl, M. Martin, *Proceedings of the Second National Passive Solar Conference*, edited by D. Prowler, I. Duncan, B. Bennett, American Section of the International Solar Energy Society, Newark, DE [1978], Vol. 2, pp. 684–686, as replotted in C. G. Granqvist, A. Hjortsberg, *J. Appl. Phys.* 52 [1981] 4205–4220. With permission.)

In other words, spectral fine-tuning is of interest within the sky window range. This is one aspect where nanoeffects will come into play and, as we will see below, SiC nanoparticles are ideal from this perspective.

Figure 7.7a demonstrated that the atmospheric radiance has a strong angular dependence, and it is worth exploring whether angular selective surfaces can enhance the cooling of an underlying material. Just as for spectral selectivity the aim is to reach as low temperatures as possible by further reducing the amount of absorbed radiation while still emitting strongly for angles where it is of value to do so. Thus, an angular selective surface devised for sky cooling should have very low reflectance within the sky window toward the zenith, but the reflectance should gradually rise as the angle of incidence goes up. This occurs for many smooth surfaces, whose reflectance increases steeply at angles of incidence above 60° and a similar dependence can take place in some doped polymer coatings. Deliberate engineering of reflectance profiles with two- or three-layer thin films is also feasible. For some situations it is useful to have an angular-spectral profile for which the absorptance is blackbody-like for zenith radiation and decreases continuously as θ_z increases across the whole Planck spectrum.

7.4.2 Spectral Selectivity and Sky Cooling: Idealized Surfaces

When surfaces are exposed to the sky, they can experience cooling due to a net loss of thermal radiation. This means that the thermal radiation given off by the surface has to be larger than that coming in from the atmosphere (and possibly other surrounding surfaces) and is absorbed, rather than being reflected or transmitted. And absorption of incoming radiation reduces the cooling potential. The incident radiation is largely beyond our control, but it is possible to control how much of it is absorbed, provided that we understand the spectral and angular distribution of the sky radiance. This was discussed in the previous section. The technologies addressed here aim to maximize the net radiative output, which is the difference between the outgoing and the absorbed part of the incoming radiation, and it is very instructive to consider outgoing and incoming radiation relative to each other when the goal is to cool to subambient temperatures. Cooling of hot objects, on the other hand, is a lot different as the incoming radiation is then much weaker, relatively speaking. A number of strategies for maximizing the difference will be introduced, but the main emphasis is on the spectral properties of the emitting surface.

Section 2.7 outlined the basic physics of thermal radiation from surfaces at finite temperatures and especially introduced the blackbody radiator, which emits with a spectral power density given by $P(\lambda,\tau)$. This

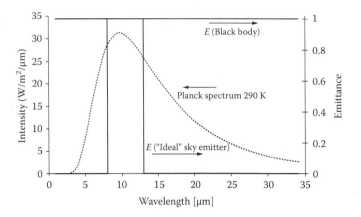

FIGURE 7.8 Spectral emittance (or absorptance) of a blackbody and ideal sky window emitter relative to the Planck radiation spectrum from an atmosphere at 290 K.

is the Planck spectral radiant exitance, which we refer to for simplicity as the "Planck spectrum." The surface temperatures τ_s of interest in sky cooling lie near or below those of the ambient atmosphere at τ_a, and then $P(\lambda, \tau_s)$ ranges on average from around 2.5 to 35 µm in wavelength. Chapter 2 also introduced the ideal spectral properties of a surface capable of reaching very low temperatures at night (cf. Figure 2.15b). Its spectral reflectance $R(\lambda)$ was not that of a blackbody, which has $R(\lambda) = 0$ over the Planck spectrum. Instead, the ideal radiator only radiates with 100% efficiency, or equivalently has $R(\lambda) = 0$, at wavelengths where the atmosphere is significantly transparent to thermal radiation, which is the segment of the Planck spectrum we called the "sky window" from 8 to 13 µm. Over the remainder of the Planck spectrum, this ideal surface reflects all of the incoming radiation, that is, it has $R(\lambda) = 1$.

Thus, the blackbody and this ideal sky window radiator represent two spectral extremes, as shown in Figure 7.8. We shall see that in practice surfaces approaching both of these limits have roles to play, because both radiate strongly within the sky window. Which is to be preferred depends mainly on how far below τ_a one wants to cool, and also on the details of the practical arrangement for cooling and its geographic location. A blackbody is better just below ambient because it then gives off a great deal more radiation than the selective emitter. It should be kept in mind, however, that a surface approaching the ideal properties of the sky window selective radiator, shown in Figure 7.8, always is able to get coldest if nonradiative heat gains from the surrounds are kept very small, as we shall soon see. Given that both spectral limits have their uses, one also needs to ask whether it is also worth considering intermediate spectral properties. The answer is that some work quite well, but not all.

7.4.3 Calculated Cooling for Ideal and Practical Materials

We will first consider various surfaces open to the sky and then surfaces under practical covers designed so as to limit convection gains but transmit most of the radiated heat. Suitable cover materials were introduced in Section 3.5. The net radiated power P_{rad} without covers is given by

$$
P_{rad} = \left[\int_0^{\pi/2} d(\sin^2 \theta_z) \int_0^\infty d\lambda P(\lambda, \tau_s) E_s(\theta_z, \lambda) \right]
$$
$$
- \left[\int_0^{\pi/2} d(\sin^2 \theta_z) \int_0^\infty d\lambda P(\lambda, \tau_a) E_a(\theta_z, \lambda) E_s(\theta_z, \lambda) \right]
\tag{7.3}
$$

It is important to understand each of the terms in this equation. The first term is the outgoing radiation from the surface, which is at temperature τ_s. The second term is the incoming atmospheric radiation that is not reflected or transmitted and thus involves both the atmospheric emittance E_a and surface emittance or absorptance E_s. If atmospheric temperature and humidity are fixed, this second term does not change, but as cooling proceeds and τ_s drops, the first term weakens until $P_{rad} = 0$. At this point τ_s is the radiative stagnation temperature $\tau_{s,min}$. If the surface happens to be colder than $\tau_{s,min}$, then it will be warmed by atmospheric radiation. The change in solid angle and intensity projection onto the surface with angle of incidence are combined in the $d(\sin^2\theta_z)$ term as introduced in Section 2.4 for general diffuse beams.

How do different surface spectral properties influence the two terms in Equation 7.3? The first term will be largest for a blackbody emitter, so why bother with a sky window emitter? The ideal sky window emitter has $E_s \sim 0.32$, so it emits a lot less radiation than the blackbody. Its value lies in the second term, which is very small for the ideal spectrum of Figure 7.7 for clear dry skies. We shall now show that ultimately this factor enables the sky window emitter to pump heat at a higher rate and achieve much lower temperatures than the blackbody. But at near-ambient temperature it does have a lower value of P_{rad}. One should keep in mind that when we get to real surfaces and systems, Equation 7.3 dictates a balance between the advantages of maximizing output (first term) and minimizing input (second term), but since the second term is fixed and the first term falls with increasing τ_s this balance gradually shifts toward minimizing the second term. Thus, the preferred practical spectrum depends on how far τ_s needs to be below ambient temperature. The humidity also plays a significant role, as expected from Figure 7.7b. Interestingly, the drier the atmosphere, the more the balance is shifted toward broadband "blackish" emitters. This may seem counter-intuitive but arises because the second term in Equation 7.3 is then weaker for all surfaces, and hence it takes a lower magnitude of τ_s for the sky window emitters to be superior. This influence of humidity is one example of a broader principle that will come up again soon, namely, external factors that act to reduce the incoming radiation enhance the relative merits of a blackbody to those of a spectrally selective emitter.

A number of results based on Equation 7.3 have been presented in the past for ideal and blackbody surfaces. The detailed results depend on the model of the atmosphere. Figure 7.9 shows data from an application of this model to both the blackbody and the ideal sky window emitting surface of Figure 7.8 for hemispherical radiation in the case of various climate zones ranging from tropical to very dry [32]. The highest cooling performance occurs for the "U.S. standard atmosphere," for which the precipitable water is as small as 1.83 mm. The plots then steadily shift down as humidity levels rise. A key result is that the ideal spectrally selective emitter can achieve, in the absence of nonradiative heat gains, much lower stagnation temperatures than the blackbody. It will get to 50°C to 60°C below ambient temperature, while the blackbody reaches only 25°C to 30°C. For temperatures above 10°C below ambient, the blackbody is pumping more heat radiatively because the output term in Equation 7.3 then dominates, but by 15°C below ambient, the spectrally selective radiator is superior because the input term is then much more important. Nonradiative heat gains, which input heat according to $P_{nr} = U_{nr}(\tau_a - \tau_s)$, where U_{nr} is a constant, limit practical stagnation. The common line in Figure 7.9 is for a very well-insulated system with $U_{nr} = 1 \text{ Wm}^{-2}\text{K}^{-1}$. The stagnation points occur when $P_{rad} = P_{nr}$. Average values of $(\tau_a - \tau_{s,min})$ for both types of surfaces are closer together when U_{nr} has a nonzero value, but the spectrally selective surface still cools further by around 8°C to 10°C. The low-cost practical systems in the following have U_{nr} values that are higher and lie at 2 to 2.5 $\text{Wm}^{-2}\text{K}^{-1}$, so stagnation temperatures around 15°C below the ambient are their limits. In this case, the spectrally selective surface still achieves a lower value of $\tau_{s,min}$, but the blackbody reaches a temperature that is only two or three degrees higher.

Infrared transparent convection covers are needed in practice for achieving a low value of U_{nr}. This cover has a moderate impact on the second term in Equation 7.3, but it does not influence the first term since the power density leaving the surface depends only on τ_s. To find the incoming power density before absorption one must replace $P(\lambda,\tau_a)E_a(\theta_z,\lambda)$ in

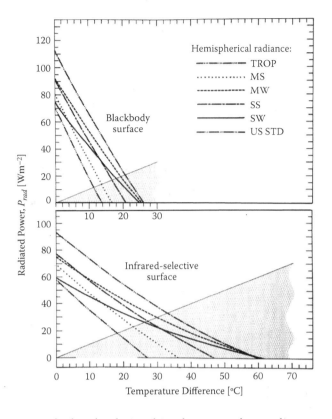

FIGURE 7.9 Calculated relationships between sky cooling power and temperature difference for infrared-selective and blackbody surfaces freely exposed to atmospheres representing six different climatic conditions specifically being tropical (TROP), midlatitude summer (MS), midlatitude winter (MW), subarctic summer (SS), subarctic winter (SW), and the U.S. standard (US STD). The shaded areas are inaccessible for a device with a nonradiative heat transfer coefficient of 1 $Wm^{-2}K^{-1}$. (From T. S. Eriksson, C. G. Granqvist, *Appl. Opt.* 21 [1982] 4381–4388. With permission.)

Equation 7.3 by three terms: (*i*) $T_c(\theta_z,\lambda)P(\lambda,\tau_a)E_a(\theta_z,\lambda)$, (*ii*) $R_c(\theta_z,\lambda)P(\lambda,\tau_s)E_s(\theta_z,\lambda)$, and (*iii*) $E_c(\theta_z,\lambda)P(\tau_a,\lambda)$. The coefficients $T_c(\theta_z,\lambda)$ and $R_c(\theta_z,\lambda)$ are the directional spectral transmittance and reflectance of the cover, respectively, and $E_c(\theta_z,\lambda) = 1 - T_c(\theta_z,\lambda) - R_c(\theta_z,\lambda)$ is the emittance of the cover. The terms (*i*) – (*iii*) are easily found either directly from data on the cover sheet or from the optical constants and thickness of the materials, normally polymers, that are used. Term (*i*) amounts to a reduction in the incoming radiation due to the cover, whereas terms (*ii*) and (*iii*) add to it. Term (*iii*) is the contribution of thermal emission from the cover. In practice, the polymer is always at a temperature very close to ambient temperature, so one can assume that any slight tendency it has to warm due to (weak) absorption of radiation emitted from the surface of interest can be neglected.

7.4.4 Some Practical Surfaces for Sky Cooling: Bulk-Type Solids

Section 3.5 introduced ionic and molecular absorption processes that can yield narrow absorption bands in the sky window range. The absorption can ensue from the natural ionic or molecular vibration modes in some materials comprised of light atoms including Si, Al, Mg, C, N, and O. Practical experiments have been conducted on thin films of materials such as silicon monoxide [31,33], silicon nitride [34], and silicon oxynitride [35,36]. Other alternatives include a number of polymer layers (with poly vinyl fluoride being especially good [15,37]) and layers of MgO and LiF [38], as well as gaseous ammonia, ethylene, and ethylene oxide [39,40]. These materials yield little absorption across the Planck spectrum lying outside the sky window range, and the desired selective reflectance profile can be approximately reached provided that the thin-film or polymer layer has an adequate thickness and is backed by a highly reflecting metal such as an aluminum film. Some selectivity can be achieved also with anodized alumina [41,42] and with alumina films [43], but the data are not as good as for the previously mentioned alternatives.

Figure 7.10a illustrates spectral reflectance for an optimized silicon oxynitride film [36]. High absorption in the sky window is due to molecular vibrations in tetrahedrally coordinated SiO_4, SiO_3N, SiO_2N_2, $SiON_3$, and SiN_4 structural units [36], and the high reflectance outside this window is caused by the underlying Al. Figure 7.10b shows spectral reflectance of a thick ceramic MgO layer backed by metal [38]; the large reflectance at $\lambda > 13$ μm is due to the Reststrahlen effect (cf. Section 3.5).

7.4.5 Nanotechnology for Optimum Sky Radiators: Computed and Measured Data

A number of surfaces and materials have been used to demonstrate sky cooling as discussed above. None of them was developed in order to benefit from the possibilities offered by nanotechnology. However, nanostructures can indeed add new sky cooling capabilities, raise performance by doping in some existing materials, and lower costs. Specifically, the phonon properties of the nanoparticles allow the infrared optical properties to be fine tuned to the sky window, and nanoparticles can serve as convenient dopants for polymers or paint binders.

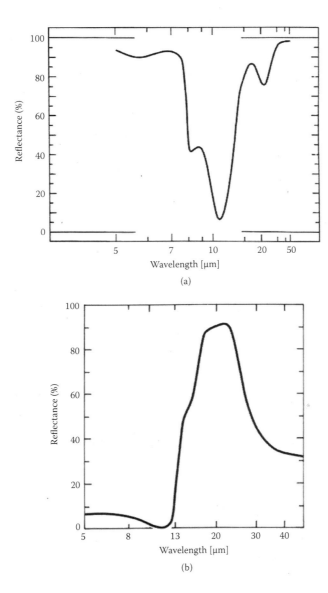

FIGURE 7.10   Spectral reflectance of a 1.2-μm-thick film of $SiO_{1.47}N_{0.54}$ made by sputtering onto Al-coated glass (a) and for a 1.1-mm-thick MgO layer on a metal backing (b). (Panel a from T. S. Eriksson, C. G. Granqvist, *J. Appl. Phys.* 60 [1986] 2081–2091. With permission; panel b from P. Berdahl, *Appl. Opt.* 23 [1984] 370–372, as replotted in C. G. Granqvist, T. S. Eriksson, *Materials Science for Solar Energy Conversion Systems*, edited by C. G. Granqvist, Pergamon, Oxford, U.K. [1991], pp. 168–203. With permission.)

Nanoparticles of SiC and SiO_2 turn out to be of special interest [44,45]. Their optical properties are governed by surface resonances in an ionic material, which has a negative dielectric constant over a limited frequency range between the transverse and longitudinal phonons at the frequencies ω_T and ω_L, respectively, as discussed in Section 3.5. The frequency range with $\varepsilon_1(\omega) < 0$ is called the Reststrahlen band. A nanosphere of an ionic material has a surface plasmon resonance centered at

$$\omega^2_{SPR} = \frac{\omega_T^2(2\varepsilon_h) + \omega_L^2\varepsilon(\infty)}{2\varepsilon_h + \varepsilon(\infty)} \tag{7.4}$$

where $\varepsilon(\infty)$ is the residue from high-frequency absorption and ε_h is the dielectric constant of the host in which the nanospheres are embedded. Equation 7.4 is obtained from the formulas given in Chapter 3, specifically Equation 3.16 for the polarizability of a sphere. The resonance occurs when the dielectric constant of the nanoparticle fulfills $\varepsilon_p = -2\varepsilon_h$, with ε_p given for ionic motion by Equation 3.14. Using the values of ω_T, ω_L and $\varepsilon(\infty)$ appropriate for SiC, reported in Table 3.3, together with ε_h = 1.0, leads to an absorption peak centered at 11.3 μm (i.e., in the middle of the sky window). The relaxation rate ω_τ is small for SiC, as also found in Table 3.3, which makes the resonance very sharp and, in fact, SiC nanoparticles have been put forward as a material where surface plasmon modes are particularly clear cut [46]. The desired optical properties are not achieved if the SiC is in the form of a smooth continuous layer or large particles; for those cases, the SiC would reflect or scatter in the resonant range. However, embedding a SiC layer in certain specific multilayers can produce useful tuning of the phonon absorption, as will be discussed below. The latter feature is analogous to what one can find in thin silver layers, as discussed in Section 4.3.4. SiO_2 nanoparticles can serve as an alternative to SiC and are also of interest. They resonate in the same way as SiC and have a narrow peak centered at 8.97 μm.

We note, in passing, that SiC nanoparticles have been widely studied by astronomers [47,48]. Such nanoparticles are pervasive in interstellar dust and leave a sharp spectral signature in the IR radiation from deep space. This spectral feature lies where the atmosphere has its largest IR transmittance, which clearly has aided ground-based IR astronomy. Nevertheless, and despite the previous interest in SiC by astronomers, it seems that applications of this material to sky cooling are only very recent [44,45].

We now turn to detailed results for nanoparticles embedded in 10-μm-thick polyethylene foil and specifically illustrate transmittance data for doping with 10 vol.% SiO_2, 10 vol.% crystalline SiC (c-SiC), and 20 vol.% amorphous SiC (a-SiC). Figure 7.11 shows calculated results for which the effect of the nanoparticles was obtained from the models

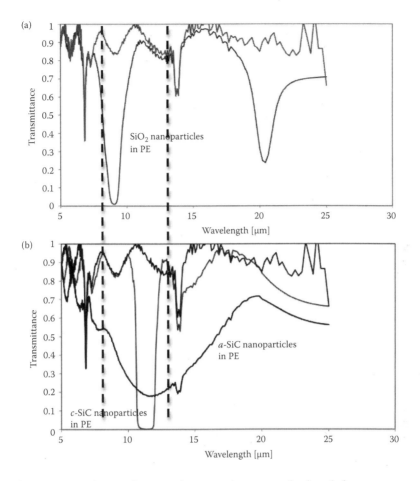

FIGURE 7.11 Spectral normal transmittance calculated for nanoparticles of SiO_2 (a), c-SiC (b), and a-SiC (b) embedded in polyethylene (PE) foil. Topmost curves refer to undoped PE. Fine structure in the curves emanate from optical interference. Vertical lines denote the sky window. (From A. R. Gentle, G. B. Smith, *Nano Lett.* 10 [2010] 373–379. With permission.)

for optical homogenization in Section 3.9 [44,45]. The optical constants of SiC and SiO_2 were obtained from the literature [49], whereas optical data for the undoped foil material were determined separately. Sharp absorption features due to surface plasmons stand out distinctly within the $8 < \lambda < 13$ μm sky window for SiO_2 and c-SiC. However, a-SiC only gives a very broad absorption feature, which is as expected, since that material does not support surface phonons. The qualitative difference between the data for SiO_2 and c-SiC on one hand and a-SiC on the other

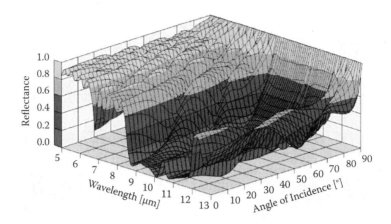

FIGURE 7.12 *A color version of this figure follows page 200.* Spectral and angular-dependent reflectance for nanoparticles of SiO_2 and *c*-SiC embedded in polyethylene foil. (From A. R. Gentle, G. B. Smith, *Nano Lett.* 10 [2010] 373–379. With permission.)

serves as a clear illustration of the importance of the Reststrahlen condition on $\varepsilon(\omega)$.

The SiO_2 doping leads to a strong absorption centered at 8.97 μm, which straddles the narrow but sharp absorption band due to atmospheric ozone. There is also a resonant peak near 20 μm, but this feature is less pronounced and lies near the tail of the Planck spectrum, so it is not a major concern here. The *c*-SiC-doped foil has one very sharp absorption peak around 11.3 μm, which lies at a wavelength where the sky window has minimum absorption. Hence, *c*-SiC nanoparticles are more efficient with regard to sky cooling than SiO_2 nanoparticles.

Angular selectivity is of importance for sky cooling and suppresses the effect of blackbody-like radiation emerging at large zenith angles, as discussed previously. Combined angular and spectral selectivity can be computed, and Figure 7.12 shows some initial results for a 10-μm-thick polyethylene foil containing 5 vol.% SiO_2 and 5 vol.% of *c*-SiC [45]. This foil is backed by nontransparent aluminum. The reflectance is given in the $5 < \lambda < 13$ μm range for the full angular interval from normal to glancing. The low reflectance in the sky window is seen to prevail for angles up to ~60°, whereas the reflectance rapidly approaches high values beyond this angle. Interference features stand out more clearly in Figure 7.12 than in Figure 7.11 as expected from the influence of the aluminum layer.

We now turn to experimental data on nanoparticle-based coatings. Figure 7.13 shows the structure of clustered 50-nm-diameter SiC nanoparticles prepared by spray coating onto Al-coated glass. Similar

FIGURE 7.13   Scanning electron micrograph of SiC nanoparticles made by spray coating onto an aluminum substrate.

coatings, on larger areas of Al sheet, were used in outdoor sky cooling tests. SiO_2 nanoparticle coatings look very much the same.

As an alternative to having a doped polymer, the SiC or SiO_2 nanoparticles can be embedded in, or positioned under, a vacuum-coated layer that absorbs by normal phonon modes in a narrow range not covered by the surface phonon resonances. Figure 7.14 refers to such a combination, with SiO_2 particles under a 264-nm-thick vacuum-coated SiO thin film backed by aluminum. The reflectance data shown indicate that phonons in SiO combine with surface phonons in SiO_2 nanoparticles to cover the sky window range [44]. The coverage would be even better for a thicker SiO film. Some scattering can be seen at short wavelengths and is due to clustering, but this effect has a very minor impact on the potential for sky cooling.

Dispersions of SiC or SiO_2 nanoparticles in appropriate silicon-based or polymer-based paint binders (such as silicones, siloxanes, or polyvinylidene fluorides) can be expected to yield results similar to those in Figure 7.14 if applied to aluminum in a sufficiently thin layer. Different mixtures may apply depending on the binder layer's contribution to the absorption in the sky window.

7.4.6 Practical Sky Cooling: Systems and Data

There are many design issues to consider when setting up practical sky cooling devices. Among those already discussed we note local atmospheric

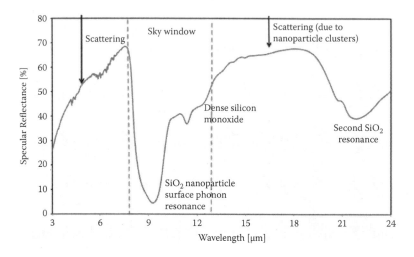

FIGURE 7.14   Spectral specular reflectance of a SiO_2 nanoparticle layer under a continuous SiO film and backed by Al. Vertical lines denote the sky window. (From A. R. Gentle, G. B. Smith, *Proc. SPIE* 7404 [2009] 74040J 1–8. With permission.)

conditions, average operating temperature relative to that of the ambient, and spectral properties of the emitting surface. But there are more issues such as the emitting area needed to provide the desired amount of heat pumping, the possible use of heat mirrors to confine the radiation angles, the housing in which the emitting surface sits, the properties of the IR transparent cover sheet (introduced in Section 7.4.3), the specific location (e.g., roof, façade, or on open ground), possible tilt to the horizontal, and the purpose of the cooling. This purpose might require that the collected coolness is used or stored away from the collecting unit, in which case a heat exchange fluid and either pumping or thermo-siphoning is needed. Of course, this fluid should not freeze and it must be in good thermal contact with the emitting surface as it passes through.

Figure 7.15 illustrates a basic arrangement for practical sky cooling. Other designs have been summarized elsewhere [26,50]. The coating indicated in Figure 7.15 may have high thermal emittance or be infrared selective and adapted to the sky window. The IR transparent cover can be UV-stabilized polyethylene. It is needed to reduce convection exchange with ambient air and, along with the nontransparent insulation, helps achieve a satisfactory nonradiative heat transfer. The particular design in Figure 7.15 also incorporates a plate with channels for a heat exchange fluid.

One factor that comes into play for outdoor experiments is water condensation, which occurs when surface temperature falls below the

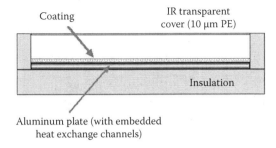

FIGURE 7.15 Schematic of a basic sky cooling system.

dew point. Condensation thus happens first on the emitter surface, which is coldest, and may sometimes occur as dew on the cover as well. Water has high thermal emittance and is hence strongly IR absorbing. It is essential that water is removed from the cover. The impact of water condensation on the infrared emitting surface, on the other hand, is usually not a problem for the cooling performance achievable in basic set-ups like the one in Figure 7.15. Not even ice formation is usually a problem. There are several strategies to reduce the effect of water drops on the cover or IR emitter, and a small tilt of the collector can lead to adequate drainage. The cooling performance is, of course, affected by the tilt, and the quantitative effect can be calculated from formulas similar to Equations 7.3 and 7.5, to come. Excessive tilt should be avoided in order to prevent thermal influx from surrounding surfaces. As discussed in Section 7.5, condensation onto surfaces undergoing sky cooling can be used for obtaining substantial amounts of water.

Figure 7.16 shows an example of some actual cooling results recorded over a continuous clear period comprising three nights and two days [51]. Data were taken on two cooling devices placed side by side, both similar to the one in Figure 7.15. One of the devices had a blackbody-like emitter with a paint surface, and the other had a surface with spectral and angular-dependent selectivity similar to the data in Figure 7.12. Both surfaces had high solar reflectance and an emitting area of 0.08 m². The devices were shaded from direct solar radiation by a wall, but they received diffuse solar radiation during the day.

The data in Figure 7.16 show many interesting features. It is evident that both surfaces are capable of significant cooling and reach temperatures below 0°C and that the selective radiator clearly outperforms the blackbody-like surface except during daytime. The selective surface quickly gets to 13°C below ambient temperature once the solar intensity begins to drop, and that temperature difference is then maintained during the night. The only deviation from this pattern is during short periods when some water vapor, enclosed between the radiator surface and

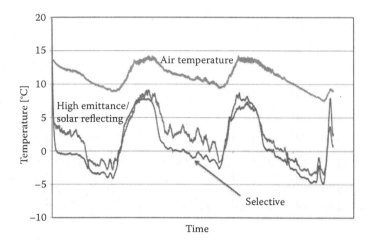

FIGURE 7.16 Cooling performance of two devices similar to the one in Figure 7.15 during a period of three nights and two days. One of the devices had a blackbody-like coating and the other a selectively emitting surface. The temperature of the ambient air is shown for reference. (From A. R. Gentle, G. B. Smith [2010] unpublished data.)

the polyethylene cover, condenses and forms ice; the cooling power is then needed for the phase change. Once the water has condensed as ice, it has no obvious effect on the cooling performance.

Figure 7.16 proves that net cooling prevails during the day. The increase in temperature after sunrise mirrors the drop in net cooling rate due to the onset of diffuse solar influx. Some sharp but short-lived temperature peaks at night arise when clouds pass over and add some incoming radiation. The cooling surface then responds rapidly to the cloud cover and finds a new steady-state temperature.

The cooling devices giving the data in Figure 7.16 had a nonradiative heat transfer as large as 2.5 to 3 $Wm^{-2}K^{-1}$. Scaling up and better insulation should enable 2 $Wm^{-2}K^{-1}$ or less, and hence even lower temperatures. Nevertheless, the transparent cover remains a significant challenge as elaborated in Box 7.4.

Much further engineering development of sky cooling systems and their interface with storage or other devices is clearly needed. However, the potential for easily implemented, low-cost cooling for a variety of applications is beyond doubt. We believe that sky cooling technology will become increasingly important and possibly emerge as a centerpiece for tackling a variety of environmental and lifestyle challenges that the world will face in coming years.

BOX 7.4 TRANSPARENT CONVECTION
SHIELDS FOR SKY COOLERS

Sky radiators reaching temperatures well below those of the ambience need good thermal insulation. This is easy for nontransparent parts of the cooling device but a challenge for the upward-facing part, which should transmit radiation at least in the $8 < \lambda < 13$ μm range while providing small non-radiative heat transfer, comprising conduction and convection. The conductive component is relatively straightforward and requires a sufficient distance between the radiating surface and the shield. At first sight, the convection might seem easy too for the case of a horizontal device, and one could hope for a stagnant gas layer that is coldest at the bottom. In reality, this is a problem, though, since wind blowing past the convection shield will move it—as long as it is a thin foil or a series of foils—so that forced convection transfers heat to the radiating surface. The convection shield hence must have mechanical rigidity.

A first attempt at developing a material with the properties required for sky cooling is illustrated in Figure B7.4.1. It shows a three-dimensional material consisting of crossed layers of vee-corrugated high-density polyethylene foils [52]. Measurements on a typical material consisting of three components, each with a height h of 1.5 cm and an apex angle θ_a of 45°, yielded $U_{nr} \approx 0.9$ Wm^{-2}K^{-1}, together with an infrared transmittance (measured with an IR-imaging instrument) of 73%. The result is encouraging for further materials development, which might involve other types of polymers in alternative configurations, as well as fiber-based reinforcement. Thus, the transparent convection shield is not an Achilles' heel for sky cooling, but it is definitely a material in need of more development!

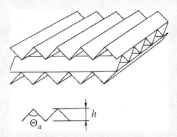

FIGURE B7.4.1 IR-transparent convection shield comprised of corrugated polymers foils characterized by height h and apex angle θ_a. (From N. A. Nilsson et al., *Solar Energy Mater.* 12 [1985] 327–333. With permission.)

One should appreciate the principal difference from a flat-plate solar collector, such as the device shown in Figure 6.2. In that case, the convection shield needs to be transparent in the $0.3 < \lambda < 3$ μm solar range, which can be accomplished with a pane of iron-free glass (cf. Figure 4.2) that has a thickness of several millimeters. The solar wavelengths are shorter than those of typical lattice vibrations, and a material serving as a transparent convection shield in the thermal infrared must be judiciously chosen to avoid lattice vibrations in the $8 < \lambda < 13$ μm range and/or have a very small mass. The solar transmittance of this shield is also of concern in many cooling applications.

7.4.7 Amplifying Sky Cooling with Heat Mirrors

Spectral and angular selectivity are not the only strategies to reduce incoming atmospheric radiation and hence amplifying the net radiative output of a surface exposed to the sky. If part of the sky vault seen by the emitter is replaced by a surface with much lower emittance than the sky in that zone, then the incoming radiation from those solid angles must be lower. The effect is as though the whole sky, on average, got a lot colder [53]. The sky is most "black" at high angles to the zenith and most transparent toward the zenith, so the obvious choice is to surround the emitting surface with a heat mirror that leaves open a reasonable-size aperture around the zenith direction. The heat mirrors must have a low emittance in order to maximize this benefit, and thus can be made of bare aluminum or a metallized surface (e.g., silver) with a thin protective oxide layer over the metal. Such surfaces can have an emittance between 0.02 and 0.1, whereas the sky at high zenith angles has an emittance approaching unity. Figure 7.17 illustrates a few practical heat mirror arrangements for boosting sky cooling. Some of these structures have features in common with solar concentrators used, for example, in solar thermal power plants.

The optical design does not have to be particularly rigorous in order to provide strong gains, and thus the structures can be low cost. But it is important that the optical design ensures that incoming near-horizontal rays are not reflected so that they reach the emitting surface. An acceptance angle, denoted $\theta_{z,max}$ for the case of a radially symmetric heat mirror, defines what radiation gets in and must exclude rays from $\theta_{z,max} < \theta_z < 90°$. All emitted radiation, apart from what the heat mirror absorbs, gets out. The incoming radiation is now strongly attenuated, so one expects net cooling gains and lower stagnation temperatures than in the absence of heat mirror arrangements.

Which type of radiator benefits most from a heat mirror arrangement: the broad-band absorber with high emittance or the sky window

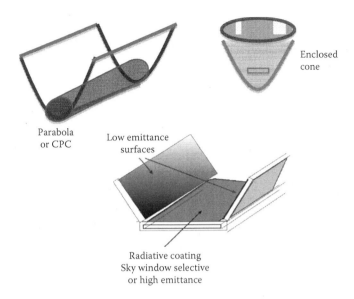

Enclosed
cone

Parabola
or CPC

Low emittance
surfaces

Radiative coating
Sky window selective
or high emittance

FIGURE 7.17   Structures with heat mirrors blocking some or all of the atmospheric radiation emerging from large zenith angles.

selective emitter with intermediate emittance? In other words, does the evaluation regarding the benefits of the two types of surfaces in Figure 7.9 change when angular restrictions are included in the analysis? It is easy to model P_{rad} with mirrors, which we will do, but some reasoning can also tell what will happen: the net output from the blackbody falls off most rapidly as τ_s goes down, as seen in Figure 7.9, because this surface absorbs most of the incoming radiation. Thus, the blackbody stands to benefit most from using heat mirrors. Calculations of P_{rad} for an aperture that is symmetric about the zenith direction can be made by replacing Equation 7.3 with

$$
P_{rad} = \left[\int_{0}^{\pi/2} d(\sin^2 \theta_z) \int_{0}^{\infty} d\lambda P(\lambda, \tau_s) E_s(\theta_z, \lambda) \right] -
$$
$$
\left[\int_{0}^{\theta_{z,max}} d(\sin^2 \theta_z) \int_{0}^{\infty} d\lambda P(\lambda, \tau_a) E_a(\theta_z, \lambda) E_s(\theta_z, \lambda) \right] - \qquad (7.5)
$$
$$
\left[\int_{\theta_{z,max}}^{\pi/2} d(\sin^2 \theta_z) \int_{0}^{\infty} d\lambda P(\lambda, \tau_a) E_m(\theta_z, \lambda) E_s(\theta_z, \lambda) \right]
$$

where E_m is the emittance of the heat mirror and taken to be wavelength independent.

Calculations based on Equation 7.5 were performed in the same way as for obtaining the data in Figure 7.9 and using the U.S. standard model atmosphere. The acceptance angle $\theta_{z,min}$ was put to 45° and 90°, where the latter value represents free exposure to the sky. Figure 7.18 shows that the data for $\theta_{z,min}$ = 45° yield a very large enhancement of P_{rad} over what was found for free exposure [53]. The enhancement is particularly large for the blackbody surface, whose net radiative power exceeds that for the infrared selective surface for $(\tau_a - \tau_s)$ being as large as 50°C. Including a realistic value of U_{nr} then leads to the conclusion that the blackbody surface is consistently superior to the infrared selective surface. For $\theta_{z,min}$ = 90°, on the other hand, the results are very similar to those earlier found in Figure 7.9. Figure 7.18 also shows two triangles, which indicate the possible performance of nonideal, though carefully chosen, surfaces; the upper triangle refers to a blackbody-like surface with an emittance of 0.9 and $\theta_{z,min}$ = 45°, and the lower triangle refers to an infrared selective surface with an emittance of 0.1 in the 8 < λ <

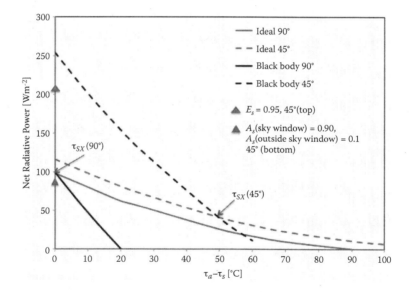

FIGURE 7.18 Net radiative cooling power for ideal blackbody and infrared selective surfaces exposed to the U.S. standard atmosphere under condition so that the acceptance angle is limited to 45° and for free exposure to the sky. The triangles on the vertical axis indicate cooling powers for surfaces with non-ideal properties as discussed in the main text. (From G. B. Smith, *Solar Energy Mater. Solar Cells* 93 [2009] 1696–1701. With permission.)

13 μm range and 0.90 outside this range along with $\theta_{z,min} = 45°$. Cooling curves emanating from these triangles will run almost in parallel with the pertinent curves for ideal surfaces.

The main result of Figure 7.18 is the prediction that a device with a blackbody-like emitter and with highly reflecting surfaces confining the radiative exchange to an angular range around the zenith direction can lead to a very large cooling power and to temperatures lying far below those of the ambience. For example, with $U_{nr} = 2$ Wm^{-2}K^{-1} one can remove heat at 20°C below ambient temperature at 60 to 70 Wm^{-2}.

7.4.8 Impact of Solar Irradiance on Sky Cooling

Sky cooling can function both night and day. Cooling during the night is most effective, but there are many possibilities to combine this with cooling also in the day (though at a lower rate than in the night), heating in the day (or during winter), daylight transmission, and photovoltaic energy generation. Thus, it is of interest to consider the impact of solar irradiance on sky cooling.

Even if night cooling is the sole concern, it is worthwhile to limit solar heating during the day. This is so because the entire cooling device might heat up, and then there is a surplus of energy that takes time to remove by sky cooling before subambient temperatures are reached. There are a number of ways to limit the solar heating: The simplest may be to totally close off the system, or to put it away if it is portable. Secondly, the cooling surface itself may combine its desirable IR properties with high solar reflectance or transmittance. A third approach, which may be the best for most practical cooling devices, is to have a convection-suppressing shield that reflects or back scatters solar radiation while it transmits in the thermal infrared. Microparticles of ZnS can be effective for the latter purpose, as we will see below in the discussion about foils for water condensation. Another option is nanosized TiO$_2$ incorporated in polyethylene [54].

It is now becoming clear that useful net heat pumping is possible during the day, provided that the impacts of daytime solar radiation are managed. It is very difficult to achieve net cooling in the presence of direct solar irradiation, though it is physically possible in principle with materials that have very small solar absorption and very large radiation output. One such material might be sheet glass in which the iron content is almost zero (cf. Figure 4.2). To have a worthwhile net cooling all day or most of the day in direct sun, a rough guide is that the solar heating of the cooling surface should be less than 40 Wm^{-2}. This means a solar absorptance less than ~8%, along with high emittance across the sky window. These properties are possible but may be difficult to accomplish in practice, and an easier approach is to combine a significant sky view with shading of the direct solar irradiance during most of the day. A recent study with façade-mounted radiators, facing away from solar incidence at midday, demonstrated useful net sky cooling in the day and almost as large cooling as for horizontal mounting in the night [51]. Continuous cooling during night and day was demonstrated also in some early work with heat mirrors, one of which also acted as a shield of direct solar radiation [41,42].

7.5 WATER CONDENSATION USING SKY COOLING

Surfaces which can pump heat at subambient temperatures can be used to condense water vapor from the atmosphere; applications of this type may have a venerable history and go back to the ancient Greeks [55]. Condensation can be made from other sources, such as the condensation part of a solar desalination cycle for extracting fresh water from salty or brackish water. The condensation rate from air depends on the relative humidity or dew point temperature τ_{dp} and the surface temperature τ_s of the radiating surface relative to the ambient temperature τ_a [27,56]. Air is saturated when $\tau_a = \tau_{dp}$. Net radiative cooling still occurs normally down to and below the dew point, but below τ_{dp} some of the pumped heat is taken up by the condensation process, thus limiting the minimum value of τ_s. A steady-state energy balance is reached at some rate of condensation R_d (in kgs^{-1}m^{-2}) such that

$$P_{rad} = U_{nr}(\tau_a - \tau_s) + R_d L_w \qquad (7.6)$$

where $L_w = 2.26 \times 10^6$ Jkg^{-1} is the latent heat of water vaporization, that is, the heat that must be pumped away each second to condense 1 kg of water. For an open surface, $R_d L_w$ depends on the convective exchange between the air away from the surface and that near the radiating surface. This is governed by differences in vapor pressures, which are higher at τ_a than at τ_s, plus the convective exchange part of U_{nr} according to relations that are well known in atmospheric science [57]. The pertinent equations must be solved iteratively together with Equation 7.6 to determine self consistently both τ_s and R_d for the particular relationship between P_{rad} and τ_s for a given experimental set-up.

Figure 7.19 shows nightly collected dew in Dodoma, Tanzania, versus relative humidity [58,59]. Both data sets are based on a very simple system comprised of a 0.4-mm-thick polyethylene foil containing 15 vol.% of ZnS pigment. The solar reflectance was ~85 %, and the emittance in the sky window was high as a consequence of the ZnS [18]. The

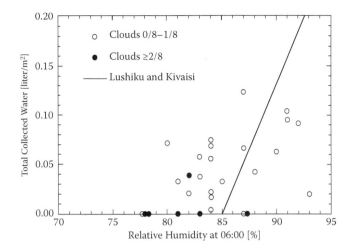

FIGURE 7.19 Water condensed on a pigmented polymer foil as a function of relative humidity overnight in Dodoma, Tanzania, during November 1993. Data refer to different fractional cloud covers. The line indicates averaged data taken during earlier measurements. (From T. Nilsson, *Solar Energy Mater. Solar Cells* 40 [1996] 23–32. With permission.)

collected water can reach ~0.1 liter per m² but is often much less. Clouds are found to deteriorate the dew collection.

Dew collection is currently being implemented for drinking water in hot and arid India [60] and is studied elsewhere too [61]. The technique is referred to as "dew-rain" and typically uses pigmented foils similar to the one used for the experiments shown in Figure 7.19. Figure 7.1 depicted a unit in France capable of producing significant amounts of water. Even simple galvanized iron roofs are capable of collecting some dew, given the proper climatic conditions [62]. The preliminary nature of the dew collection performed thus far should be emphasized, and several of the sky cooling systems discussed above could lead to superior results.

The performance of simple foil collectors does not only depend on the radiative properties of the materials but also on the rate at which water runs off the collecting surface (i.e., on its hydrophobic or hydrophilic character). A good approach might be to let the water condense inside a channel in a polymer or under a cooling plate within a heat mirror aperture system. If the vapor is forced by a fan or by natural convection through a channel under a plate, or through channels in open IR-transparent polymer systems where the polymer is doped with surface plasmon resonant nanoparticles such as SiC, then there will be a thermal gradient across the channel as the vapor progressively falls off. $R_d L_w$ will vary with location and flow speed.

Dew formation can be beneficial but it can also be a nuisance, particularly if it is accompanied by frost formation. Thus, modern fenestration with excellent thermal insulation can make it possible for the outer window pane to drop below τ_{dp} so that water is condensed on surfaces exposed to the clear sky, especially during early autumn mornings [63,64]. A striking illustration of this effect is given below in Figure 8.8. The water usually disappears shortly after sunrise, so the problem is seldom a very serious one.

An analogous situation is well known for cars that are parked outdoors during clear winter nights in cold climates: their windscreens are covered with adherent frost in the morning [65]. The condensation is effectively removed under most, though not all, weather conditions if the glass surface has a transparent low-emittance coating that diminishes the sky cooling. Pyrolytically produced SnO_2:F, discussed in Section 4.4, is an excellent coating for this purpose. Another way to make the water condensation inconspicuous, as long as the water is liquid, is to have a thin photocatalytic and super-hydrophilic coating of TiO_2 on the outer glass surface (cf. Section 8.2).

7.6 A ROLE FOR COOLING AND WASTE HEAT IN ELECTRIC POWER GENERATION

Today's power stations place great thermal and water resource burdens on the environment. Cooling is an essential aspect of thermally driven generation of electricity, and cooling towers are stark visual reminders of this need. If local lakes and rivers are employed for cooling, the consequence is degradation of the environment and reduction of natural water flows for irrigation or human use. It is a fundamental aspect of thermodynamics, as outlined in Chapter 2, that for engines utilizing heat input at temperature τ_h the useful work available to drive a turbine and produce power depends on the temperature τ_c at which the waste heat is exhausted. The maximum available power is given by the Carnot efficiency $[1 - \tau_c/\tau_h]$. Practical, or irreversible, heat engines have more realistically efficiencies of around $[1 - (\tau_c/\tau_h)^{1/2}]$ [66], with the rate at which heat can be exhausted by cooling having an important influence on this efficiency. Thus, for every GW of normal electric power with τ_c lying near the ambient temperature, around 1.8 GW of heat has to be exhausted, often as saturated air at around 35°C.

Many emerging concepts for large-scale, low-cost renewable power exploit vast thermal resources with lower temperatures than those for the usual power stations. When differences between τ_h and τ_c are small—as for power generation in OTEC (cf. Section 2.10), salt gradient solar ponds and solar chimneys—low efficiencies for conversion of heat to mechanical work must thus result. But if the natural energy resource is large enough and free at both input and output, these systems never-theless are very attractive. Furthermore, the source of coolness is differ-ent from the local ambient, which is a significant aspect. Any thermal power system operating at τ_h below ~170°C can benefit significantly in efficiency by having τ_c fall from, say, τ_c > 25°C, which would be typi-cal for cooling by air or water, to a temperature of 5°C to 10°C. Such a decrease of τ_c is achievable, and a number of power systems with such characteristics are now projected.

Waste heat from conventional power stations and industries, using high-temperature processes, can also be tapped for additional power. Power from waste heat requires turbines working at low temperatures, so cooling can add significantly to the power output. A viable and con-servative goal for the world could be up to 1,000 TWh of power per annum from a fraction of such resources operating at 10 to 15% con-version efficiency. Costs will depend on the specific type of source but should undercut solar thermal power generation if sufficient efforts are put into their design. Savings in CO_2 emissions would be up to 1 Gt per annum. Thus, designers of new large-scale power stations, found-ries, kilns, and steel plants should considering incorporating waste heat capture into plant design for subsequent use in power production. Such features are easier and much cheaper if they are put it in at the outset rather than as add-ons later. The associated carbon credits would also help offset the carbon liabilities for normal operation and hence add to the economic viability of such additions.

Cooling, if sufficiently cheap and simple, can also be a useful adjunct for boosting the output from renewable power systems, especially ther-mal systems operating at low temperatures. Solar thermal power gen-eration with an input temperature τ_h in the range 80°C to 100°C could have a boost in efficiency of 50% to 100% if the engine condensation cycle is done with coolness collected via sky cooling. This idea, which could provide very low-cost power, is considered among our specula-tions in Chapter 9. And cooling may also be an asset for some electronic power systems, whose efficiency depends on junction temperatures or thermal gradients. Large-scale photovoltaic generation systems do not produce electricity at night, but they are commonly located in near-per-fect locations for night sky cooling under clear skies and in dry air. Such generation systems would benefit from the additional cooling by use of night-cooled fluids, which may be able to decrease the temperatures of the solar cells by ~5°C or more over what is possible with regular air cooling during daytime. Stirling cycle engines, used in small solar ther-mal systems, could also have their efficiencies raised and their environ-mental impact diminished by using stored coolness, but the gains are smaller than for solar cells.

Interest in thermoelectric power is growing. It can be solar driven and would be much improved with overnight-generated coolness. Some developments in this area involve new nanotechnology, and it has the possibility of becoming a key source of highly efficient refrigeration in the future, so it is addressed in detail next.

7.7 ELECTRONIC COOLING AND NANOTECHNOLOGY

Cooling as well as electric power generation can be achieved by using the mobile electrons in two different conducting solids connected at one end and arranged with a common temperature difference $\Delta \tau$ across each. Thermocouples using two different metals work on this principle as do thermoelectric devices using n- and p-doped semiconductors as the two materials. Bismuth telluride has been the best semiconductor so far and has been available for this kind of devices for some time.

Solid-state energy conversion of heat to electricity, and using electricity to pump heat, would be ideal, but it is not efficient enough at present for replacing compressor-based cooling systems or to use in solar thermal systems for power generation. However, the area of electronic cooling is one where future developments utilizing known nanoscience could have a major impact and reduce the problems associated with leaking refrigerants, which are very bad greenhouse gases. A practical breakthrough could revolutionize many aspects of how we use and generate energy and lead to vast reductions in CO_2 emissions. Transferring scientific concepts into commercial devices appears to be the major hurdle. Figure 7.20 shows a standard commercial device configuration for thermoelectric cooling [67].

There are three factors that reduce efficiency when electrons are driven by an external voltage to carry heat from a cold junction to a hot junction for cooling applications: First, the electrons lose energy by normal resistive losses as they traverse the solid and, second, heat can flow back through the device from the hot side. The third factor affecting efficiency is intrinsic to the materials, namely the Seebeck coefficient S (i.e., the voltage per unit of temperature difference for each arm in the thermoelectric device). These three factors are lumped together into a performance factor Z, which should be as large as possible. It is defined by

$$Z = \sigma S^2 / k_\tau \qquad (7.7)$$

where σ is electrical conductivity and k_τ is thermal conductivity. A widely used figure of merit is then obtained by multiplying Z with the

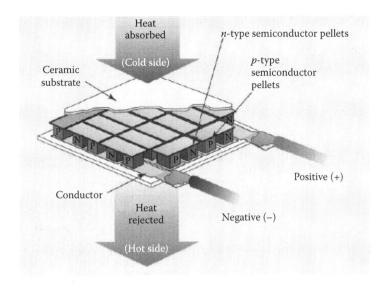

FIGURE 7.20   Standard configuration for commercial *p-n* junction thermoelectric cooling modules. The two white layers are ceramic. (From L. E. Bell, *Science* 321 [2008] 1457–1461. With permission.)

average temperature (in K). This quantity is labeled $Z\tau$ and is about 1.0 in the best currently available commercial modules. This is much too small to match conventional cooling methods, but the compactness of the thermoelectric devices has led to a number of valuable niche markets. A $Z\tau$ value of ~2 would open up the markets for domestic heating and cooling.

There are many variations on the simple system in Figure 7.20 that work better or cost less, and nanostructures can be used to boost the systems [67]. Nanostructures such as quantum wells can in principle help by raising S and reducing k_τ, and various promising related developments have occurred, but these ideas remain to be made commercially attractive.

Power generation needs higher values of $Z\tau$ than those available with present materials, and the requirements for cooling are even more stringent. Cooling also needs high values of $\Delta\tau$. A $Z\tau$ of around 1.25 will open opportunities for widespread use of waste heat and solar radiation for power. In practice, Z depends on τ, and lamination of materials with different values of Z may be the best option [68].

It is very difficult to keep the cold side of a thermoelectric generator at air temperature if the hot side is at ~400°C, say, as is appropriate in a system operating with concentrated solar radiation. Clearly, cold-side warming will reduce the power output a lot, so the application of stored coolness from overnight cooling should raise the efficiency.

Very high-efficiency thermoelectric conversion may be possible if two conditions are fulfilled, that is,

- The electrons are transported ballistically from cold to hot (i.e., without the collisions which lead to zigzag paths and cause resistance.)
- The physical structure itself limits heat flow.

Ballistic transport is possible, in principle, by thermionic emission or tunneling across a vacuum gap that is less than 10 nm in width, and this can lead to power generation with close to Carnot efficiency, because thermal transport from hot to cold is limited to radiation [69]. It is a very difficult technical challenge to maintain such a gap uniformly over useful areas in the presence of thermal expansion and mechanical stress, but serious commercial efforts are being made. Arrays of nanopores in insulators and stacks of layers may be the solution. Keeping the heat transport down then becomes the key issue. Another important aspect is the ease of getting electrons to leave a surface. The potential barrier they must overcome—that is, the work function—can be reduced by surface nanostructure [70].

Cooling has been the dominant focus to date for applications of thermoelectric devices, and power generation has been confined mainly to difficult locations, including in deep space where the devices are driven by the heat supplied by small radioactive sources. Thermoelectric generators have proven their durability also in other harsh environments, such as at the earth's poles. The enhancements in $Z\tau$ for making power generation competitive are not as large as those needed for general cooling applications, so it is possible that solar thermal plants using thermoelectrics could be one of the first renewables to benefit from improvements in this technology, while energy efficiency will benefit through waste heat becoming a viable source of extra power. A solar thermal plant not needing fluids or steam would be much easier and cheaper to run and maintain than current plants of this type, so solar thermal electricity could drop sufficiently in price to undercut coal- and nuclear-fired plants—if only the thermoelectric components were not too expensive. An expanding use at present is in hybrid and normal cars to provide extra electric power from their hot exhaust systems.

In summary, electronic cooling and electronic thermal power generation is a tantalizing field with regard to potential. It is growing in impact, but its true potential will only be realized if advances in nanotechnology and nanoscience open up ways to practical mass production of materials and systems that exhibit a significant jump in S, $Z\tau$, and overall efficiency relative to today's (2010) values. The required materials also must be sufficiently abundant.

7.8 WHITHER COOLING?

7.8.1 Some Environmental and Health-Related Benefits

New cooling techniques can reduce and offset CO_2 emissions, as discussed above, but there are other major environmental and health benefits as well. Thus, they can have an impact on water supply and water quality for agriculture and human consumption, and in this context one should note the necessity to have fresh water in order to diminish diarrheal diseases and other severe health threats in the third world. Other outcomes of the new cooling techniques can be to allow long-term storage of food and to cut down emissions of potent greenhouse gases.

Water condensation directly via sky cooling was discussed above, but there are numerous other water-related applications. For example, there are many areas in the world where it is necessary to intercept salty groundwater before it enters major rivers, which are needed for irrigation and human use. This salty water is typically pumped into evaporation basins so both water and salt are wasted. But fresh water can be extracted cheaply using the sky cooling from the air just above and around such basins, or even from select estuaries, because this air has elevated humidity whereas the sky above can be clear. Crops can be grown starting with salt water and using special greenhouses. Extraction of useable bacteria from water can be aided by using low-cost sky cooling together with doped polymers or coated metals.

Consider keeping food fresh. Field-based, low-cost sky cooling systems, operating both day and night, can provide the temperature drop that is needed to preserve fruit and vegetables long enough to avoid spoilage. Cooling hen houses cheaply in warm countries, perhaps mainly in the third world, can enable higher egg and chick yields. And there are no doubt many other niche examples.

Today's refrigeration technologies use compressed gases that are worse greenhouse gases than CO_2, and these gases may end up in the atmosphere unless the scrapping of old refrigeration systems is done with care. If the projected growth in conventional air conditioning by 2030 was to occur, then refrigerant gases may eventually pose an equal or a worse threat than that from fossil fuels. Thus, widespread use of sky cooling and/or electronic cooling, along with conscious building design to reduce cooling loads, should be a matter of urgency if this risk is to be averted.

7.8.2 Cooling Plus

Cooling technologies—whether based on electronic advances, radiation to the night sky, or hybrids of the latter with existing methods—could mean that we are at the threshold of a new technical era in cooling and refrigeration, but much more research and development is clearly needed. And this is a matter of urgency.

In a BAU scenario, the cooling demand is set to soon escalate into a dominant problem for power supply and for the environment. And these projections may be underestimates, as proceeding global warming will lead to increasing demands on cooling. Seen in this perspective, it is fortunate that there is an untapped and vast natural low-cost cooling resource—the clear night sky—overhead. But sky cooling must be combined with other measures in order to meet the growing cooling demand with less environmental impact than at present: these measures include energy-efficient building structures with much less pumping of heat into local air, cooler urban microclimates from raised albedos via new roof coatings and new road surfaces, and more urban "greening." Implementing these measures will make the urbanscapes—indoors and outdoors—much more pleasant places for the billions who are destined to reside in them in the years ahead.

There are other aspects of cooling that we have barely touched because studies have yet to begin. In particular, spectrally selective transmittance may enable cooling in the daytime jointly with various uses of the transmitted solar energy, as noted earlier. The adjunct ability to generate electric power in new low-cost ways with the aid of sky cooling could ultimately also have a major influence. We might even go as far as to say that it is not unreasonable to imagine a world where clean power sources, using some combination of solar energy and sky cooling, become the backbone of a low-carbon economy. The prospect then is not only less pollution but, in due course, lower power cost. Further comments on this issue follow in Chapter 9.

REFERENCES

1. M. Bojić, F. Yik, Cooling energy evaluation for high-rise residential buildings in Hong Kong, *Energy Buildings* 37 (2005) 345–351.
2. K. W. J. Barnham, M. Mazzer, B. Clive, Resolving the energy crisis: Nuclear or photovoltaics? *Nature Mater.* 5 (2006) 161–164.
3. A. Jäger-Waldau, editor, *REF-SYST Status Report 2004*, EUR 21297 EN, Joint Research Center, Ispra, Italy (2004).
4. D. Brockett, D. Fridley, J.-M. Lin, J. Jin, A tale of five cities: The China residential energy consumption survey, in *Human and Social Dimensions of Energy Use: Understanding Markets and Demands*, ACEEE Summer Study on Building Energy Efficiency, 2002, pp. 8.29–8.40.
5. www.opur.fr.

6. H. Akbari, S. Menon, A. Rosenfeld, Global cooling: Increasing world-wide urban albedos to offset CO_2, *Climatic Change* 94 (2009) 275–286.

7. G. B. Smith, M. B. Riley, Solar absorption cycle cooling installations in Sydney, Australia, in *Proceedings of the International Solar Energy Society Silver Jubilee Congress*, edited by K. W. Böer, B. H. Glenn, Pergamon, Elmsford, NY, 1979, Vol. 1, pp. 729–733.

8. J. C. Hedstrom, H. S. Murray, J. D. Balcomb, Solar heating and cooling results for the Los Alamos study center, in *Proceedings of the Conference on Solar Heating and Cooling Systems Operational Results*, Colorado Springs, CO, November 28–December 1, 1978; Los Alamos Scientific Laboratories LA-UR-78-2588, 1978, pp. 1–7.

9. http://www.un.org/esa/population/publications/sixbillion/sixbil-part1.pdf.

10. http://www.urbanheatislands.com.

11. M. Kolokotroni, Y. Zhang, R. Watkins, The London heat island and building cooling design, *Solar Energy* 81 (2007) 102–110.

12. H. Taha, Episodic performance and sensitivity of the urbanized MM5 (uMM5) to perturbations is surface properties in Houston Texas, *Boundary-Layer Meterol.* 127 (2008) 193–218.

13. H. Taha, Urban surface modification as a potential ozone air-quality improvement strategy in California: A mesoscale modeling study, *Boundary-Layer Meteorol.* 127 (2008) 219–239.

14. Vattenfall AB (2007), Global Mapping of Greenhouse Gas Abatement Opportunities up to 2030: Building Sector Deep-Dive, June 2007, www.vattenfall.com/www/ccc/ccc/Gemeinsame_Inhalte/DOCUMENT/567263vattenfall/P0272861.pdf.

15. A. Addeo, E. Monza, M. Peraldo, B. Bartoli, B. Coluzzi, V. Silvestrini, G. Troise, Selective covers for natural cooling devices, *Nuovo Cimento* C 1 (1978) 419–29.

16. A. Addeo, L. Nicolais, G. Romeo, B. Bartoli, B. Coluzzi, V. Silvestrini, Light selective structures for large scale natural air conditioning, *Solar Energy* 24 (1980) 93–98.

17. T. M. J. Nilsson, G. A. Niklasson, Optimization of optical properties of pigmented foils for radiative cooling applications: Model calculations, *Proc. Soc. Photo-Opt. Instrum. Engr.* 1536 (1991) 169–182.

18. T. M. J. Nilsson, G. A. Niklasson, Radiative cooling during the day: Simulations and experiments on pigmented polyethylene cover foils, *Solar Energy Mater. Solar Cells* 37 (1995) 93–118.

19. J. Bell, R. Lehman, G. Smith, Advanced roof coatings: Materials and their applications, in *Proceedings CD and Summary Book of the CIB International Conference on Smart and Sustainable Built Environment*, Brisbane, Australia, November 19–21, 2003 (ISBN: 1-74107-040-6), Paper T606.

20. SkyCool Pty Ltd, Sydney, Australia (2007), private communication.
21. B. Schuler, Innovation for construction and buildings by BASF, in *Proceedings of Convation06*, Sydney, Australia, 2006 (unpublished).
22. A. Synnefa, M. Santamouris, K. Apostolakis, On the development, optical properties and thermal performance of cool colored coatings for the urban environment, *Solar Energy* 81 (2007) 488–497.
23. G. B. Smith, A. Gentle, P. D. Swift, A. Earp, N. Mronga, Coloured paints based on coated flakes of metal as the pigment, for enhanced solar reflectance and cooler interiors: Description and theory, *Solar Energy Mater. Solar Cells* 79 (2003) 163–177.
24. G. B. Smith, A. Gentle, P. Swift, A. Earp, N. Mronga, Coloured paints based on iron oxide and silicon oxide coated flakes of aluminium as the pigment, for energy efficient paint: Optical and thermal experiments, *Solar Energy Mater. Solar Cells* 79 (2003) 179–197.
25. M. R. Kuehnle, H. Statz, Encapsulated nanoparticles for the absorption of electromagnetic energy, US patent publication US 2005/0074611 A1 (2005).
26. C. G. Granqvist, T. S. Eriksson, Materials for radiative cooling to low temperatures, in *Materials Science for Solar Energy Conversion Systems*, edited by C. G. Granqvist, Pergamon, Oxford, U.K., 1991, pp. 168–203.
27. T. M. J. Nilsson, Optical Scattering Properties of Pigmented Foils for Radiative Cooling and Water Condensation: Theory and Experiment, Ph.D. thesis, Department of Physics, Chalmers University of Technology, Göteborg, Sweden (1994).
28. X. Berger, J. Bathiebo, Directional spectral emissivities of clear skies, *Renewable Energy* 28 (2003) 1925–1933.
29. P. Berdahl, R. Fromberg, The thermal radiance of clear skies, *Solar Energy* 29 (1982) 299–314.
30. P. Berdahl, M. Martin, The resource for radiative cooling, in *Proceedings of the Second National Passive Solar Conference*, edited by D. Prowler, I. Duncan, B. Bennett, American Section of the International Solar Energy Society, Newark, DE, 1978, Vol. 2, pp. 684–686.
31. C. G. Granqvist, A. Hjortsberg, Radiative cooling to low temperatures: General considerations and application to selectively emitting SiO films, *J. Appl. Phys.* 52 (1981) 4205–4220.
32. T. S. Eriksson, C. G. Granqvist, Radiative cooling computed for model atmospheres, *Appl. Opt.* 21 (1982) 4381–4388.
33. M. Tazawa, H. Kakiuchida, G. Xu, P. Jin, H. Arwin, Optical constants of vacuum evaporated SiO film and an application, *J. Electroceram.* 16 (2006) 511–515.

34. Z. Liang, H. Shen, J. Li, N. Xu, Microstructure and optical properties of silicon nitride thin films as radiative cooling materials, *Solar Energy* 72 (2002) 505–510.

35. T. S. Eriksson, C. G. Granqvist, Infrared optical properties of electron-beam evaporated silicon oxynitride films, *Appl. Opt.* 22 (1983) 3204–3206.

36. T. S. Eriksson, C. G. Granqvist, Infrared optical properties of silicon oxynitride films: Experimental data and theoretical interpretation, *J. Appl. Phys.* 60 (1986) 2081–2091.

37. P. T. Tsilingiris, The total infrared transmittance of polymerized vinyl fluoride films for a wide range of radiant source temperature, *Renewable Energy* 28 (2003) 887–900.

38. P. Berdahl, Radiative cooling with MgO and/or LiF layers, *Appl. Opt.* 23 (1984) 370–372.

39. E. M. Lushiku, A. Hjortsberg, C. G. Granqvist, Radiative cooling with selectively infrared-emitting ammonia gas, *J. Appl. Phys.* 53 (1982) 5526–5530.

40. E. M. Lushiku, C. G. Granqvist, Radiative cooling with selectively infrared-emitting gases, *Appl. Opt.* 23 (1984) 1835–1843.

41. F. Trombe, A. Lè Phat Vinh, M. Lè Phat Vinh, Description des expériences sur le refroidissement des corps terrestres, *J. Rech. CNRS* 65 (1964) 563–580.

42. F. Trombe, Perspectives sur l'utilization des rayonnements solaires et terrestres dans certaines régions du monde, *Rev. Gén. Therm.* 6 (1967) 1285–1314.

43. T. S. Eriksson, A. Hjortsberg, C. G. Granqvist, Solar absorptance and thermal emittance of Al_2O_3 films on Al: A theoretical assessment, *Solar Energy Mater.* 6 (1982) 191–199.

44. A. R. Gentle, G. B. Smith, Angular selectivity: Impact on optimised coatings for night sky radiative cooling, *Proc. SPIE* 7404 (2009) 74040J 1–8.

45. A. R. Gentle, G. B. Smith, Radiative heat pumping from the earth using surface phonon resonant nanoparticles, *Nano Lett.* 10 (2010) 373–379.

46. C. F. Bohren, D. R. Huffman, *Absorption and Scattering of Light by Small Particles*, John Wiley & Sons, New York, 1983.

47. M. Dkaki, L. Calcagno, A. M. Makthari, V. Raineri, Infrared spectroscopy and transmission electron microscopy of polycrystalline silicon carbide, *Mater. Sci. Semicond. Proc.* 4 (2001) 201–204.

48. T. Henning, H. Mutschke, Formation and spectroscopy of carbides, *Spectrochim. Acta, Part A* 57 (2001) 815–824.

49. E. D. Palik, *Handbook of Optical Constants of Solids*, Academic, Orlando, FL, 1985.

50. T. S. Eriksson, Surface Coatings for Radiative Cooling to Low Temperatures, Ph.D. thesis, Department of Physics, Chalmers University of Technology, Göteborg, Sweden, 1985.

51. A. R. Gentle, G. B. Smith (2010) unpublished data.

52. N. A. Nilsson, T. S. Eriksson, C. G. Granqvist, Infrared-transparent convection shields for radiative cooling: Initial results on corrugated polyethylene foils, *Solar Energy Mater.* 12 (1985) 327–333.

53. G. B. Smith, Amplified radiative cooling via optimised combinations of aperture geometry and spectral emittance profiles of surfaces and the atmosphere, *Solar Energy Mater. Solar Cells* 93 (2009) 1696–1701.

54. Y. Mastai, Y. Diamant, S. T. Aruna, A. Zaban, TiO_2 nanocrystalline pigmented polyethylene foils for radiative cooling applications: Synthesis and characterization, *Langmuir* 17 (2001) 7118–7123.

55. V. S. Nikolayev, D. Beysens, A. Gioda, I. Milimouk, E. Katiushin, J.-P. Morel, Water recovery from dew, *J. Hydrology* 182 (1996) 19–35.

56. D. Beysens, The formation of dew, *Atmospheric Res.* 39 (1995) 215–237.

57. J. L. Monteith, M. H. Unsworth, *Principles of Environmental Physics*, 3rd edition, Academic, Burlington, MA, 2008.

58. E. M. Lushiku, R. T. Kivaisi, Applications of radiative cooling for condensation irrigation, *Proc. Soc. Photo-Opt. Instrum. Engr.* 1149 (1989) 111–113.

59. T. Nilsson, Initial experiments on dew collection in Sweden and Tanzania, *Solar Energy Mater. Solar Cells* 40 (1996) 23–32.

60. G. Sharan, H. Prakash, Dew condensation on greenhouse roof at Kothara (Kutch), *J. Agriculture Engr.* 40 (4) (2003) 75–76.

61. M. Muselli, D. Beysens, J. Marcillat, I. Milimouk, T. Nilsson, A. Louche, Dew water collector for potable water in Ajaccio (Corsica Island, France), *Atmospheric Res.* 64 (2002) 297–312.

62. G. Sharan, D. Beysens, I. Milimouk-Melnytchouk, A study of dew water yields on galvanized iron roofs in Kothara (north-west India), *J. Arid Environments* 69 (2007) 256–269.

63. C. G. Granqvist, Transparent conductors as solar energy materials: A panoramic review, *Solar Energy Mater. Solar Cells* 91 (2007) 1529–1598.

64. A. Werner, A. Roos, Simulations of coatings to avoid external condensation on low *U*-value windows, *Opt. Mater.* 30 (2008) 968–978.

65. I. Hamberg, J. S. E. M. Svensson, T. S. Eriksson, C. G. Granqvist, P. Arrenius, F. Norin, Radiative cooling and frost formation on surfaces with different thermal emittance: Theoretical analysis and practical experience, *Appl. Opt.* 26 (1987) 2131–2136.

66. F. L. Curzon, B. Ahlborn, Efficiency of a Carnot engine at maximum power output, *Am. J. Phys.* 43 (1975) 22–24.

67. L. E. Bell, Cooling, heating, generating power, and recovering waste heat with thermoelectric systems, *Science* 321 (2008) 1457–1461.

68. G. J. Snyder, Application of the compatibility factor to the design of segmented and cascaded thermoelectric generators, *Appl. Phys. Lett.* 84 (2004) 2436–2438.

69. Y. Hishinuma, T. H. Geballe, B. Y. Moyzhes, T. W. Kenny, Refrigeration by combined tunneling and thermionic emission in vacuum: Use of nanometer scale design, *Appl. Phys. Lett.* 78 (2001) 2572–2574.

70. A. Tavkhelidze, V. Svanidze, I. Noselidze, Fermi gas energetics in low-dimensional metals of special geometry, *J. Vac. Sci. Technol.* B 25 (2007) 1270–1275.

CHAPTER 8

Supporting
Nanotechnologies
Air Sensing and Cleaning,
Thermal Insulation, and
Electrical Storage

This chapter brings together a range of important issues that impact on energy use in general and, in particular, on the quality of spaces within buildings. Specifically, we discuss sensing and purification of indoor air, self-cleaning surfaces, thermal insulation, and storage of electricity. These are huge topics in their own right, and our treatment of each of them will take the form of an introduction, overview of key aspects, and some recent progress. The reader who is interested in more detail could start with the references provided. Figure 8.1 shows one of the materials we will discuss below [1]. It is an aerogel comprised of silica nanoparticles and has a porosity that can be as high as 99% or even above. This makes silica aerogel the lightest solid material known today, and its density can be only about three times that of air. Such "solid gases" are interesting for transparent thermal insulation, as further discussed in Section 8.3.

8.1 AIR QUALITY AND AIR SENSING

8.1.1 The Sick Building Syndrome

People in industrialized countries typically spend as much as 80 to 90% of their time indoors, in buildings and vehicles [2]. Hence, it is obvious that the indoor air quality is important for health and well-being and for work productivity [3,4]. But the air quality experienced is often not as good, and the "sick building syndrome" (SBS) [5] is a frequently used term to describe situations in which the users of buildings experience

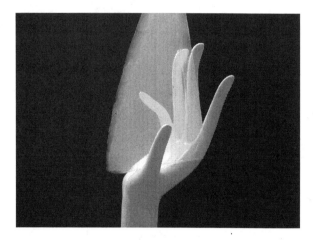

FIGURE 8.1 *A color version of this figure follows page 200.* Silica aerogel tile. (From http://www.airglass.se. With permission.)

acute health effects and discomfort seemingly connected with the time spent in the building while no specific illness or cause can be pinned down. As much as 30% of all new and refurbished buildings worldwide may be the subject of complaints connected with indoor air quality. The SBS is usually nonspecific with symptoms such as headache, eye, nose or throat irritation, dry cough, itchy or dry skin, dizziness and nausea, difficulty in concentrating, fatigue, and sensitivity to odors. The cause of these symptoms is normally not known, but the syndrome is no longer felt by persons who have left the "sick" building.

There are three main causes of the SBS:

- *Inadequate ventilation* can cause the SBS, which implies that efforts to improve energy efficiency by cutting down on ventilation may backfire and lead to health problems.
- *Chemical contaminants* are another cause and can originate from indoor sources such as adhesives, carpeting and upholstery, and manufactured wood products. All of these sources may emit volatile organic compounds (VOCs), such as formaldehyde (HCHO), some of which are known carcinogens. Tobacco smoke should be noted as it produces high levels of VOCs, other toxic compounds, and respirable particulate matter. Inadequately vented heaters, stoves, and fireplaces can lead to harmful combustion products such as carbon monoxide, nitrogen dioxide, and respirable particles.
- *Biological contaminants* are the third major cause of SBS and include bacteria, moulds, pollen, and viruses, which contribute

microbial volatile organic compounds (MVOCs) and emanate from stagnant water in ducts, drains, and humidifiers, and through water ingress by way of leaks. The indoor bacterium *Legionella* is a particularly well-known problem.

These three causes may act in combination and may supplement other perceived problems such as inadequate lighting, temperature, or humidity, or excessive noise. Radon and asbestos are not included in the causes of the SBS. There may be a gender aspect in SBS, and women tend to be more vulnerable than men.

There are many VOCs—such as decane, toluene, xylene, formaldehyde, tetrachlororethylene, ozone, and nitrogen dioxide—each with specific threshold levels for causing problems [6]. Formaldehyde can be particularly problematical, especially in new buildings [7], and has a human sensory irritation threshold of 0.1 mg/m^3 and an odor threshold of 1 mg/m^3. Inhaled formaldehyde may affect learning and memory, as found from animal studies [8]. The concentrations of the VOCs are often higher indoors than outdoors [9], and the same is true for MVOCs [10].

It is clearly important to be able to measure and survey the air quality inside buildings. This requires good sensors, which must be low cost and energy efficient in order to be of practical interest. It will be shown in the following text that nanostructured oxides can be used for this purpose. Removal of VOCs and MVOCs is also of great relevance and, as discussed in Section 8.2, this can be done by photocatalysis using, again, nanostructured oxides. Interestingly, the very same transition metal oxides are relevant for sensing and purification of air, and these oxides are of central interest also for electrochromic window coatings (cf. Section 4.11) [11] and solar reflective high emittance paints (cf. Section 7.3). Gas sensors enable demand controlled ventilation, and photocatalytic air cleaning makes it possible to keep the air quality at a desired level without excessive and energy-consuming air exchange. The energy efficiency will be particularly good if the photocatalysis is driven by direct solar irradiation.

8.1.2 Gas Sensing with Nanoporous Metal Oxides: General

It has long been known that semiconducting metal oxides can be used for gas sensing. The underlying mechanism is chemical interaction of the gas molecules with the semiconductor's surface leading to changes in its electrical conductivity [12–15]. Molecules in the gas phase, which can serve as electron donors or acceptors, adsorb on the oxide and form surface states, which then can exchange electrons with the oxide. If the molecule is an acceptor it will extract electrons from the oxide and hence decrease its conductivity (in the case of an *n*-type conductor); if the molecule is a donor it will increase the conductivity. A space charge layer is thus formed with a conductivity that is a function of the concentration and type of the gas, and this chemically induced effect can be transformed into an electrical signal via an appropriate electrode arrangement.

The influence of the gas absorption will be particularly strong if the oxide is nanoporous and comprises nanoparticles that are weakly linked to one another. Such a structure is shown in the upper part of Figure 8.2. Also shown there is a simplified model of the electronic bands, illustrating

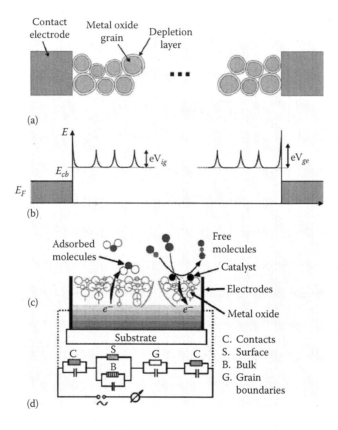

FIGURE 8.2 Physical structure of a gas sensor with oxide grains between electrodes (a) and a one-dimensional electronic structure showing inter-grain and grain-electrode band bending (by eV_{ig} and eV_{ge}, respectively), as well as the lower limit of the conduction band of the oxide (E_{cb}) and its Fermi level (E_F) (b). Also shown are processes upon gas adsorption in the presence and absence of a catalyst with e^- denoting electrons (c) and an equivalent circuit of the various contributions to the overall conductivity (d). (After M. Graf et al., *J. Nanoparticle Res.* 8 [2006] 823–839. With permission.)

that inter-grain band bending leads to energy barriers whose height can be influenced by gas adsorption. The lower part of Figure 8.2 shows how adsorbed molecules provide charge to the oxide. This process is usually not very species selective, but it can be made so if the metal oxide is doped with a suitable catalyst, as also shown in Figure 8.2. The bottom part of the figure contains a simple equivalent circuit of the sensor with separate contributions due to the surface, bulk, and grain boundaries of the nanoparticles and of the electrodes.

A good gas sensor is characterized by four main features, namely

- *Sensitivity*, so that the conductivity is strongly changed upon gas exposure
- *Selectivity*, meaning that the sensitivity is large only for the desired gas or gases
- *Operating temperature*, which typically is 200°C to 400°C and requires a heating device that in practice should be small (known as a "micro-hotplate")
- *Long-term stability*

These features have a number of consequences for the implementation of gas sensors in buildings. Of course, the sensors must be able to reliably and unambiguously detect particularly relevant gases such as formaldehyde. Furthermore, it is desirable to allow operation at room temperature in order to avoid excessive energy consumption during long-term operation, and a fire hazard.

8.1.3 Gas Sensing with Nanoporous Metal Oxides: Illustrative Examples

Highly sensitive detection of gases is possible by use of a variety of oxides, with SnO_2 being a particularly well-known example [16]. WO_3, TiO_2, In_2O_3, Ga_2O_3, $SrTiO_3$, and many other oxides have been extensively investigated as well. The films should be nanoporous in order to allow a large contact area to the gas; such structures can be made with a variety of the thin film techniques as discussed in Appendix 1. Chemically prepared nanoparticulate layers [17], "nanosponges" [18], and hybrid layers consisting of oxide and carbon nanotubes [19] are some recently studied options.

Box 8.1 discusses gas sensing with WO_3-based layers composed of well-characterized nanoparticles with and without the addition of catalyst nanoparticles. The data show that high sensitivity can be combined with selectivity and room-temperature operation.

BOX 8.1   GAS SENSING WITH NANOPARTICULATE WO₃-BASED THIN FILMS

Figure B8.1.1 shows the structure of two nanoparticulate WO₃-based layers made by the advanced gas deposition (AGD) technique introduced in Appendix 1 [20]. These films are somewhat substoichiometric and contain particles with a tetragonal structure. The images indicate agglomerations of nanoparticles with narrow size distributions. The WO₃ particles in Figure B8.1.1a are shown in as-deposited state, whereas those in Figure B8.1.1b—referring to a WO₃:Pd mixed-particle film with a Pd:W atomic ratio of 0.5%—have been sintered at 600°C in air for 1 hour. Clearly, the sintering has resulted in grain growth with preserved nanoporosity. The AGD technique is unusual in that it separates nanoparticle nucleation and growth from film growth—with both processes occurring in a clean ambience—which enables the interactions between neighboring particles to be fine tuned.

FIGURE B8.1.1   Scanning electron micrographs of an as-deposited film of WO₃ (a) and of a WO₃:Pd film sintered at 600°C (b). The horizontal bars are 100 nm. (From A. Hoel et al., *Sensors Actuators* B 105 [2005] 283–289. With permission.)

Figure B8.1.2 shows changes in the conductivity when unheated films such as those in Figure B8.1.1 were exposed to different amounts of H₂S gas in air. These changes are expressed in terms of a sensitivity Σ defined by $\Sigma = \sigma_{gas}/\sigma_{air}$, where σ_{gas} is the electrical conductivity of the sensor exposed to the pertinent gas and σ_{air} refers to pure, dry, synthetic air, respectively. The H₂S was introduced after 10 minutes and led to a gradual increase of the conductivity in proportion to the H₂S content, as shown in Figure B8.1.2b for the WO₃:Pd film. After 10 minutes of exposure, the films were subjected to a 10-minutes-long heating pulse at 200°C, which restored the original sensitivity. The sensitivity, recorded after 10 minutes of gas exposure, depends strongly on the H_2S concentration, but readily measurable effects can be seen at the 500 ppb level, and Figure B1.2b indicates that $\Sigma \approx 10$ then applies to the WO_3:Pd film. For the undoped WO_3 film, the straight line indicates that the data were consistent with $\Sigma \approx A\ C_{gas}{}^B$, where C_{gas} is the gas concentration and A and B are constants.

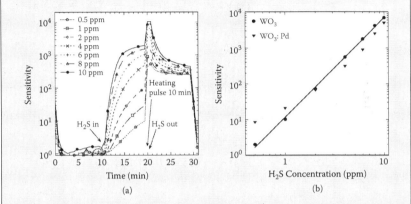

FIGURE B8.1.2 Room-temperature sensitivity versus time of a nanoporous WO_3:Pd film exposed to different concentrations of H_2S in air (a), and room-temperature sensitivity versus H_2S concentration for sensors comprised of WO_3 and WO_3:Pd (b). The latter data for WO_3 fall on a straight line in the log–log plot. (From A. Hoel et al., *Sensors Actuators* B 105 [2005] 283–289. With permission.)

Gas specificity is of obvious interest. This can be accomplished in a number of different ways [21], the most straightforward being by use of an array of sensors with each element optimized for a specific gaseous species. This is only one option, though. Nanoporous oxides are usually sensitive to a range of different gases and, for example, WO_3-based films can detect NO_2 and CO for the former gas down to ~1 ppm [22]. The sensitivity is peaked at different operating temperatures for the various gases, and Figure B8.1.3 shows large differences for H_2S, NO_2, and CO for the case of WO_3:Al. If one allows sensor operation at an elevated temperature, then it is clearly possible to use this effect to accomplish specificity. Temperature-modulated resistance together with statistical analysis using pattern recognition algorithms make it possible to reliably detect as little as 20 ppb of H_2S and 200 ppb of ethanol [23,24].

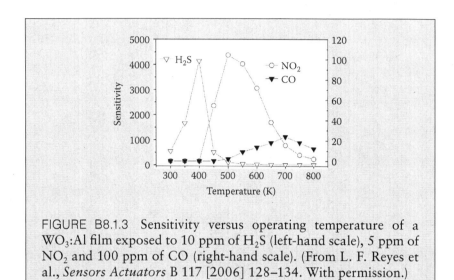

FIGURE B8.1.3 Sensitivity versus operating temperature of a WO₃:Al film exposed to 10 ppm of H₂S (left-hand scale), 5 ppm of NO₂ and 100 ppm of CO (right-hand scale). (From L. F. Reyes et al., *Sensors Actuators* B 117 [2006] 128–134. With permission.)

Electrical conductivity, as discussed in Box 8.1, is one of the many properties that can be used for sensing purposes. Electromagnetic fluctuation spectroscopy ("noise") is another alternative that has attracted attention recently [25]. This technique is interesting for buildings since it can detect several gases simultaneously with a single sensor and since it is not vulnerable to long-term drift of the dc electrical properties; on the negative side, the noise detection requires rather advanced electronics.

Indoor air often contains formaldehyde, which is one of the prime contaminants associated with the SBS, as pointed out above. Detecting this gas is hence an important issue, but detector development is still in its infancy. Sputter deposition of NiO (a p-type material) under conditions giving submicrometer grains has been demonstrated to give films that respond to formaldehyde exposure at levels of a few ppm when the film is kept at 300°C [26]. Specifically, a change in the gas content by 5 ppm altered the resistance by ~1.5%. Recent work on NiO nanoparticle films made by advanced gas deposition (cf. Appendix 1) showed good sensor properties for H₂S and NO₂, particularly when the operating temperatures were ~150 and ~100°C, respectively [27].

8.2 PHOTOCATALYSIS FOR CLEANING

8.2.1 General

Photoexcitation can occur when semiconductor nanoparticles are exposed to light with energy above the band gap energy. Figure 8.3 shows what

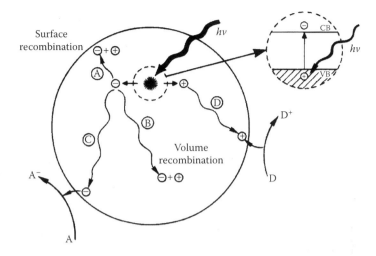

FIGURE 8.3 Photoexcitation in a semiconductor particle by a photon with energy $h\nu$ lifts an electron from the valence band (VB) to the conduction band (CB). The electron-hole pair can then undergo several different processes. (From A. L. Linsebigler et al., *Chem. Rev.* 95 [1995] 735–758. With permission.)

goes on [28]: Photon absorption produces an exciton followed by charge separation so that an electron-hole pair is created. Charge transport to the particle surface along the paths C and D then leads to reduction and oxidation processes at the surface, respectively. Other processes, denoted A and B, end up with surface recombination and bulk recombination. The processes clearly show some resemblance to those in the dye sensitized and organic solar cells discussed in Section 6.2.

The detailed physical and chemical processes occurring at the surfaces of the nanoparticles are surprisingly poorly understood. A discussion of them would take us deeply into surface science even for the most widely investigated photocatalyst, which is TiO_2 [29], and this is beyond the scope of this book. However, there is now convincing evidence of an intimate coupling between nanoparticle morphology, exposed crystalline phases, and particle reactivity [30]. TiO_2 has a number of crystalline structures, with the anatase and rutile being the most common, and each of these can form nanoparticles of different shapes. Figure 8.4 depicts five different shapes of TiO_2 nanoparticles present in samples made by solution-based chemical preparation. The particle sizes are in the 10 to 70 nm range. Each of the faces corresponds to a specific crystallographic index, as shown.

Figure 8.5 shows the normalized surface concentration as a function of time for HCOOH and HCOO molecules absorbed on five different

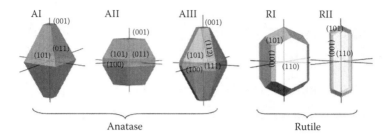

FIGURE 8.4   Compilation of most probable shapes of TiO_2 nanoparticles with anatase (A) and rutile (R) structures. The crystallographic face indices are indicated.

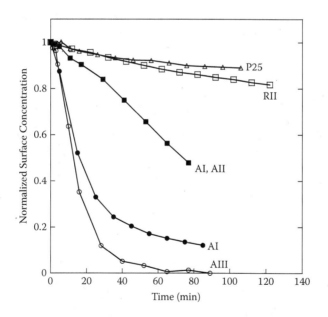

FIGURE 8.5   Normalized surface concentration of adsorbed HCOOH and HCOO molecules on anatase (A) and rutile (R) nanocrystals versus time for irradiation by simulated sunlight. Data are shown also for a commercial photocatalyst (P25).

samples comprised of TiO_2 nanocrystals under simulated solar irradiation. The data were obtained from in situ infrared spectroscopy [30]. The samples are dominated by the nanoparticle shapes shown in Figure 8.4, and results are given also for a material designated P25 which is used extensively as a base line for work on photochemical and photoelectric effects of TiO_2-based materials (cf. Figure 6.21). The important conclusion to be drawn from Figure 8.5 is that photocatalytic reaction rates depend strongly on the type of nanoparticle that dominates the material. Hence, one can boost the photocatalysis by controlling the surface structure on the nanometer scale. The data in Figure 8.5 should be regarded as indicative for a practical situation, and the presence of water vapor, the temperature, and other parameters are important for the performance of the photocatalyst. Another important aspect is that residuals can build up on the surface during the photocatalysis and cause its deactivation.

Photocatalysis is well known in oxides based on Cd, Ce, Fe, Sb, Sn, Ti, W, Zn, and Zr. The most widely studied material is TiO_2 with anatase-type crystal structure [28,31,32]. Thin films of this material are absorbing in the UV, whereas they are nonabsorbing in the luminous and NIR ranges. Thus, only a few percent of the solar radiation is useful for photocatalysis, and it is of interest to widen the active range so as to include longer wavelengths, too (cf. Box 6.3). This widening can be accomplished by doping the TiO_2 with N, C, S, F, various metals, or combinations of these dopants. Nitrogen doping to form $TiO_{2-x}N_x$ has been investigated in particular detail [33–36]. Figure 8.6 shows that nitrogen doping extends the absorption to the $0.4 < \lambda < 0.5$ µm range. Large research activities are currently (2009) pursued in order to explore how this correlates with photocatalytic activity and the general mechanism outlined in Figure 8.3.

TiO_2 has a high refractive index, and microparticles of this material can backscatter strongly and hence be efficient in high-albedo paints (cf. Section 7.3). Doping leads to some coloration, but this effect is confined to a minor part of the solar spectrum so that $TiO_{2-x}N_x$ pigments still make it possible to combine overheat protection with efficient solar photocatalysis, as long as the paint binder does not cover the pigment surfaces.

Highly reactive species are usually created during the processes following photoexcitation, as shown in Figure 8.3. They can be used for breakdown of organic molecules or microorganisms in contact with the semiconductor surface. These processes can be used for numerous applications [32], such as

- Self-cleaning surfaces
- Air purification
- Self-sterilization

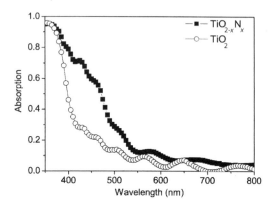

FIGURE 8.6   Spectral optical absorption for films of TiO_2 and $TiO_{2-x}N_x$. (From G. Romualdo Torres et al., *J. Phys. Chem. B* 108 [2004] 5995–6003. With permission.)

- Antifogging
- Heat transfer and heat dissipation
- Water purification
- Corrosion protection

A number of these applications are discussed next.

8.2.2 Self-Cleaning Surfaces

TiO_2-coated ceramic tiles, cements, plastic foils, membrane materials (for tents, etc.), and glass panes have been used for several years, especially for outdoor applications [31,32]. The self-cleaning is by far most efficient in conjunction with water flow, such as natural rainfall, and is associated with the superhydrophilic character of the TiO_2 surface. This surface is structurally modified and metastable, and its nature does not seem to be fully understood [37]; it is not simply a very clean surface. The water penetrates the molecular-level space between this surface and the stain so that the latter can be washed away. Thus, cleaning can occur even if the UV irradiation is not strong. Drying artifacts normally do not remain after this cleaning. The cleaning may not be perfect, but the superhydrophilicity increases the intervals between periodic cleanings and hence cuts down maintenance costs. So "easy-clean" may be a better designation than "self-clean." A number of striking illustrations of the cleaning of objects that have been partly covered with TiO_2 can be found in the literature [31,32]. TiO_2-based films also can serve as anticorrosion treatment, especially if they are deposited onto a layer of phosphotungstic acid [38].

Figure 8.7 illustrates a number of features associated with super-hydrophilic TiO_2 surfaces [37]. The middle part shows the difference between the wetting of an unirradiated film and of an UV irradiated, superhydrophilic one. The wetting is characterized by a water contact angle θ_{wc}, defined as the angle between the underlying surface and the tangent of the liquid phase at the interface of the solid–liquid–gas phase as shown in the upper part of Figure 8.7. Figure 8.7a shows that UV irra-diation by 1.1 mW/cm² makes θ_{wc} decrease gradually as a function of the irradiation time, and $\theta_{wc} \approx 0$ if this time is ~2 h, that is, the water then

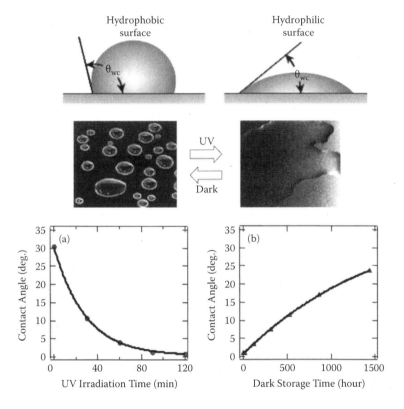

FIGURE 8.7 The upper part shows the definition of a water contact angle θ_{wc}, which is large for a hydrophobic surface and small for a hydro-philic surface. The middle part contains photographs of water coverage on TiO_2 surfaces kept in the dark and after UV irradiation. The lower part shows water contact angle after UV irradiation (a) and after subse-quent dark storage (b). (From K. Hashimoto et al., *Jpn. J. Appl. Phys.* 44 [2005] 8269–8285. With permission.)

covers the surface as a uniform thin layer. The superhydrophilicity is not permanent, though, and storing the TiO_2 in the dark for a few months brings back the original wetting properties, as seen from Figure 8.7b. Doped TiO_2 films can show improved hydrophilicity [39].

Self-cleaning TiO_2-based surfaces have been deposited onto float glass in commercial products for several years; the coatings have been made by spray pyrolysis as well as by reactive dc magnetron sputtering. There are several issues with such products; one of them has to do with the high refractive index of TiO_2, which leads to higher reflectance and lower transmittance than for uncoated glass, and the other concerns the loss of self-cleaning ability in the absence of significant solar irradiation. As elaborated in Box 8.2, films consisting of TiO_2 and SiO_2 can improve both of these issues.

The superhydrophilic surfaces spread water condensation so that light scattering from droplets is avoided, that is, the surfaces are experienced as fog free. This is of obvious interest for windows and mirrors. Figure 8.8 is a striking illustration and shows a thermally well-insulated window in a private house in Uppsala, Sweden [43]. The data were taken just after sunrise during a clear day in the fall. The right-hand part of the window is covered by scattering droplets caused by dew formation, which is associated with significant sky cooling (cf. Section 7.5). The left-hand part of the same window has a commercial self-cleaning and superhydrophilic TiO_2-based coating on its outer surface. This coating spreads the droplets and renders them invisible.

Antifogging can have wider applications than those discussed above and can be useful to enhance boiling and evaporation heat transfer [44]. One application in this context is for cooling exterior walls of buildings via falling water films in order to combat the urban heat island effect (cf. Section 7.2) [37]. Another application of superhydrophilic surfaces is in air coolers in which water bridges can be avoided between adjacent cooling fins and therefore heat exchange efficiency can be maintained, as well as air throughput. Superhydrophilicity is also useful for surfaces employed in water supply by atmospheric water condensation via night sky cooling.

Self-sterilization is a variety of self-cleaning and is of particular interest for hospitals and facilities for elder care. Tiles covered with TiO_2:Cu and irradiated with UV have been shown to function well for these applications [45].

Self-cleaning can be achieved not only by superhydrophilicity but also by the reverse effect, superhydrophobicity. This is usually referred to as the "lotus effect" and implies that water drops roll off the surface of interest and then carry undesirable particulates with them. Superhydrophobicity is at hand for $\theta_{wc} > 150°$ and leads to water removal from surfaces at tilt angles that are only a few degrees from the

BOX 8.2   ENHANCED SUPERHYDROPHILICITY IN FILMS COMPRISING TiO$_2$ AND SiO$_2$

Figure B8.2.1 shows spectral transmittance of glass coated with a TiO$_2$-SiO$_2$ composite film in which submicrometer-size SiO$_2$ and nanosize TiO$_2$ particles were applied by stepwise electrostatic deposition. These films serve as efficient antireflection coatings, and it is seen that the transmittance can exceed 99% in the luminous range, while the self-cleaning properties are maintained [40]. Similar properties can be achieved in analogous double-layer TiO$_2$-SiO$_2$ films [41].

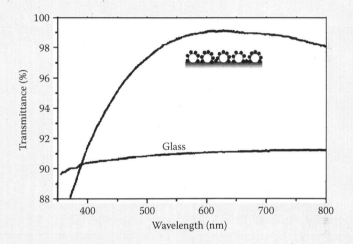

FIGURE B8.2.1   Spectral transmittance of glass with and without a coating consisting of SiO$_2$ particles covered with TiO$_2$ nanoparticles as indicated in the inset sketch. (From X.-T. Zhang et al., *Chem. Mater.* 17 [2005] 696–700. With permission.)

TiO$_2$-SiO$_2$ films made by sol-gel technology can display natural superhydrophilicity [42]. Figure B8.2.2 shows that θ_{wc} can remain small for months in the dark, which is qualitatively different from the behavior of pure SiO$_2$ and TiO$_2$ (cf. Figure 8.7). Much research in self-cleaning oxides is ongoing, and the last word on this subject is far from spoken.

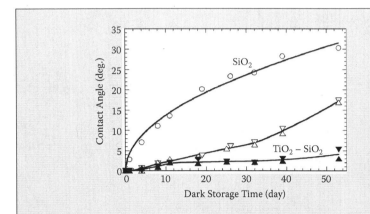

FIGURE B8.2.2   Water contact angle for films of SiO₂ and four different TiO₂-SiO₂ composites after dark storage. (From M. Houmard et al., *Surface Sci.* 602 [2008] 3364–3374. With permission.)

FIGURE 8.8   Photograph of a window of a private house in Uppsala, Sweden. Both parts of the window are triple glazed and have a center-of-glass thermal conductance of 0.65 Wm⁻²K⁻¹. The window is in a northeast facing façade with an open view toward the sky. The left-hand part of the window has an outer glass with a self-cleaning and super-hydrophilic TiO₂-based coating, and the right-hand part has standard uncoated glass.

horizontal. Detailed calculations have shown that θ_{wc} depends strongly on the root-mean-square surface roughness amplitude but is rather insensitive to other geometrical features [46]. Low tilt angles are associated with heterogeneous wetting, meaning that the fluid does not penetrate into regions between the protrusions [47]. Surfaces meeting the conditions for superhydrophobicity can be constructed in a variety of ways and can consist of many materials, with "sponge-like" nanostructured TiO_2 being one example of many [48].

8.2.3 Air Purification

An important application of photocatalysis is for decontamination, deodorization, and disinfection of indoor air. The photocatalyst can be attached to surfaces on walls, roofs, and floors, as well as to other surfaces such as on textiles or nonwoven fabrics. The photocatalytic systems are different from traditional filter-based air cleaners in that they do not accumulate the pollutants and hence they avoid the risk of secondary pollution.

There is a large literature on air cleaning using devices incorporating UV sources. However, the natural solar radiation, as well as residual UV light from fluorescent lights, can be used and offers advantages with regard to energy use. Figure 8.9 shows the breakdown of acetone under simulated sunlight in a reaction chamber containing surfaces of four different commercial catalysts based on anatase TiO_2 [49]. Clearly, some of these catalysts can bring down the acetone concentration from 10 to 20% of the original content during 20 minutes of exposure.

One should note here that a regular window can be arranged as a device for air cleaning if the outer glass is UV transparent (i.e., iron free as discussed in Section 4.1) and a glass pane with a photocatalytic surface is positioned so that it is irradiated by UV light while there are provisions allowing the room air to circulate past the photocatalytic surface [50]. Pigments of photocatalysts operating with visible light might be used as paint materials on walls and ceilings in buildings equipped with windows having normal float glass. However, one should be aware of the fact that photocatalytic breakdown of VOCs may lead to the creation of noxious intermediate compounds such as formaldehyde and acetaldehyde [51], and these aspects need further investigation.

Cementitious building materials are in themselves nanocomposites [52]. They can be made photocatalytic by TiO_2 admixture [53,54], and such materials can be efficient for purification not only of indoor air but also outdoors. This effect has been demonstrated convincingly with regard to the removal of NO_x caused by automobile exhausts. The purification can be particularly strong with concrete containing glass cullets

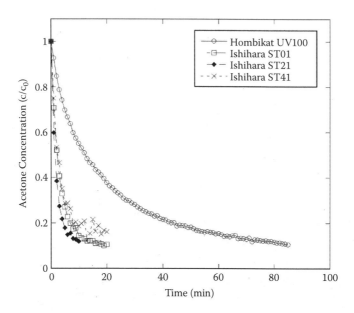

FIGURE 8.9   Relative acetone concentration versus time as measured in a reaction chamber containing four different commercial catalysts and irradiated by artificial sunlight. The catalysts contain TiO_2 anatase nanoparticles with sizes in the 5 to 200 nm range and surface areas between 300 and 10 m^2g^{-1}. (From L. Österlund [2009], private communication. With permission.)

[55]. Air cleaning also can be achieved by inclusion of TiO_2 in roadway surfaces [32] and pavements [56]. The ensuing improvement in the air quality will not only have beneficial health effects for persons being outside buildings, but the improved quality of the air that is let into the building will diminish the need for indoor air treatment and hence be an energy boon. Furthermore, the high refractive index of TiO_2 can lead to an increase of the albedo over that for common materials for buildings and roads, thereby offsetting global warming, as further discussed in Section 7.1.

8.3   THERMAL INSULATION WITH NANOMATERIALS

Thermal insulation materials for walls, ceilings, and roofs might be thought of as based on mature technologies and hence are rather "low tech." However, the science behind even the most traditional porous insulators is relatively complex, and the image of the thermal insulator is about to change. Indeed, a leap forward is imminent and will involve nanostructures and vacuum. Most insulation is used in situations where there is no light transmittance, but there are also applications where insulation materials with good light transparency or translucency are of interest. Clearly, transparent insulation is important for windows and luminaires discussed in Chapters 4 and 5, and this issue is considered separately at the end of this section.

8.3.1 Thermal Conductance of Porous and Nanoporous Materials

Good thermal insulation is essential for buildings in both cold and warm climates and for refrigeration. It also plays an important role in the efficiency of solar hot water collectors, high performance sky cooling systems, hot water storage tanks, cold storage tanks, and for transport and storage of fresh food and some medicines. Improvements and wider use of good insulation can lead to very large energy savings, greater comfort, and less food and medicine spoilage. Codes for energy-efficient buildings continue to require ever-higher standards of thermal insulation, and this means ever-increasing thicknesses as long as one is bound to traditional insulating materials. But larger thickness can be awkward, and in many structures impossible, especially for retrofitting of old buildings. Thus, the aim of the emerging insulation materials addressed here is to achieve higher levels of performance in thinner layers. The material property of interest is usually called thermal conductivity or "k_τ-value" [$Wm^{-1}K^{-1}$]. The use of the term thermal conductivity, however, hides the fact that most good insulating materials—because they are porous—transport heat by three mechanisms: conduction, convection, and radiation. Thus, k_τ is, strictly speaking, an effective thermal conductivity.

Thermal conductance U is given by $U = k_\tau/d_i$ [$Wm^{-2}K^{-1}$], where d_i is the thickness of the insulating layer, and the heat flow P across a temperature drop $\Delta\tau$ is then $P = U\Delta\tau$ [Wm^{-2}]. Many products are defined in terms of their R-value or thermal resistance given by $R = 1/U$ [$W^{-1}m^2$]. As a reference point for insulating ability, recall from Section 4.2 that multiglazed windows with good thermal insulation generally had a thermal conductance between 1 and 2 $Wm^{-2}K^{-1}$, though the best achieved well below 1 $Wm^{-2}K^{-1}$ and, under special conditions, down to 0.2 $Wm^{-2}K^{-1}$ (the window in Figure 8.8 has 0.65 $Wm^{-2}K^{-1}$). How thick would a typical mineral wool insulating panel need to be to achieve such U values? With $k_\tau \sim 0.04$ $Wm^{-1}K^{-1}$, a thickness of around 4 cm would provide $U = 1$ $Wm^{-2}K^{-1}$. However, most modern building codes presently require that U be around 0.2 $Wm^{-2}K^{-1}$ or less, which means for the traditional products thicknesses of 20 cm or more. To put this in perspective we note

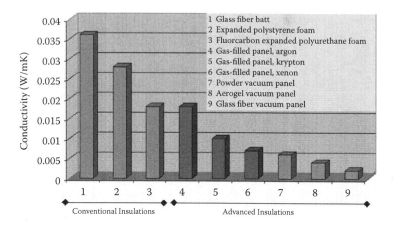

FIGURE 8.10 Thermal conductivity for conventional and advanced materials. (From http://gfp.lbl.gov/.)

that for a 20 m² area of wall or ceiling and, say, $\Delta\tau = 15°C$ this means the heat gain or heat loss falls from 300 to 60 W.

Conventional insulating materials are compared with one another and with emerging ones in Figure 8.10 [57]. Traditional materials based on glass and expanded polymer foam have $0.18 < k_\tau < 0.35$ Wm⁻¹K⁻¹. The best materials, on the other hand, utilize nanostructured aerogels and vacuum and can achieve a k_τ-value as low as 0.002 Wm⁻¹K⁻¹. These two types of materials will be considered next.

The advantages of the nanoporous materials, such as inorganic aerogels and related polymeric materials, ensues from their favorable convective heat transfer. Convection arises because density gradients in a gas exist across a space in which there is a thermal difference (i.e., molecules spread out more in warmer gases). This fact, in addition to gravity, causes hot gases to rise and cold to fall, and heat flow is then established. To eliminate convection, one must ensure that there is not enough gas present to generate such flows. There are two ways of doing this:

- To have vacuum
- To make the gas-filled pores small enough that they contain very little gas even at normal pressure

A nanopore fulfills the second requirement, and convection is impossible when the mean free path between molecule collisions becomes much larger than the pore size. Nitrogen gas at room temperature, to take an example, has a mean free path of around 100 nm.

Figure 8.11 illustrates convection prevention as the pore size is decreased [58]. Once convection is under way, the molecular mass comes

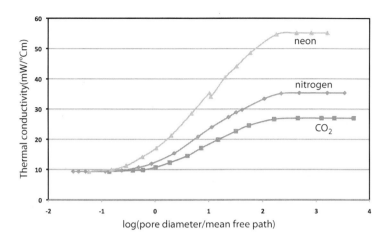

FIGURE 8.11 Pore size versus onset of convection in nanoporous and microporous insulating material incorporating three different gases. (Replotted from data in J. C. Harper, A. F. El Sahrigi, *Ind. Engr. Chem. Fundamentals* 3 [1964] 318–324.)

into play and lighter molecules are more able to transfer heat, as seen in Figure 8.11. The residual thermal conductivity $k_\tau \sim 0.009$ Wm^{-1}K^{-1} is due to conduction along the thin skeleton material plus thermal radiation across the pores.

8.3.2 Vacuum Insulation Panels

Vacuum insulation panels (VIPs) have become of interest in recent years and can achieve k_τ-values from 0.005 to 0.002 Wm^{-1}K^{-1} (i.e., a U-value as low as 0.1 Wm^{-2}K^{-1}). This would need a thickness of only 2 to 5 cm, which is suitable for retrofitting into many current wall cavities or onto the internal wall surfaces without much loss of room space [59]. Figure 8.12 compares the thicknesses needed for the same level of insulation between a VIP and a traditional insulating material [60]. A 20 m^2 wall area with $\Delta\tau = 15$°C would pass less than 30 W of heat, and such panels would also be very useful for sky cooling applications.

The VIP is normally based on a microporous material which is "open" (i.e., without enclosed cavities), and vacuum is maintained by aluminum foils which seal off the internal structure. Obviously, the VIPs must be handled so that this seal is not damaged. Another practical aspect is that VIPs may decrease the acoustic insulation. VIPs can be used within door and window frames, on floors, under flat roofs, and on terraces and internal floors. The insulating performance of doors, which

FIGURE 8.12   Thicknesses for traditional glass wool (left) and a VIP (right), both having the same U-value. (From R. Baetens et al., *Energy Buildings* 42 [2010] 147–172. With permission.)

is a weak spot in many well-insulated buildings, is typically increased by 50% [59].

Air and moisture ingress slowly degrade the vacuum in VIPs, and hence their U-value will increase over time. This loss of performance may eventually be manageable with gettering and desiccants, but that remains for future study. Thus establishing the relative merits of VIPs and nanofoams, characterized by closed nanocavities, must await further development and testing. The main hold-up for the widespread use of VIPs seems to be the current manufacturing methods, which are labor intensive.

8.3.3 Silica Aerogel

Silica aerogel is a nanoporous translucent material that can be used in fenestration of various types [61–63]. A photo of this material was shown in Figure 8.1. It consists of an open structure of interconnected SiO_2 nanoparticles with diameters down to 1 nm. The pore size is 1 to 100 nm, and the overall porosity can exceed 99%. The material is mechanically fragile, though it can be strong enough to be load bearing

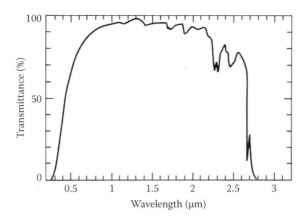

FIGURE 8.13 Spectral transmittance through a 4-mm-thick silica aerogel tile. (From C. G. Granqvist, *Materials Science for Solar Energy Conversion Systems*, edited by C. G. Granqvist, Pergamon, Oxford, U.K., 1991, pp. 106–167. With permission.)

in evacuated devices for transparent insulation. It is normally produced as tiles or granules via supercritical drying.

Figure 8.13 shows the spectral transmittance through an aerogel tile [64,65]. The transmittance lies above 80% for 0.6 < λ < 2.2 μm but drops sharply at shorter wavelengths as a result of diffuse scattering due to density fluctuations. This scattering leads to a bluish appearance of thick aerogel tiles, as apparent from Figure 8.1. The aerogels can be used in VIPs and yield superior thermal insulation, especially if radiative transfer across the aerogel is prevented by low-emittance surfaces.

8.4 GREEN ENERGY STORAGE

8.4.1 Energy Storage: Survey of a "Missing Link"

Energy storage is a vital aspect of systems for renewable energy and improved energy efficiency. It is sometimes referred to as the "missing link" for such systems [66]. Storage is essential if the renewables component of our energy supply is to be capable of providing base-load power; without this capability the nonhydro contribution will be capped, probably well below 50%. One attraction of renewable power sources to some people, especially those in remote locations, is that they can create an independence from external grid-based supplies or diesel fuel deliveries. For many of the millions who make up the world's nonurban poor, these sources can liberate them from the grinding task of collecting waning supplies of firewood.

It is also attractive from an energy security viewpoint to have local supplies, backed by moderate-size storage, in addition to external supplies. This means, for example, that power is available both during a blackout and during an extended lack of local solar energy due to bad weather. Energy security is also a national issue as outside resources become scarce, too expensive, or are cut off as a result of political, industrial, or other events. Thus, fires in gas processing plants, or pipeline problems, have been known to severely reduce gas or oil supplies in some regions for many months on end. Large local and community-scale storage of energy will also become increasingly attractive if severe weather events become more common, which is predicted to happen as the impact of global warming grows. Thus, renewables combined with advanced storage technologies will not only help to reduce the extent of global warming but they also hold much promise as measures for adaptation to global warming.

Energy can be stored in many forms: gravitational, electrical, magnetic, thermal, chemical, and mechanical. Volume, weight, cost, leakage, internal loss, and ease and speed of charging and discharging must all be considered. The more energy one can easily pack into and remove from a given volume or mass the better. Mechanical energy storage is not just in kinetic form, as in a flywheel, but also in stationary form as elastic strain or compressed air. Coal, oil, and natural gas are stores of chemical energy laid down by natural processes over millennia. The sun is a store of nuclear fusion fuel. Each form has a major role in renewable energy systems, including dams (gravitational), hydrogen storage (chemical splitting of water), hot water tanks, massive building components, deep underground rocks heated by natural nuclear decay, phase change materials (thermal and geothermal), and superconducting coils (magnetic). Thermal storage is important and a natural adjunct to solar thermal systems and night sky cooling. The key aspect is what material to use, and this depends on the temperature at which the thermal store operates. For cooling purposes one needs to store energy from ~0 to ~20°C, for domestic hot water from around 60 to 80°C, and for solar thermal power at hundreds of degrees. Water, water-glycol mixtures, and phase change materials are ideal for temperatures from 100°C and down to about 0°C. For very high temperatures, one can use molten salts, solid rocks, or large blocks of graphitic carbon. Hot rock and carbon block storage is, in essence, man-made geothermal storage.

The focus for the discussion below is on advanced electric energy storage in batteries and capacitors utilizing nanostructured materials. The likely eventual shift to an all-electric or hybrid-electric vehicle fleet seems to be the main driving force at present in the quest for improvements in electrical storage and electrical power management, with batteries and capacitors being the key components. Better batteries are also needed for remote and stand-alone renewable energy systems. High-performance capacitors, which can charge and discharge quickly, are very important for vehicle acceleration and power switching, as well as for information, communication, and display technologies.

Nearly all of the major advances in compact electrical storage are being driven by progress in nanomaterials, and the main attraction is in their exceptionally large surface-to-volume ratio. Surfaces and interfaces are where the electrical energy is stored in advanced capacitors, and it is also where charging and discharging take place in batteries.

8.4.2 Electrical Storage Using Electrochemistry

Most of us carry around batteries in a variety of small devices such as mobile phones, laptops, digital cameras, calculators, etc. The majority of these devices are rechargeable, sometimes by use of small solar cells. For larger power, lead acid car batteries have long been the stalwarts of vehicle electrical systems and of solar power storage. These batteries are heavy, include poisonous or otherwise undesired materials, and often occupy too much space to be convenient for solar power systems.

Today's electronic systems can also rely on small, or even tiny, devices for temporary storage of energy (i.e., capacitors). Battery and capacitor technologies have evolved extensively in recent years, and further large improvements are expected. Many of these are due to nanoscience. Developments over the last two decades have begun to fill in missing parts of the "electrical storage landscape," which traditional batteries and capacitors did not cover. We first need to become familiar with this "landscape" via its "map" so as to fully appreciate what nanoscience has recently added and has to offer in the future.

Batteries and capacitors have to both store charge and convert their stored energy into electrical power. The latter is a time-dependent issue. If a lot of energy has to go in or out in a short time, then power flow is the main issue. And if a lot of energy has to be stored, then high energy capacity is needed. Capacitors are good at the former, batteries at the latter. Storage capacity is important for systems which deliver power steadily, such as solar systems for homes or buildings that are not grid connected. Having both high power flow and high energy capacity in one device would be ideal for electric cars, but it is not easily accomplished. Cars need a lot of storage and also high power for acceleration, and ideally high power flow for fast charging ability. The latter aspect is the least developed one, but some nanostructured batteries make it a genuine possibility to fully charge a car at a "battery power station" in the same time it takes to fill today's car with petrol at a gas station. At the moment, though, an electric car needs all-night charging, or "most-of-the-day" or "while-at-work-or-shopping" charging. An alternative might be a quick change-over battery pack from specialized "battery stations," and such facilities are starting to appear to service the growing electrical vehicle fleet.

The power delivered per unit mass in W/kg is called the *power density*, and the energy stored per unit mass in Wh/kg or J/kg is called the *energy density*. The "electrical storage landscape" involves a "map" where these two functions are plotted against each other in a Ragone chart such as the one shown in Figure 8.14 [67].

Traditional capacitors can deliver 10^4 to 10^7 W/kg but can only store 0.01 to 0.05 Wh/kg. Lead acid batteries can deliver, at best, a much more leisurely 100 W/kg but have high storage, between 20 Wh/kg to 8 Wh/kg. What about the storage "territory" between 0.05 and 10 Wh/kg with good power? Can we exceed 20 Wh/kg storage and/or achieve much better power rates in batteries other than lead-acid ones?

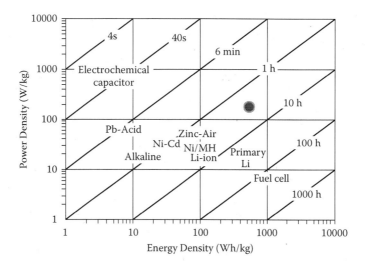

FIGURE 8.14   Ragone chart of power density versus energy density for electrochemical capacitors, batteries, and fuel cells. MH denotes metal hydride. Traditional capacitors are not included as they would require an extension of the chart by two orders of magnitude down for the x-axis and three orders up for the y-axis. Diagonal lines represent charge/discharge times. (From Y. Zhang et al., *Int. J. Hydrogen Energy* 34 [2009] 4889–4899. With permission.) The solid circle has been added to show the measured performance of a nanostructured virus-based battery in order to indicate what is becoming possible (cf. Box 8.3 for details).

Definitely! The question now is how far we can go to cover the map in Figure 8.14. A combination of nanoscience and advances in electrochemistry have allowed the filling in of many gaps with batteries based on lithium chemistry, Ni-Cd, and Ni metal hydrides as well as with more recent devices such as fuel cells and electrochemical capacitors similar to the one discussed shortly.

With the advent of very large scale photovoltaic systems, a totally new class of batteries is being considered for stationary applications; they need both very large storage capacity and ability for ultra large current flows upon charging and discharging. These batteries utilize two liquid metals separated by a liquid salt, which means they have to be maintained at ~700°C. The underlying idea is that ions can move much more easily in liquids than in solids. This work has only just begun, but is of high interest [68]. It builds upon the well-known high-temperature electrochemical process for producing aluminum metal in which aluminum oxide is first dissolved in molten salt cryolite (Na_3AlF_6) at 900°C to enable extraction and reduction of the Al^{3+} ions to Al metal.

8.4.3 Electrochemical Super- and Ultracapacitors

The structure and chemistry of electrodes and electrolytes determine power flows and can also influence storage capacity. Electrochemical capacitors include super- and ultracapacitors, and their scientific designation "electrical double-layer capacitors" indicates how they work. Here, a charge that is opposite to that on the electrode builds up alongside it in the electrolyte, not on the opposite electrode as in a conventional capacitor. The thickness d_{dc} of this double charge layer is only about 0.5 to 1 nm. Using the basic relation for capacitance, $C = \varepsilon_0 \varepsilon A / d_{dc}$ where A is the area, now yields that C is 10 to 20 microfarad per cm^2 or, in mass units, of the order of 100 farad per gram of electrode (for nanoporous carbon electrodes). The voltage U ranges from around 5 to 20 V, and the stored energy, given by $\frac{1}{2}CU^2$, is about 10^{-3} J/cm^2. No chemical reaction occurs at the electrode, only charge separation, and this means that many more charge–discharge cycles can occur than in a battery over the lifetime of the device.

The capacitance is very high in supercapacitors and ultracapacitors, and ranges from ~ 0.1 to more than 20 F, which implies that the active electrode area lies between ~1 and 200 m^2. How can such a large area occur in a compact device? The reason is that the conducting electrodes are highly nanoporous. The lower size limit of the pores, which then dictates the maximum effective area, is set by the requirement that ions in the solution should be able to enter them and is estimated to be about 2 nm [67]. Nanoporous carbon, which exists in many forms, has been

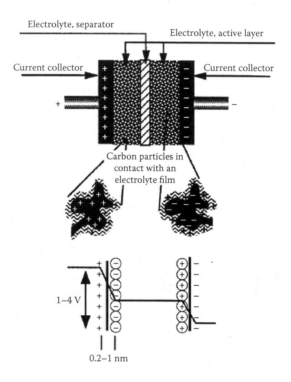

FIGURE 8.15 Basic double-layer electrochemical cell using carbon particles packed around each electrode. An inert, porous separator serves as a pure ion conductor and prevents any flow of electrons from one side to the other. The carbon particles are charged oppositely on the two sides. The lower plot shows the double-layer on each particle and the associated voltage profile on pairs of particles on opposite sides of the separator. (From R. Kötz, M. Carlen, *Electrochim. Acta* 45 [2000] 2483–2498. With permission.)

the mainstay of this technology, and Figure 8.15 outlines the principles and basic structure of electrochemical capacitors based on this material [69]. Devices can involve packed nanoparticle powders, as well as activated carbon, which is a highly porous form that is usually prepared by pyrolysis. Carbon nanotubes and nanofibers have been used recently [67,70] and lead to fast charge/discharge rates though rather low energy density. Mesoporous and nanoporous gold is also of interest for electrochemical capacitors and enables moderate storage with fast discharge at high power [71].

Recent interest in electrochemical storage has been directed at the new highly conducting carbon-based nanomaterial known as "graphene." It essentially comprises single layers or sheets of carbon atoms as in graphite, and the structure was shown in the right-hand part of Figure 4.27 [72]. In a study of their potential as ultracapacitors, the graphene sheets agglomerated partly into particles but had sections with many free individual sheets [73]. Capacitance values as large as 135 and 99 F/g were measured with aqueous and organic electrolytes, respectively. Given their scope for improvements, these nanomaterials may even ultimately play a role in energy storage rivaling some batteries. In the future, commercial carbon-based electrochemical capacitors may store 20 to 40 Wh/kg [70]. Looking at Figure 8.14, this puts them in the domain of many of today's batteries, but with longer lifetimes and higher powers if required.

The need for both power and sufficient storage has led to hybrid systems. These are based on a variety of the electrochemical capacitors, called pseudo-capacitors, in which the electrode interacts chemically to some extent with the electrolyte (i.e., it is part capacitor and part battery). The electrodes in this case are conducting oxides or conducting polymers whose functionality again relies on nanoporosity. One of the cheapest options is porous MnO_2, which can supply up to 260 F/g at 0.8 V. More complex and expensive oxide and polymer materials under development can achieve up to ~1000 F/g at around 1 V [67,74].

8.4.4 Nanomaterials for Advanced Batteries

Developments in nanostructured materials for battery electrodes and electrolytes are both needed if major advances are to occur, and the main focus is currently on Li^+ ion-based batteries. The goals are faster charging and discharging and better storage, along with extended lifetime.

Anode materials may involve nanostructured alloys and composites which store Li^+—that is, intercalation hosts for Li^+ such as mixtures of Li and TiO_2 nanowires that can accommodate enough lithium to form $Li_{0.91}TiO_2$ [74]. Various oxides with internal nanostructures that let them take up Li^+ are other alternatives. Rod-shaped nanoparticles of lithium iron phosphate coated with thin layers of a conducting polymer such as polypyrrole, and a mix of polypyrrole and polyethylene glycol are of current interest for electric cars but need further development. The aim of coating the nanoparticles with a polymer conductor is to minimize the charge transfer resistance between the $LiFePO_4$ particles and the polymer electrolyte [75]. Recently, $LiFePO_4$ nanoparticles, devised for battery electrodes, with slightly shifted stoichiometry have been found to conduct ions along the particle surface at high rates. A full battery discharge in 10 to 20 s seems to be achievable with electrodes based on these nanoparticles mixed with carbon. The addition of carbon is needed to allow fast electron flow in order to match the fast ion flow

[76]. Clearly, much new science is needed for the processes that occur at nanoelectrodes of different materials before one can proceed with confidence to commercial systems in this field. It is an area of much technical promise that needs urgent and accelerated research and development.

Nanocathodes are less developed than nanoanodes. The reaction with the electrolytes can be facilitated with nanoparticles, but this may also lead to overheating and safety issues. Nanopillars of various oxides are of concern for cathodes, and such pillars of V_2O_5 inside porous alumina are of special interest. The V_2O_5 can also be in nanoporous form. There is much interesting science to sort out in order to understand how the physical and lattice structures of these and other cathode materials interact with the electrolyte and how they intercalate lithium ions [74].

Nanocomposites of ceramic nanoparticles embedded in solid polymeric ion conductors—such as polyethylene oxide (PEO) combined with a lithium salt—show much promise for future improvement in the performance of electrolytes. They have already led to large enhancements compared to the starting materials. Significant ionic conductivity in polymeric conductors was only thought to happen at high temperatures, above the glass transition temperature where the polymer chains become disordered. But crystalline polymer compounds have now been found in which the Li^+ ions conduct and reside in tunnels formed among the polymer chains, as seen in Figure 8.16. These structures facilitate fast ion transport. The same electrolyte materials are also of value for supercapacitors.

We end with two examples of batteries devised according to principles rooted in biology. Thus, recent work on an all-polymer paper-based battery-employed cellulose fibers of algal origin (*Cladophora*) coated with 50-nm-thick layers of polypyrrole [77]. The specific surface area of this cellulose is exceptionally large—much larger than for the cellulose typically used in the paper industry—which allows large charging capacity at a high rate. It is interesting to note that *Cladophora* is usually considered an environmental pollutant, which here it is put to good use via nanotechnology.

Another example of a new nanomaterials-based approach to high-performance batteries touches on an area of materials development that is in its infancy but has much to offer. It uses genetic engineering, which now has reached a point at which it is feasible to implement nanoscale electrical wiring based on biological principles. This is further elaborated in Box 8.3.

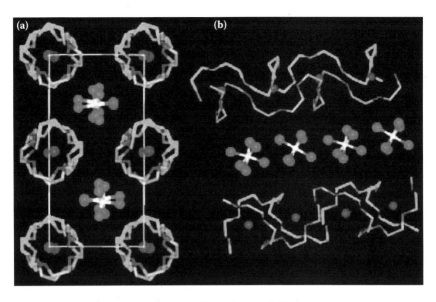

FIGURE 8.16 *A color version of this figure follows page 200.* Structure of the nanostructured polymer ion conductor $PEO_6:LiAsF_6$ as seen looking into the chains and tunnels (a) and along them (b). The Li^+ sits inside the tunnels formed by the PEO. The atoms are fluorine, carbon, and oxygen. (From A. S. Aricò et al., *Nature Mater.* 4 [2005] 366–377. With permission.)

BOX 8.3   A VIRUS-BASED NANOBATTERY

Multifunctional viruses can be used as scaffolds for the synthesis of high-performance nanostructured electrodes in batteries [78]. A two-gene virus enabled linking of amorphous $a\text{-}FePO_4 \cdot H_2O$ nanowires to single-wall carbon nanotubes (SWNTs) and made it possible to engineer very high performance nanoelectrodes that combine a large storage capacity with high power delivery, as apparent from the Ragone chart in Figure B8.3.1a. The cycling stability was good, which can be inferred from Figure B8.3.1b. The nanostructure resulting from the two-gene tethering of SWNTs to $a\text{-}FePO_4 \cdot H_2O$ is depicted in Figure B8.3.2 at high and medium magnifications. The performance of these batteries is remarkable, as can be seen by comparison to other battery technologies in Figure 8.14, and points at the opportunities that nanomaterials will provide in order to meet future needs for electrical storage.

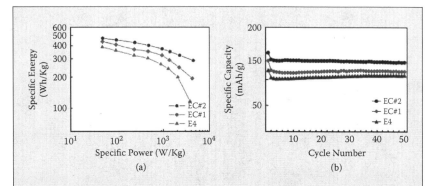

FIGURE B8.3.1 Ragone chart of energy storage versus power capability for batteries based on one-gene (E4) and two-gene (EC#1 and EC#2) tethering (a), and cycling stability of these systems (b). (Modified image from Y. J. Lee et al., *Science* 324 [2009] 1051–1055. With permission.)

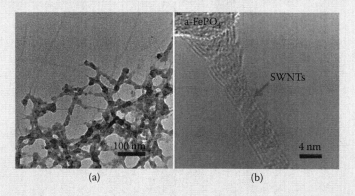

FIGURE B8.3.2 Nanostructure of *a*-FePO$_4$ grown on the two-gene virus system EC#2 and tethered to single-walled carbon nanotubes (SWNTs). Panel (a) shows a low-magnification transmission electron micrograph (30,000x) with faint images of SWNTs in the upper part and *a*-FePO$_4$ attached to EC#2 in the lower part. Panel (b) is a very high magnification image (800,000x) showing tethering of a single SWNT to an *a*-FePO$_4$ particle. (From Y. J. Lee et al., *Science* 324 [2009] 1051–1055. With permission.)

REFERENCES

1. http://www.airglass.se.
2. J. A. Leech, W. C. Nelson, R. T. Burnett, A. Aaron, M. E. Raizenne, It's about time: A comparison of Canadian and American time-activity patterns, *J. Exposure Anal. Environm. Epidem.* 12 (2002) 427–432.
3. T. Godish, *Indoor Environmental Quality*, CRC Press, Boca Raton, FL, 2001.
4. M. J. Mendell, W. J. Fisk, K. Kreiss, H. Levin, D. Alexander, W. S. Cain, J. R. Girman, C. J. Hines, P. A. Jensen, D. K. Milton, L. P. Rexroat, K. M. Wallingford, Improving the health of workers in indoor environments: Priority research needs for a national occupational research agenda, *Am. J. Public Health* 92 (2002) 1430–1440.
5. M. Murphy, *Sick Building Syndrome and the Problem of Uncertainty: Environmental Politics, Technoscience, and Women Workers*, Duke University Press, Durham, NC, 2006.
6. P. Wolkoff, C. K. Wilkins, P. A. Clausen, G. D. Nielsen, Organic compounds in office environments: Sensory irritation, odor, measurements and the role of reactive chemistry, *Indoor Air* 16 (2006) 7–19.
7. H. Guo, N. H. Kwok, H. R. Cheng, S. C. Lee, W. T. Hung, Y. S. Li, Formaldehyde and volatile organic compounds in Hong Kong homes: Concentrations and impact factors, *Indoor Air* 19 (2009) 206–217.
8. Z. Lu, C. M. Li, Y. Qiao, Y. Yan, X. Yang, Effect of inhaled formaldehyde on learning and memory of mice, *Indoor Air* 18 (2008) 77–83.
9. B. M. Eklund, S. Burkes, P. Morris, L. Mosconi, Spatial and temporal variability in VOC levels within a commercial retail building, *Indoor Air* 18 (2008) 365–374.
10. B. Wessén, K.-O. Schoeps, Microbial volatile organic compounds: What substances can be found in sick buildings? *Analyst* 121 (1996) 1203–1205.
11. C. G. Granqvist, A. Azens, P. Heszler, L. B. Kish, L. Österlund, Nanomaterials for benign indoor environments: Electrochromics for "smart windows," sensors for air quality, and photo-catalysis for air cleaning, *Solar Energy Mater. Solar Cells* 91 (2007) 355–365.
12. G. Eranna, B. C. Joshi, D. P. Runthala, R. P. Gupta, Oxide materials for development of integrated gas sensors: A comprehensive review, *Crit. Rev. Solid State Mater. Sci.* 29 (2004) 111–188.
13. M. Graf, A. Gurlo, N. Bârsan, U. Weimar, A. Hierlemann, Microfabricated gas sensor systems with sensitive nanocrystalline metal-oxide films, *J. Nanoparticle Res.* 8 (2006) 823–839.
14. N. Barsan, D. Koziej, U. Weimar, Metal oxide-based gas sensor research: How to? *Sensors Actuators B* 121 (2007) 18–35.

15. M. Tiemann, Porous metal oxides as gas sensors, *Chem. Eur. J.* 13 (2007) 8376–8388.
16. N. Barsan, M. Schweizer-Berberich, W. Göpel, Fundamental and practical aspects in the design of nanoscaled SnO_2 gas sensors: A status report, *Fresenius J. Anal. Chem.* 365 (1999) 287–304.
17. S. Pokhrel, C. E. Simion, V. S. Teodorescu, N. Barsan, U. Weimar, Synthesis, mechanism, and gas-sensing application of surfactant tailored tungsten oxide nanostructures, *Adv. Functional Mater.* 19 (2009) 1767–1774.
18. A. S. Zuruzi, N. C. MacDonald, M. Moskovits, A. Kolmakov, Metal oxide "nanosponges" as chemical sensors: Highly sensitive detection of hydrogen using nanosponge titania, *Angew. Chem. Int. Ed.* 46 (2007) 4298–4301.
19. E. H. Espinosa, R. Ionescu, E. Llobet, A. Felten, C. Bittencourt, E. Sotter, Z. Topalian, P. Heszler, C. G. Granqvist, J. J. Pireaux, X. Correig, Highly selective NO_2 gas sensors made of MWCNTs and WO_3 hybrid layers, *J. Electrochem. Soc.* 154 (2007) J141–J149.
20. A. Hoel, L. F. Reyes, S. Saukko, P. Heszler, V. Lantto, C. G. Granqvist, Gas sensing with films of nanocrystalline WO_3 and Pd made by advanced reactive gas deposition, *Sensors Actuators B* 105 (2005) 283–289.
21. P. Heszler, R. Ionescu, E. Llobet, L. F. Reyes, J. M. Smulko, L. B. Kish, C. G. Granqvist, On the selectivity of nanostructured semiconductor gas sensors, *Phys. Stat. Sol. B* 244 (2007) 4331–4335.
22. L. F. Reyes, A. Hoel, S. Saukko, P. Heszler, V. Lantto, C. G. Granqvist, Gas sensor response of pure and activated WO_3 nanoparticle films made by advanced reactive gas deposition, *Sensors Actuators B* 117 (2006) 128–134.
23. R. Ionescu, A. Hoel, C. G. Granqvist, E. Llobet, P. Heszler, Ethanol and H_2S gas detection in air and in reducing and oxidising ambience: Application of pattern recognition to analyse the output from temperature-modulated nanoparticulate WO_3 gas sensors, *Sensors Actuators B* 104 (2005) 124–131.
24. R. Ionescu, A. Hoel, C. G. Granqvist, E. Llobet, P. Heszler, Low-level detection of ethanol and H_2S with temperature-modulated WO_3 nanoparticle gas sensors, *Sensors Actuators B* 104 (2005) 132–139.
25. L. B. Kish, Y. Li, J. L. Solis, W. H. Marlow, R. Vajtai, C. G. Granqvist, V. Lantto, J. M. Smulko, G. Schmera, Detecting harmful gases using fluctuation-enhanced sensing with Taguchi sensors, *IEEE Sensors J.* 5 (2005) 671–676.
26. C.-Y. Lee, C.-M. Chiang, Y.-H. Wang, R.-H. Ma, A self-heating gas sensor with integrated NiO thin-film for formaldehyde detection, *Sensors Actuators B* 122 (2007) 503–510.

27. C. Luyo, R. Ionescu, L. F. Reyes, Z. Topalian, W. Estrada, E. Llobet, C. G. Granqvist, P. Heszler, Gas sensing response of NiO nanoparticle films made by reactive gas deposition, *Sensors Actuators B* 138 (2009) 14–20.

28. A. L. Linsebigler, G. Lu, J. T. Yates, Jr., Photocatalysis on TiO_2 surfaces: Principles, mechanisms, and selected results, *Chem. Rev.* 95 (1995) 735–758.

29. T. L. Thompson, J. T. Yates, Jr., Surface science studies of the photoactivation of TiO_2: New photochemical processes, *Chem. Rev.* 106 (2006) 4428–4453.

30. L. Österlund, Structure-reactivity relationships of anatase and rutile TiO_2 nanocrystals measured by in situ vibrational spectroscopy, in *Solid-State Chemistry and Photocatalysis of Titanium Dioxide*, edited by M. K. Nowotny and J. Nowotny, Trans Tech Publ., Stafa-Zürich, Switzerland, 2010, *Solid State Phenomena*, 162, 203–219.

31. A. Fujishima, K. Hashimoto, T. Watanabe, *TiO_2 Photocatalysis: Fundamentals and Applications*, Bkc, Tokyo, Japan, 1999.

32. A. Fujishima, X. Zhang, D. A. Tryk, TiO_2 photocatalysis and related surface phenomena, *Surf. Sci. Rep.* 63 (2008) 515–582.

33. R. Asahi, T. Morikawa, T. Ohwaki, K. Aoki, Y. Taga, Visible-light photocatalysis in nitrogen-doped titanium oxides, *Science* 293 (2001) 269–271.

34. T. Lindgren, J. M. Mwabora, E. Avendaño, J. Jonsson, A. Hoel, C. G. Granqvist, S.-E. Lindquist, Photoelectrochemical and optical properties of nitrogen doped titanium dioxide films prepared by reactive DC magnetron sputtering, *J. Phys. Chem. B* 107 (2003) 5709–5716.

35. G. Romualdo Torres, T. Lindgren, J. Lu, C. G. Granqvist, S.-E. Lindquist, Photoelectrochemical study of nitrogen-doped titanium dioxide for water oxidation, *J. Phys. Chem.* 108 (2004) 5995–6003.

36. J. M. Mwabora, T. Lindgren, E. Avendaño, T. F. Jaramillo, J. Lu, S.-E. Lindquist, C. G. Granqvist, Structure, composition, and morphology of photoelectrochemically active $TiO_{2-x}N_x$ thin films deposited by reactive DC magnetron sputtering, *J. Phys. Chem. B* 108 (2004) 20193–20198.

37. K. Hashimoto, H. Irie, A. Fujishima, TiO_2 photocatalysis: A historical overview and future prospects, *Jpn. J. Appl. Phys.* 44 (2005) 8269–8285.

38. P. Ngaotrakanwiwat, S. Saitoh, Y. Ohko, T. Tatsuma, A. Fujishima, TiO_2-phosphotungstic acid photocatalyst systems with an energy storage ability, *J. Electrochem. Soc.* 150 (2003) A1405–A1407.

39. Y. W. Sakai, K. Obata, K. Hashimoto, H. Irie, Enhancement of visible light-induced hydrophilicity on nitrogen and sulfur-doped TiO_2 thin films, *Vacuum* 83 (2009) 683–687.

40. X.-T. Zhang, O. Sato, M. Taguchi, Y. Einaga, T. Murakami, A. Fujishima, Self-cleaning particle coating with antireflection properties, *Chem. Mater.* 17 (2005) 696–700.

41. X.-T. Zhang, A. Fujishima, M. Jin, A. V. Emeline, T. Murakami, Double-layered TiO_2-SiO_2 nanostructured films with self-cleaning and antireflective properties, *J. Phys. Chem. B* 110 (2006) 25142–25148.

42. M. Houmard, D. Riassetto, F. Roussel, A. Bourgeois, G. Berthomé, J. C. Joud, M. Langlet, Enhanced persistence of natural super-hydrophilicity in TiO_2-SiO_2 composite thin films deposited via sol-gel route, *Surface Sci.* 602 (2008) 3364–3374.

43. A. Roos, A. Werner, Formation of external condensation on low U-value windows with and without hydrophilic coatings, in *Proceedings 7th International Conference on Coatings on Glass and Plastics*, edited by C. I. M. A. Spee, J. L. B. de Groot, H. A. Meinema, J. Pütz, TNO Science and Industry, Eindhoven, the Netherlands, 2008; pp. 295–298.

44. Y. Takata, S. Hidaka, J. M. Cao, T. Nakamura, H. Yamamoto, M. Masuda, T. Ito, Effect of surface wettability on boiling and evaporation, *Energy* 30 (2005) 209–220.

45. T. N. Rao, D. A. Tryk, A. Fujishima, Applications of TiO_2 photocatalysis, in *Encyclopedia of Electrochemistry: Semiconductor Electrodes and Photoelectrochemistry*, edited by A. J. Bard, M. Stratmann, S. Licht, Wiley-VCH, Weinheim, Germany, 2002; Vol. 6, pp. 536–561.

46. C. Yang, U. Tartaglino, B. N. J. Persson, Influence of surface roughness on superhydrophobicity, *Phys. Rev. Lett.* 97 (2006) 116103 1–4.

47. A. Marmur, The Lotus effect: Superhydrophobicity and metastability, *Langmuir* 20 (2004) 3517–3519.

48. Y. Lai, C. Lin, J. Huang, H. Zhuang, L. Sun, T. Nguyen, Markedly controllable adhesion of superhydrophobic spongelike nanostructure TiO_2 films, *Langmuir* 24 (2008) 3867–3873.

49. L. Österlund (2009), private communication.

50. C.G. Granqvist, Materials for good day-lighting and clean air: New vistas in electrochromism and photocatalysis, in *Society of Vacuum Coaters 50th Annual Technical Conference Proceedings*, Society of Vacuum Coaters, Albuquerque, NM, 2007; pp. 561–567; also in *Spring Bulletin*, Society of Vacuum Coaters, Albuquerque, NM, 2008; pp. 48–53.

51. A. T. Hodgson, H. Destaillats, D. P. Sullivan, W. J. Fisk, Performance of ultraviolet photocatalytic oxidation for indoor air cleaning applications, *Indoor Air* 17 (2007) 305–316.

52. Z. Bittnar, P. M. J. Bartos, J. Němeček, V. Šmilauer, J. Zeman, Eds., *Nanotechnology in Construction: Proceedings of the NICOM3*, Springer, Berlin, Germany, 2009.

53. L. Cassar, Photocatalysis of cementitious materials: Clean buildings and clean air, *MRS Bull.* (May) (2004) 328–331.

54. B. Ruot, A. Plassais, F. Olive, L. Guillot, L. Bonafous, TiO_2-containing cement pastes and mortars: Measurements of the photocatalytic efficiency using a rhodamine B-based colourimetric test, *Solar Energy* 83 (2009) 1794–1801.

55. J. Chen, C.-S. Poon, Photocatalytic activity of titanium dioxide modified concrete materials: Influence of utilizing recycled glass cullets as aggregates, *J. Environ. Management* 90 (2009) 3436–3442.

56. M. M. Ballari, M. Hunger, G. Hüsken, H. J. H. Brouwers, Heterogeneous photocatalysis applied to concrete pavement for air remediation, in *Nanotechnology in Construction: Proceedings of the NICOM3*, edited by Z. Bittnar, P. M. J. Bartos, J. Němeček, V. Šmilauer, J. Zeman, Springer, Berlin, Germany, 2009; pp.409–414.

57. http://gfp.lbl.gov/.

58. J. C. Harper, A. F. El Sahrigi, Thermal conductivities of gas-filled porous solids, *Ind. Engr. Chem. Fundamentals* 3 (1964) 318–324.

59. R. Baetens, B. P. Jelle, J. V. Thue, M. J. Tenpierik, S. Grynning, S. Uvsløkk, A. Gustavsen, Vacuum insulation panels for building applications: A review and beyond, *Energy Buildings* 42 (2010) 147–172.

60. H. Cauberg, M. Tenpierik, Vacuümisolatiepanelen en andere noviteiten, in *Praktijkhandboek Duurzaam Bouwen*, edited by D. W. Dicke, E. M. Haas, Weka Publishers, Amsterdam, the Netherlands; pp. 4.1 VAC 1–19.

61. J. Fricke, Ed., *Aerogels*, Springer Proceedings in Physics, Vol. 6, Springer, Berlin, Germany, 1986.

62. M. Reim, W. Körner, J. Manara, S. Korder, M. Arduini-Schuster, H.-P. Ebert, J. Fricke, Silica aerogel granulate material for thermal insulation and daylighting, *Solar Energy* 79 (2005) 131–139.

63. A. Soleimani Dorcheh, M. H. Abbasi, Silica aerogel: Synthesis, properties and characterization, *J. Mater. Proc. Technol.* 199 (2008) 10–26.

64. P. H. Tewari, A. J. Hunt, J. G. Lieber, K. Lofftus, Microstructural properties of transparent silica aerogels, in *Aerogels*, edited by J. Fricke, *Springer Proceedings in Physics*, Springer, Berlin, Germany, 1986; Vol. 6, pp. 142–147.

65. C. G. Granqvist, Energy-efficient windows: Present and forthcoming technology, in *Materials Science for Solar Energy Conversion Systems*, edited by C. G. Granqvist, Pergamon, Oxford, U.K., 1991, pp.106–167.

66. G. Marsh, RE storage, the missing link, *Refocus* 3(2) (2002) 38–42.

67. Y. Zhang, H. Feng, X. Wu, L. Wang, A. Zhang, T. Xia, H. Dong, X. Li, L. Zhang, Progress of electrochemical capacitor electrode materials: A review, *Int. J. Hydrogen Energy* 34 (2009) 4889–4899.

68. D. Sadoway, Liquid-battery: Research in energy storage (2009); http://mit.edu/dsadoway/www/.

69. R. Kötz, M. Carlen, Principles and applications of electrochemical capacitors, *Electrochim. Acta* 45 (2000) 2483–2498.

70. P. Simon, A. Burke, Nanostructured carbons: Double-layer capacitance and more, *Electrochem. Soc. Interface* (Spring) (2008) 38–43.

71. M. B. Cortie, A. I. Maaroof, G. B. Smith, Electrochemical capacitance of mesoporous gold, *Gold Bull.* 38 (2005) 14–22.

72. A. K. Geim, K. S. Novoselov, The rise of graphene, *Nature Mater.* 6 (2007) 183–191.

73. M. D. Stoller, S. Park, Y. Zhu, J. An, R.S. Ruoff, Graphene-based ultracapacitors, *Nano Lett.* 8 (2008) 3498–3502.

74. A. S. Aricò, P. Bruce, B. Scrosati, J.-M. Tarascon, W. van Schalkwijk, Nanostructured materials for advanced energy conversion and storage devices, *Nature Mater.* 4 (2005) 366–377.

75. A. Fedorková, H.-D.Wiemhöfer, R. Oriňáková, A. Oriňák, M. C. Stan, M. Winter, D. Kaniansky, A. N. Alejos, Improved lithium exchange at LiFePO$_4$ cathode particles by coating with composite polypyrrole–polyethylene glycol layers, *J. Solid State Electrochem.* 13 (2009) 1867–1872.

76. B. Kang, G. Ceder, Battery materials for ultrafast charging and discharging, *Nature* 458 (2009) 190–193.

77. G. Nyström, A. Razak, M. Strømme, L. Nyholm, A. Mihranyan, Ultrafast all-polymer paper-based batteries, *Nano Lett.* 9 (2009) 3635–3639.

78. Y. J. Lee, H. Yi, W.-J. Kim, K. Kang, D. S. Yun, M. S. Strano, G. Ceder, A. M. Belcher, Fabricating genetically engineered high-power lithium-ion batteries using multiple virus genes, *Science* 324 (2009) 1051–1055.

CHAPTER 9

Conclusions
Nanotechnologies for a Sustainable Future

9.1 ENERGY AND THE FUTURE

A transition is under way unlike any other in human history. We are faced with an urgent agenda: first to stabilize and then to rejuvenate the global environment, while coping with resource scarcity and a growing population. Many factors will determine how, when, and in what shape earth and humanity will emerge from this current unsettling period. Will technological developments provide the means for dealing with the environmental forces unleashed by our past and continuing impacts, and eventually for mitigating them? Will we have a sustainable resource base for the needs of this new world?

Let's be optimistic and assume humanity takes on the challenge and eventually emerges into a more stable and less uncertain future. In this concluding chapter we will assume that many of the technological opportunities raised in this book, and those that will develop in the future from emergent nanoscience and other scientific advances, will shape that future world. This is a world with affordable technologies, in harmony with the environment and not degrading it, available to and used by its citizens. One of the benefits will be an improved quality of life for the world's poorest, which in turn will help stabilize the global population. The challenges to attaining this future world are international, so global implementation of the solutions will be central to success.

What will this new world look like? Perhaps one of the most visible and obvious features will be more buildings looking, in principle, like the ones in Figure 9.1 [1]. We will discuss such buildings at the end of

(a)   (b)

FIGURE 9.1   External (a) and internal (b) view of The Eden Project located in Cornwall, U.K. The membrane structure consists of ETFE foil. It is discussed further below.

this chapter. But first we would like to take a broader look at how energy and nanotechnologies may interact in the future world.

The technologies needed for a sustainable world—as we saw in Chapters 2 and 3—will emerge from two key scientific advancements: a wider and deeper understanding of nature's energy flows and the integration of that understanding with knowledge of materials' responses to external stimuli. Tuning materials responses to the environment will efficiently supply or diminish our needs of energy and other resources. This tuning, or harmonizing, is at the heart of nature's natural selection and has been going on for billions of years, ever since life emerged on earth, and has led to ever more sophisticated nanostructures, from cells to seashells. We will both mimic and surpass nature's achievements in nanotechnology in the quest for better materials. It is ironic that, to restore balance in nature, we will have to outdo in decades nature's own efforts in "natural nanotechnology" over billions of years.

Energy use and energy supply are the core problems. It is thus useful at this point—as we approach the end of this book—to have a fresh look at "energy" and "energy technology" from a fundamental perspective; this is elaborated in Box 9.1. In one sense all energy is stored, and energy technology is about tapping safely into these various "stores." Every time we drive a car, turn on a computer, or sow a wheat crop, we draw on an energy resource to carry out the desired task. In almost all cases, tasks require energy in some form other than that in the "store," so the energy must be converted.

Daylight for vision is an interesting exception to the generalization about energy conversion. One reason for its high value is that it does not require any conversion until it reaches our retinas, just occasionally some redirection, as discussed in Chapter 5. Plant growth based on solar energy is another example of a process in which energy conversion is not

BOX 9.1   ENERGY IN TIME AND SPACE

From a scientific perspective, the universe is a bundle of energy that is evolving and changing forms. To fully describe the changes that occur requires the introduction of an additional concept: "geometry" or, in simple language, time and space. We know that the universe is expanding, and hence its energy density is falling. Changes in geometry often accompany changes in energy forms, but geometric restrictions limit the extent of these changes, and time is a meaningless concept in the absence of evolution in energy systems.

Technology is in essence human intervention in the process of change, in order to force it in directions aligned to our needs in a well-controlled manner. In the quantum world even the "vacuum"— which is an integral part of the universe—contains energy. Most energy is not even accounted for today (2010) and is ascribed to mysterious entities called "dark matter" and "dark energy."

All energy was present and localized at the time of the Big Bang some 13.7 billion years ago, but its forms have evolved continuously ever since, from

- Elementary particles (which includes photons with energies across the radiation spectrum), probably entities responsible for gravity, and antiparticles, which can annihilate particles and release energy,
- To protons, neutrons, nuclei, atoms, and molecules, and finally
- To liquids and solids.

The types of motion that can occur inside this "universal soup," within the confines of time and space, include thermal energy, which is, in effect, an outcome of random motion in many-component systems. Energy technology is about tapping into one of these "stored" forms of energy so as to convert it into a form needed for some function.

It is interesting that, as energy technology has progressed, our perspectives on time and space have evolved in tandem. Think of typical villagers in the Middle Ages. They might have spent their entire lives within a few square kilometers; their typical work tasks or journeys involved periods of hours, days, or months and used human or animal power, or fire. But now, in the 21st century, humans can travel almost anywhere in the world in a day. Microsecond and nanosecond processes in computer and electronic chips underpin much manufacturing, and the advanced machines they control allow production volumes with high engineering precision at rates unimaginable to preindustrial people.

Our brains adapted long ago to function at the speed needed to collect, process, and store information simultaneously, as well as direct actions as required by the many incoming signals from our senses. Analogously, sensing is an integral part of advanced technology and increasingly dictates the way we control our use of energy. This control will help save a great deal of energy.

demanded, although this is a complex case as other forms of energy— such as organic nutrients for associated biochemical reactions—are also needed. But these two examples are exceptions, and in most cases energy must be transformed into a different form before it can be used.

As discussed in Chapter 2, thermodynamics prescribes that conversion from one form of energy to another leads to a little or a lot of the initial resource being lost. Energy technology is about accessing the "store" with practical and controllable conversion processes and, as we are well aware of today, a minimum of associated waste.

Some energy stores such as hydrogen and deuterium nuclei (which power the sun) might be accessed on earth via nuclear fusion, but a controlled conversion process is yet to be perfected. The energy stored in some heavy nuclei such as uranium, on the other hand, is available because basic nuclear science showed how to both initiate and control nuclear fission. Controlled mini-explosions power our cars. Other basic "stores" of energy may one day be accessible with processes we have yet to invent or stumble upon. Whatever these processes turn out to be, control and safety issues will define them. Releasing and storing energy via any energy-related technology can be a very dangerous business and a source of environmental disasters. Managing risk is integral to energy use. To mention one obvious example, vast volumes of highly inflammable petroleum are handled every day by millions of people with relatively few direct problems. An attraction of most renewables is that they are comparatively benign and safe.

9.2 NEW TECHNOLOGIES AND GROWING UPTAKE OF PROVEN TECHNOLOGIES

Some nanotechnologies and nanomaterials are already mature and yield close to optimum performance. Among these current products one finds, for example, multilayer thin films for solar control glazing and sputtered nanocomposite layers for solar selective absorbers. This does not mean these technologies will go unchallenged in the future by lower-cost alternatives, use of different materials and production methods, added functionality, or entirely new approaches. Examples discussed in this book that have such prospects to challenge current technologies include nanoparticle-doped glazing for lower cost and chromogenic glazing for added functionality. These last two are examples of technologies that are just starting on their path to maturity.

Future advances in the technologies covered in this book will emerge from new optical, thermal, electrical, and materials science associated with nanostructures. Science tells what is feasible, but converting these possibilities into real technological developments is another matter. Choices and options abound, many with strong and vociferous advocates, and there is a sense of urgency. But investment dollars are finite and need to be spent well. Dealing wisely with all of these issues requires enhanced, and properly focused, R&D programs in parallel with immediate efforts to commercialize and implement the best of current alternatives. Furthermore, the new policy initiatives must be flexible and open to change as R&D and commercialization outcomes unfold. New policies are needed to facilitate the coming transitions in areas such as energy supply, urban planning, manufacturing, transport, water, and agriculture. In addition, and importantly, changes reflecting the transitions in technology must be made in education and vocational training related to the built environment.

Hype and disappointment are natural elements in a world of change, so a word of caution is in order. Many projections and R&D breakthroughs never come to fruition, or technologies may take very much longer than anticipated to yield benefits. This is always the case and serves as a filtering process that ultimately helps achieve the best outcomes. Well-known examples in energy-related work—where high initial hopes have given way to long, hard, unrewarded slogging—include controlled fusion, room-temperature superconductivity, high-efficiency thermoelectrics, and TiO_2 that absorbs visible light but also retains good electrical properties. Still, nanotechnology might someday lead to results even in such stalled cases.

9.3 TOWARD A "NANOWORLD"

"Nano" is upon us. It is used as a prefix in "nanobuilding," "nanohouse," "nanooffice," and "nanocar." These terms turn up at least tens of thousands hits with a Google search today (2010). But what do these terms mean, and what are their attributes? Will they lead to the "green" world we so earnestly need? The terms are certainly not about housing estates built on the head of a pin. They are about buildings and vehicles whose structural materials, surface coatings, and designs fully exploit the unique capabilities that nanostructured materials provide. These nanomaterials-based buildings will function in ways that will change how we live, work, and play because the materials allow controlled and optimized properties and wider design options. These structures can, if so desired, look much the same as today's buildings, though they will function differently. But radically different architectural expressions and styles will also be possible. The materials used in construction of these nano-buildings will be produced at commodity scales and largely be used in megacities. They will have little embodied energy (also called embedded energy) and so minimize the drawdown on stored resources of minerals and fossil fuels. During production they will add benign chemicals and gases to the environment. Heat and waste gases used in materials processing and power production will be increasingly captured and reused with the aid of nanostructured surfaces or materials. Waste heat can be channeled into space via the sky window, using optimized sky cooling devices.

Nanoproduction techniques such as self-assembly, nanoimprinting, controlled nanoetching, and nanocomposite formation will become everyday processes in manufacturing. Engineering and structural materials will be lighter and stronger, and some will use growth techniques and structures borrowed from the bioworld. How far one can go in replacing steel, aluminum, glass, and concrete remains to be seen, but the initial energy intensity in buildings, roads, and vehicles will decrease.

Can we be more specific about the coming nanoworld? It is futile to try to predict the future, and attempts to do so have often, and for good reason, been met with ridicule within a few years or decades. So we will not predict! However, when we imagine a scenario in which the world becomes more environmentally friendly and sustainable, that scenario includes certain essential elements related to energy and nanotechnology. Thus, in our scenario:

- *Energy savings will become more important in general (1).* The best energy is the energy we do not need, and techniques and practices for energy conservation are coming ever more to the forefront. The major driving force today (2010) seems to be an economic one, and several studies have delineated pathways to a low-carbon economy in which energy savings in the built environment are shown to reconcile a large greenhouse gas abatement potential with a large and negative cost for this abatement. In other words, there are technologies capable of saving energy and money at the same time [2]. Particularly benign technologies include the introduction of LED lighting, improvements of the building envelope by insulation, and better residential appliances and electronics. As we have seen, these are technologies with important nanoaspects. The economic driving force also acts more indirectly. For example, a recent study of market transactions in the United States [3] showed that "green" buildings command significantly higher rental rates and selling prices than comparable nongreen buildings.

- *More focus will be directed toward the built environment (2).* It is customary to speak of energy for buildings, industry, and transport. The portion used for buildings is huge and amounts to some 30 to 40% of the primary energy worldwide [4]. The buildings-related use of electricity varies among countries and can amount to more than 70%. The potential for substantial decreases—which has been neglected for many years—is larger for buildings than for the other sectors, both in percent and in absolute numbers. Savings of 70% by the year 2030 and perhaps 90% long term have been put forward in a recent influential study for the United States [5]. Obviously, these dramatic changes require new technologies.

- *Priority will be given to solutions with little embodied energy (3).* Clearly, it is important to have materials and devices which in themselves do not have large energy-related footprints. This means doing a lot with a little, which brings us to the heart of "nano." Materials that are nanosized in one dimension (thin films), two dimensions (nanowires), and three dimensions (nanoparticles) are essential. The small amounts that are used allow the application of nanomaterials based on elements that are comparatively rare in the earth's crust, so the options are much greater than for traditional materials.

- *Thin-film technologies will play an essential role (4).* A block of, say, aluminum that is so small that it can easily be carried by hand can be evaporated or sputter-deposited so that it covers a square kilometer. In the process, the material also changes its surface properties from low to high reflectance and from high to low emittance. The practical techniques for doing this are available today, and higher deposition rates, even for complex and temperature-sensitive materials, are under rapid development. Thus nano-thin surface coatings will be even more affordable and important in the future than they are presently.

- *Pigments and nanoparticles will also be essential (5).* Pigments, which can be applied with well-established painting technologies, are part of our future scenario. Some can scatter solar energy and prevent overheating, even if just moderately colored. Other pigments can be used for efficient solar absorption and hot fluid production. Nanoparticles can do even more, absorbing and emitting radiation in well-defined wavelength ranges as discussed at various places in this book. Carbon nanostructures (nanotubes and graphene sheaths) may also lead to technology breakthroughs, though these are yet to be seen.

- *Foil- and fabric-based solutions and membrane architecture will become more widespread (6).* Thin films, as well as pigments and nanoparticles, need a backing or a medium to be dispersed in. Foils based on PET are particularly important as thin-film substrates, and coextruded multilayer structures can have high reflectance in well-defined wavelength ranges. ETFE is of particular interest in our scenario since it has a documented durability over several decades of outdoors exposure, which opens possibilities for large-scale membrane structures [6,7]. Durable foils with high transmittance in the $8 < \lambda < 13$ µm sky window are needed for cooling applications. Foils can be made and coated by continuous processes, and can then be cut to size for specific applications, which allows low-cost manufacturing.

- *Sky cooling will take its proper place (7).* Solar energy warms the earth, and sky cooling creates a comfortable temperature. Solar energy is already widely used today for daylighting, heating via solar collectors, and electricity generating via solar cells. But the other side of the equation, that for cooling, has received embarrassingly little attention. The energy flows toward the clear (night) sky can be large; these flows represent an untapped resource that could be used in sky cooling devices, as discussed at length in Chapter 7. Such devices could add significant efficiency gains to various solar power generators. Some of the most exciting cooling devices have nanoparticles with special vibrational properties. Global warming will accentuate the need for cooling, which makes sky cooling even more interesting.

- *New light sources will take over (8).* Less than half the energy used by today's compact fluorescent lamps is needed to supply the same lighting levels with LEDs, OLEDs, and other novel sources, all of which contain some nanofeatures. The improvement is even greater if we compare standard incandescent lights with the new sources. The new lamps will last up to 10 years, will not need mercury vapor, and will be ideally suited for integration with solar cells. In the future, it will also be possible to control these new lamps with sensor-based dimming devices, so that they give the ideal amount of light in combination with daylighting. Thus, a room lit with these lamps will automatically adjust to provide proper lighting of various types, from mood lighting to lighting for computer work. Novel light fixture materials and light distribution systems will help maximize out-put, eliminate glare and visual hotspots and provide unique, diverse, and variable lighting design options.

- *Chromogenic solutions will become better established (9).* Ambient conditions are always changing, and buildings have to adjust to make the most of nature's energy flows. Materials that change with temperature or irradiation level are of much interest and can be comprised of films, pigments, and nanoparticles. Devices that allow properties such as the transmittance of visible light and solar radiation to be adjusted—not only to external conditions but to human desires and needs—have even wider applicability, especially if they are foil based.

- *Natural light and visual contact will be more appreciated (10).* Natural light and views from a room are amenities that technologists have shamefully neglected in the past. Now we know how important they are for human well-being and for learning and work productivity. Clearly, there are economic incentives that can be harnessed to provide better indoor environments, yet still not much is happening. In our scenario, new nano-based approaches to mirror light pipes and fiber-based light guides will lead to new vistas in daylighting, and such devices will operate in concert with chromogenic fenestration to provide optimized lighting solutions.

- *Self-cleaning surfaces and air-cleaning materials will be common (11).* Today, we have a number of TiO_2-based nanoparticulate surfaces for a variety of cleaning applications. Although we know that the nanofeatures are essential, we do not yet know how they function in detail, so we do not know their performance limits. In the future we envision, these materials are further developed and applied to allow cleaning of the air in built environments and to reduce the need for chemical cleaning of building surfaces.

- *Sensors will be an integral part in a building's energy system (12).* Sensors are used in modern buildings today to control temperature, light level, ventilation, and many other functions. Still much remains to be done in the future, both on the technical side to make sure that the various subsystems work well and in concert, and on the human side to guarantee that the person—not the machine—is in charge.

- *Safety issues will be researched and addressed (13).* Nanoparticles can be severe health hazards, as we know from asbestos. We do not know enough about today's nanomaterials. The nanoparticles' influence may be indirect, and photocatalytic breakdown of undesired "smelly" molecules may lead to harmful intermediates. More studies about environmental and health impacts will be needed as the technologies develop to ensure that nanosolutions do not lead to unforeseen problems.

- *Multi-functionality will flourish (14).* A photovoltaic panel can also generate hot water. A sky cooling device can also produce electricity. In our scenario, clever combined devices that serve multiple roles will be common in built environments. Windows will control the light level so that it is well adapted to the users of a room. These windows will be almost as thermally insulated as the adjacent walls; they will provide adequate solar heating or prevent excessive heating; they will serve as photocatalytic cleaners of indoor and outdoor air; they will be superhydrophilic and haze free ... We are limited by only imagination and fantasy.

- *New ways with water will emerge, in greenhouses and elsewhere (15).* Present ways to source, use, and reuse water demand growing amounts of energy. In our scenario, the processes will be increasingly renewable. Salty and brackish water, polluted water, and atmospheric water will be put to use after appropriate capture and treatment with nanostructured filters, UV photocatalysts, and solar energy. Cooling systems will use much less water than now. Greenhouse roofs may be doped with nanoparticles for controlled light inlet, and may use the remaining solar energy to extract water for its plants from seawater [8]. Sky cooling may produce "dew-rain" for drip irrigation [9].

- *Aesthetics and function will merge (16).* Think of a solar collector or a solar cell panel on a rooftop today. More often than not it is ugly and renders a cold machine look to an otherwise beautiful house. Color-matched solar collectors and solar cells will no doubt widen the acceptability of these environmentally benign devices, and this change can be created with only a minimum loss in efficiency. And why must float glass be atomically smooth on both of its surfaces? Having one surface modestly structured can give a "live" impression—the one we lost when drawn glass vanished.

- *Nature will be regained via biomimetics (17).* Nature has "experimented" for billions of years and created wondrous structures. Remember the butterfly wing in Figure 1.3! Some of the underlying nanobiology may be hard or impossible to mimic. But we can learn from nature. Thus, in this book we have mentioned microbially fabricated, solar-absorbing surfaces and electrical batteries based on algae and viruses. We can envision many more developments along such lines.

A world that includes these elements will look different from the one of today. Imagine a bird's eye view of future town and country or, more realistically, the view from an aircraft window. Cities and rural areas will not appear quite the same as they do now. Rooftops and roads will have man-made high-albedo surfaces—largely based on paints and on bitumenous and cementitious materials of novel kinds—to mitigate global and local warming. At suitable locations, perhaps artificial salt pans and reflective polymers will be used for the same purpose. Renewable energy sources may be building integrated, as long as they do not give too much ambient heating. Hybrid and all-electric cars will move on roads that are less dark than today. Power plant chillers will be rare; instead there will be solar thermal power generators surrounded by reflector fields. Cities will have less smog. New forests and new crops will emerge, as well as advanced algae farms, aided by nano-based bioscientific techniques at cellular and protein level. These farms may be at select seaside locations for sequestering CO_2 and supplying biofuels, and light-spreading methods—such as those outlined in this book—can help the algae farms prosper.

Then, finally, let's look at some of the buildings in this future world. Many of them appear much the same as today. A building is not forever, but if well built it lasts for many decades or for centuries. It may be improved and refurbished, though, through implementation of the scenario elements outlined above. But there will also be new buildings, some probably looking like the one at the beginning of this chapter, membrane structures based on ETFE [6,7] capable of encapsulating many of the same elements. In the future, these membranes will be able to regulate the "indoor climate" by controlling flows of light and solar energy via chromogenic technology, preferably electrochromic foils, such as the one discussed in Box 4.6 [10], which offer the greatest possibilities for finely tuned control. Membrane structures can be huge enough to contain entire buildings, as seen in Figure 9.1b, so they serve as climate shells, blurring the distinction between "indoors" and "outdoors."

In fact, membranes are already well known in architecture [11,12] and have been explored by visionary building engineers and architects such as Buckminster Fuller [13] and Frei Otto [14] for more than 50 years. And Buckminster Fuller, the engineer, is also the person whose geodesic domes are memorialized in the C_{60} buckminsterfullerene molecule that we discussed briefly in Section 4.5.2. It is interesting to remember that simpler membranes in the form of tents, teepees, and yurts have been used by human beings for thousands of years. So, here we may see a circle closing, with the buildings of future "green architecture" sharing some features with the portable dwellings of our ancestors. Yet these buildings could also have the strength of today's structures at a fraction of their weight, and with the modularity and easy scalability needed for large cities. With visionary changes in building practice, such structures could become the basis for green and healthy megacities for future generations.

So much from things so little! The powers of nanoscience and nanotechnology point to a brighter future for the children and grandchildren of our world. Transforming the future of the three great entities on earth—the atmosphere, the oceans, and the land—by starting from an understanding of new structures a few nanometers in size is not fanciful; it is the way ahead. This book, we hope, will inspire and aid in this grand task.

REFERENCES

1. The Eden Project (2009); http://upload.wikimedia.org/wikipedia/en/7/75/Eden_project.JPG.
2. McKinsey&Company, Pathways to a Low-Carbon Economy: Version 2 of the Global Greenhouse Gas Abatement Cost Curve (2009); http://solutions.mckinsey.com/climatedesk/cms/getfile.aspx?uid=39a1aa7f-f342-440a-a187-b89e6e907377&fp=design%2fClimate+Change+Center%2fPathwayTo LowCarbonEconomy_FullReportA.pdf&ru=default%2fen-us%2fhidden%2fduplicatedownload.aspx.
3. P. Eichholtz, N. Kok, J. M. Quigley, Doing well by going good? Green office buildings, Center for the Study of Energy Markets, Berkeley, CA, Working Paper CSEM WP-192 (2009); http://www.ucei.berkeley.edu/PDF/csemwp192.pdf.
4. UNEP, *Buildings and Climate Change: Status, Challenges and Opportunities*, United Nations Environment Programme, Paris, France, 2007.
5. B. Richter, D. Goldston, G. Crabtree, L. Glicksman, D. Goldstein, D. Greene, D. Kammen, M. Levine, M. Lubell, M. Sawitz, D. Sperling, F. Schlachter, J. Scofield, D. Dawson, How America can look within to achieve energy security and reduce global warming, *Rev. Mod. Phys.* 80 (2008) S1–S107; http://www.aps.org/energyefficiencyreport/.
6. J. E. Fernández, Materials for aesthetic, energy-efficient, and self-diagnostic buildings, *Science* 315 (2007) 1807–1815.
7. A. LeCuyer, *ETFE: Technology and Design*, Birkhäuser, Basel, Switzerland, 2008.
8. K. H. Strauch, C. von Zabeltitz, Closed system greenhouses with integrated solar desalination for arid regions, in *Energy Conservation and Solar Energy Utilization in Horticultural Engineering*, edited by K. V. Garzoli, ISHC Acta Hort. Vol. 257, pp. 115–126, 1989; http://www.actahort.org/books/257/257_14.htm.

9. W. E. Alnaser, A. Barakat, Use of condensed water vapour from the atmosphere for irrigation in Bahrain, *Appl. Energy* 65 (2000) 3–18.

10. ChromoGenics, www.chromogenics.com, 2009.

11. K. Ishii, Ed., *Membrane Designs and Structures in the World*, Shinkenchikusa-sha, Tokyo, Japan, 1999.

12. K.-M. Koch, Ed., *Membrane Structures*, Prestel, Munich, Germany, 2004.

13. M. J. Gorman, *Buckminster Fuller: Designing for Mobility*, Skira, Milan, Italy, 2005.

14. W. Nerdinger, Ed., *Frei Otto: Complete Works*, Birkhäuser, Basel, Switzerland, 2005.

Appendix 1:
Thin-Film Deposition

Technologies for making thin films are of great importance for most green nanotechnologies. As discussed extensively in this book, such films are used to absorb solar energy for heating purposes, to transmit daylight through windows with superior thermal insulation, to generate electricity efficiently in solar cells which require a minimum of energy to produce, and for a multitude of other applications. The films may be metallic, semiconducting, or dielectric and deposited onto plates of metal, plastic, or glass and onto foils of metal or plastic. Film thicknesses are typically between 10 nm and 10 μm.

Thin-film science and technology are huge fields and are of very great importance for almost all modern technologies. Not surprisingly there are numerous books and tutorial texts on the subject. We therefore keep the discussion brief below and refer the reader to the literature for details [1–8]. In the following text we give a brief survey of the most important technologies and then present some details of a specific technology of particular relevance known as sputtering. Nano-aspects—and there are many of these [9]—are the focus throughout our discussion.

A1.1 OVERVIEW OF MAJOR THIN-FILM TECHNOLOGIES

Table A1.1 summarizes the most important thin-film technologies. They are classified according to the depositing species being atomistic (or molecular), particulate, or in bulk form, or whether the surface of a material is modified in order to produce a layer with distinctly different properties. Atomistic deposition is most commonly used for the technologies discussed in this book.

Evaporation is a well-known technique for making thin films and is in constant use in research laboratories all over the world. It is also in widespread use industrially, mainly for metallizing. Here, the raw material of the film is heated in vacuum so that a vapor comprised of atoms or molecules transfers material to the substrate at a sufficient rate [10,11]. The energy of the impinging species is typically a fraction of an electron volt. The heating can be produced by drawing current through a

TABLE A1.1 Survey of Thin-Film Deposition Technologies

Atomistic Deposition	Particulate Deposition	Bulk Coating	Surface Modification
Vacuum environment • Evaporation • Molecular beam epitaxy • Ion beam deposition Plasma environment • Sputter deposition • Ion plating • Plasma polymerization • Glow discharge deposition Electrolytic environment • Electroplating • Electroless deposition Chemical vapor environment • Chemical vapor deposition • Spray pyrolysis Liquid phase epitaxy	Thermal spraying • Plasma spraying • Flame spraying • Detonation gun Fusion coating • Enameling • Electrophoresis	Wetting processes • Printing • Dip coating • Spin coating Printing Cladding • Explosive • Roll-binding Weld coating	Chemical conversion • Anodic oxidation • Nitridation Leaching Thermal surface treatment Ion implantation Laser glazing

resistive coil or boat in contact with the material to be evaporated or by themionic emission from a wire and focusing of the electron beam onto the material to be evaporated from a water-cooled "electron gun." The latter technique is referred to as e-beam evaporation.

Sputter deposition is very generally used to make uniform coatings on glass, polymers, metals, etc. Essentially, a plasma is set up in a low pressure of inert and/or reactive gases, and energetic ions in the plasma dislodge material from a solid plate or cylinder of the raw material of the film (known as the target) and deposit these atoms as a uniform film on an adjacent surface (the substrate) [12–16]. The sputter plasma can be inert—typically consisting of Ar ions, in which case the target and the thin film have the same composition—or it can be reactive and contain, for example, oxygen so that an oxide film can be formed by sputtering from a metallic target. The plasma is normally confined to the target area by magnets, which is called magnetron sputtering. The deposition species typically have energies of some electron volts, which are large enough to remove contaminants from the substrate. This removal usually leads to good adherence between substrate and film, which is an asset for sputtering when compared with evaporation as a thin-film technology.

Figure A1.1(a) shows the principle for coating sheet glass by this technique; the panes are transported in and out of deposition chambers via load locks so that the process is continuous [17]. Metal-based films produced in this manner were discussed in detail in Section 4.3. Sputtering and evaporation are often referred to jointly as *physical vapor deposition* or PVD.

Other technologies can be applied without recourse to vacuum. For example, coatings can be obtained by dipping a substrate in a solution containing the species to be used in the film, withdrawing at a controlled speed, and heating to remove volatile components in the solution; an alternative technique suitable for coating small objects is to apply the solution in the form of drops, then spin the substrate at a controlled speed in order to make an even coating, and finally heat treat. In either case the deposition can be repeated in order to make a thicker film. As an alternative to dipping and spinning, the chemical solution can be applied by spraying. Film creation via dipping, spinning, and spraying is often referred to as *sol-gel deposition* [18,19]. Figure A1.2 illustrates a cross section of a three-layer coating made by dip coating [20]. It has two layers of Ni particles in Al_2O_3, with different compositions, and a top layer of SiO_2. This particular coating was employed for highly efficient conversion of solar energy into heat; it is discussed further in Section 6.1.

Chemical vapor deposition (CVD) uses heat to decompose the vapor of a "precursor" chemical to make a thin film of a desired composition [21,22]. This technique can be made more efficient by combining it with

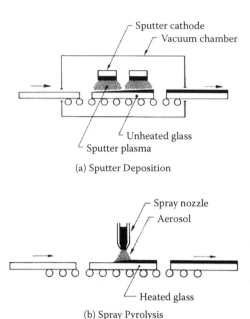

(a) Sputter Deposition

(b) Spray Pyrolysis

FIGURE A1.1　Principles for sputter deposition (a) and spray pyrolysis (b) to coat surfaces of glass transported as indicated by the horizontal arrows. (From C. G. Granqvist, in *Materials Science for Solar Energy Conversion Systems,* edited by C. G. Granqvist, Pergamon, Oxford, U.K., 1991, chap. 5, pp. 106–167. With permission.)

plasma treatment in what is known as plasma-enhanced CVD. A variety of the CVD technique is referred to as spray pyrolysis; a fluid containing the precursor is then sprayed onto a hot substrate. This method is used on a very large scale for deposition of tin oxide-based films on hot glass, either in a separate process as indicated in Figure A1.1b or continuously in conjunction with float glass production.

Electrochemical techniques include cathodic electroplating from a chemical solution [23] and anodic conversion of a metallic surface—especially of Al—to form a porous oxide [24]. Numerous alternative techniques exist as well. Anodization of aluminum can be done according to different strategies. Thus "mild" anodization can lead to a self-ordered pore structure on the nanoscale, but this technique is slow and confined to a limited set of process parameters; "hard" anodization, on the other hand, is a fast and industrially viable process leading to thick layers with a disordered pore arrangement. As realized recently, a combination of these technologies in "pulse" anodization can lead to particularly interesting nanostructures, such as those illustrated in Figure A1.3 [25] showing alternate layers with well-developed nanostructures.

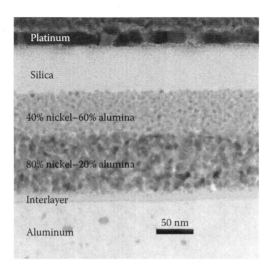

FIGURE A1.2 Scanning electron micrograph of the cross section of a sol-gel-produced multilayer coating of Ni-Al$_2$O$_3$ and SiO$_2$ deposited onto an Al substrate (with an interlayer). The top layer of Pt was applied by sputtering in order to allow the imaging. (From T. Boström, private communication, 2009.)

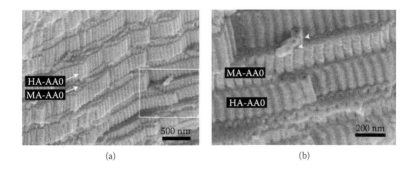

FIGURE A1.3 Panel (a) shows alternate layers of anodic aluminum oxide (AAO) prepared by "pulse" anodization; the layers are representative of "hard" anodization (HA) and "mild" anodization (MA). Panel (b) is a magnification of the displayed area. (From W. Lee et al., *Nature Nanotechnol.* 3 [2008] 234–239. With permission.)

FIGURE A1.4   Photo of a manufacturing plant for making multilayer coatings on full-size glass panes.

The cost of making thin films is often of the greatest importance for assessing whether they can be used in green nanotechnology or for other practical applications. The cost of thin-film manufacturing is a complicated issue and depends critically on the production scale. In the scientific literature it is common to read statements that sol-gel deposition is "cheap" because inherently expensive vacuum equipment is not needed. However, the necessary thermal treatments of the sol-gel coatings lead to slow manufacturing, which may be disastrous for mass fabrication. Physical vapor deposition is often the preferred technology for large-scale manufacturing. Figure A1.4 illustrates a coating plant for making thin films on window glass by use of magnetron sputtering.

Web coating of flexible substrates can be used to make films on very large surfaces [26]. Figure A1.5 illustrates one variety in which the web is transferred to a chilled drum where the deposition takes place by sputtering or any other suitable technique, and where the coated web is then collected on a take-up roll. The whole process can take place inside a vacuum chamber.

For the PVD techniques discussed above it has been tacitly implied that the incidence of the deposition species is more or less perpendicular to the substrate. If this is not the case, it is possible to build up coatings with inclined nanostructures and, if a rotation of the substrate is invoked as well, one can achieve an entire zoo of nanostructures. A few of the possibilities are illustrated in Figure A1.6 [27]. Oblique-angle deposition can be applied on an industrial scale [28] and such techniques are of interest for applications requiring angular selective optical properties, as discussed in Section 4.7.

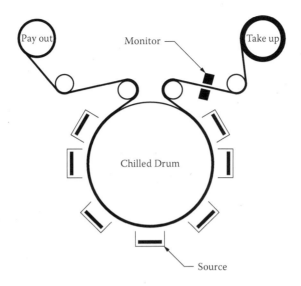

FIGURE A1.5   Schematic diagram of the internal components of a roll-to-roll coater with several sputter cathodes. (From S. F. Meyer, *J. Vac. Sci. Technol. A* 7 [1989] 1432–1435. With permission.)

So far the techniques that have been described are best suited for coating nonpatterned surfaces. However, masking is possible in order to produce patterns; alternatively etching or some other *subtractive* technique can be made to obtain a desired configuration. Rather than obscuring or subtracting material it is also possible to use an *additive* process, such as printing with an appropriate ink containing (nano)particles, normally followed by heat treatment to remove unwanted binder residues. Recent advances in printing technology, as well as the great amount of contemporary work on large-scale fabrication of nanoparticles, makes it likely that printing-related techniques will gain increased popularity in the future.

A1.2 SPUTTER DEPOSITION FOR MAKING NANOSTRUCTURED THIN FILMS

Sputter deposition, especially under reactive conditions and employing magnetron targets, can be applied to many materials and used for producing a great range of nanostructures. The nanostructure of the film depends critically on the deposition parameters, especially on the Ar pressure and the substrate temperature, and Figure A1.7 illustrates a "Thornton diagram" depicting what sputter-deposited films look like under an electron microscope as a function of these parameters [29].

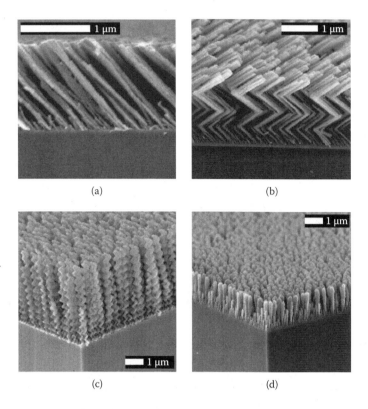

(a)   (b)

(c)   (d)

FIGURE A1.6   Nanostructured thin films made by glancing angle deposition and, in parts (b) and (c), simultaneous rotation of the substrate in order to make "nanochevrons" and a helical nanostructure ("nanotortiglioni"). (From J. Steele and M. J. Brett, *J. Mater. Sci: Mater. Electron.* 18 [2007] 367–379. With permission.)

Many applications of films require high durability so that compact films are desired, and historically sputter technology was developed to prepare films that were more durable than those made by evaporation. Parameters leading to "zone T" films are then preferred. But for other applications it is desirable to make films with a carefully chosen nanoporosity, and then one should use low substrate temperature and high Ar pressure in order to reach "zone 1." Films of the latter kind are required for example for electrochromic films and for gas sensors and photocatalytic surfaces (cf. Sections 4.11, 8.1, and 8.2); some specific nanostructures are shown in connection with the discussion of such applications.

Multilayer films are readily made by sequential sputtering from more than one target, and a full-scale production unit—such as the

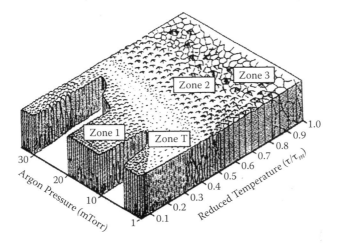

FIGURE A1.7 Schematic diagram showing nanostructures of thin films made by sputtering at different Ar pressures and substrate temperatures. The melting point of the material is denoted τ_m. (From J. A. Thornton, *J. Vacuum Sci. Technol.* 11 (1974) 666–670. With permission.)

one shown in Figure A1.4—can use a large number of targets under which the substrate is transported in a more or less continuous process. Composite films can be made by sputtering from one target of a compound material or via simultaneous sputtering from one or more targets. Mixed metal-dielectric films can be deposited reactively— for example, in Ar blended with a small amount of oxygen so that the deposited film comprises a random mixture of metallic and oxidized parts; the underlying processes can be accurately modeled [30]. Deposition of alloys can use alloy targets or separate metal targets. Chemical etching of some alloys can yield highly nanoporous conducting layers [31] with possible applications as catalysts, supercapacitors, and hydrogen storage media.

It is even possible to use sputter deposition to make films whose composition varies in a highly controlled manner over the cross section [32]. This latter possibility can be accomplished as illustrated in Figure A1.8, showing continuous deposition onto a long metallic ribbon. As the band moves past the magnetron cathode, the initial deposition is in Ar so that the film is metallic. Closer to the asymmetrically positioned oxygen inlet, the films get increasingly oxidized and—with properly adjusted parameters—the top layer can be almost purely an oxide so as to antireflect the underlying material. This novel technology has been used to make sputter-deposited surfaces for efficient conversion of solar energy into heat, as discussed in Section 6.1.

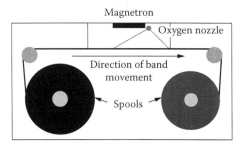

FIGURE A1.8 Schematic diagram of a roll-to-roll coating unit for continuous production of sputter deposited films with graded cross-sectional composition. (From S. Zhao, *Spectrally Selective Solar Absorbing Coatings Prepared by dc Magnetron Sputtering,* Ph.D. thesis, Uppsala University, Sweden, 2007.)

A1.3 NANOPARTICLE-BASED COATINGS

Vacuum coating methods, as discussed above, had an important historical role in making films based on nanoparticles. In fact, vacuum-based techniques provided some of the first insights into approaches to make nanoparticles under controlled conditions, specifically in the production of "gold blacks" for darkening of thermocouples by gold nanoparticles prepared by evaporation at pressures high enough to yield nanoparticle nucleation and growth in the gas phase rather than at a substrate [33].

A major step forward was taken when it was realized that the mean particle diameter and the size distribution could be understood and accurately determined, provided that the vapor source had accurate temperature control [34]. This led to the "advanced gas deposition" (AGD) technique, which is now used to mass produce nanoparticles that can be collected for later use or for coating directly onto substrates. Figure A1.9 illustrates an AGD unit; it is arranged for tungsten oxide nanoparticle production [35], but the technique can be used reactively or nonreactively to make nanoparticles of a large variety of pure metals, oxides, nitrides, etc. Evaporation takes place in the lower chamber into a laminar gas flow surrounding the vapor source. The vaporized species is then cooled via collisions with gas molecules so that it forms nuclei that subsequently grow in the gas flow. A thin transfer pipe collects nanoparticles in a zone at a controlled distance from the vapor source and transports them in a gas stream that ends in the upper deposition chamber, which is maintained at good vacuum. A separate evacuation pipe removes nanoparticles dispersed outside the growth zone. The nanoparticles are then deposited, via a nozzle in order to gain momentum, onto a substrate that can be moved so that they form a uniform film comprised

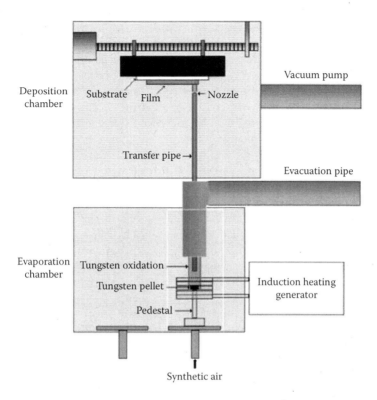

FIGURE A1.9 Schematic picture of a unit for advanced gas deposition arranged for making tungsten oxide nanoparticles. (From L. F. Reyes et al., *J. Eur. Ceram. Soc.* 24 [2004] 1415–1419. With permission.)

of nanoparticles. The technique can be implemented with multiple vapor sources and transfer pipes in order to prepare materials consisting of mixed or layered nanoparticles. A distinctive advantage of the AGD technique is that it separates nanoparticle formation and growth from film growth. This makes it possible to fine-tune particle interaction within the film, at least to some extent, which is beneficial for devices that require well-controlled electrical contact between adjacent nanoparticles such as conductometric gas sensors (cf. Section 8.1). Figure P2 showed one of the authors (CGG) in front of an AGD plant.

Gas phase synthesis of nanoparticles also can use high-temperature processes because particles can form in a flame or plasma. The reader is referred to a recent book for extended coverage of this subject [36]. Carbon nanotubes are usually made with specialized high-temperature processes; they can be used as transparent electrical conductors (cf. Section 4.5.2), in battery technology (cf. Section 8.4) and for many other applications.

Chemical approaches are now widely used to grow and precipitate metallic, inorganic, and semiconducting nanoparticles from solution and have been refined so as to limit size ranges, create elongated particles as well as spheres, and to overcoat nanoparticles or microparticles with nanoshells which enable new or improved functionality. Useful reviews and books are available [37]. Layers or coatings containing previously prepared nanoparticles are usually made by first dispersing them in a paint binder, a polymer coating solution, or a monomer solution prior to polymerization. The latter process can be used to produce master batches of concentrated nanoparticles in resin, which can subsequently be mixed with clear resin to obtain thin polymer foils doped with nanoparticles by extrusion. These foils can then be stuck onto surfaces or positioned between clear sheets. Nanoparticle-doped polymer sheets and other shapes of plastic can also be made by injection molding and extrusion from suitable resins. If dilute coatings are needed, it is important to ensure that the particles are dispersed, which usually requires a surfactant (i.e., a soap-type molecule) on their surface to ensure they do not stick together.

Our final example of nanoparticle production for functional coatings is one where work has only recently begun and for which the potential is great, though not easy to assess: biomimetic preparation. Figure A1.10 illustrates the growth of silver nanoparticles inside

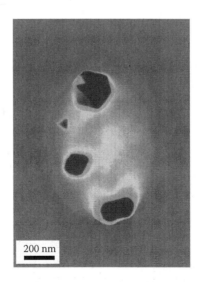

FIGURE A1.10 Transmission electron micrograph showing silver-based crystalline nanoparticles grown inside *Pseudomonas* cells. (From T. Klaus-Joerger et al., *Trends Biotechnol.* 19 [2001] 15–20. With permission.)

bacteria of *Pseudomonas stutzeri* [38,39]. Remarkably, particles of this kind can be single crystalline and can be used, for example, for making selectively solar absorbing coatings (cf. Section 6.1) [40]. In fact, there seems to a large number of organisms that can serve as ecofriendly nano-factories and produce inorganic nanoparticles either intra- or extracellularly, including magnetotactic bacteria, diatoms, fungi, and others [41].

REFERENCES

1. L. I. Maissel, R. Glang, Eds., *Handbook of Thin Film Technology*, McGraw-Hill, New York, 1970.
2. J. L. Vossen, W. Kern, Eds., *Thin Film Processes*, Academic, New York, 1978.
3. J. L. Vossen, W. Kern, Eds., *Thin Film Processes II*, Academic, New York, 1991.
4. R. F. Bunshah, J. M. Blocher, Jr., T. D. Bonifield, J. G. Fish, P. B. Ghate, B. E. Jacobson, D. M. Mattox, G. E. McGuire, M. Schwartz, J. A. Thornton, R. C. Tucker, Jr., *Deposition Technologies for Films and Coatings*, Noyes, Park Ridge, NJ, 1982.
5. D. L. Smith, *Thin-Film Deposition*, McGraw-Hill, New York, 1995.
6. H. K. Pulker, *Coatings on Glass*, 2nd edition, Elsevier, Amsterdam, the Netherlands, 1999.
7. H. J. Gläser, *Large Area Glass Coating*, von Ardenne Anlagentechnik GmbH, Dresden, Germany, 2000.
8. J. E. Mahan, *Physical Vapor Deposition of Thin Films*, JohnWiley & Sons, New York, 2000.
9. R. Messier, The nano-world of thin films, *J. Nanophotonics* 2 (2008) 021995 1–21.
10. L. Holland, *Vacuum Deposition of Thin Films*, Chapman & Hall, London, 1956.
11. R. Glang, Vacuum evaporation, in *Handbook of Thin Film Technology*, edited by L. I. Maissel, R. Glang, McGraw-Hill, New York, 1970, chap. 1, pp. 1.3–1.130.
12. B. Chapman, *Glow Discharge Processes*, Wiley, New York, 1980.
13. J. J. Cuomo, S. M. Rossnagel, H. R. Kaufman, Eds., *Handbook of Ion Beam Process Technology*, Noyes, Park Ridge, NJ, 1989.
14. M. Konuma, *Film Deposition by Plasma Techniques*, Springer, Berlin, Germany, 1992.
15. K. Wasa, S. Hayakawa, *Handbook of Sputter Deposition Technology*, Noyes, Park Ridge, NJ, 1992.
16. D. Depla, S. Mahieu, Eds., *Reactive Sputter Deposition*, Springer Series in Materials Science, Vol. 109, Springer, Berlin, 2008.

17. C. G. Granqvist, Energy efficient windows: Present and forthcoming technology, in *Materials Science for Solar Energy Conversion Systems*, edited by C. G. Granqvist, Pergamon, Oxford, U.K., 1991, chap. 5, pp. 106–167.

18. L. C. Klein, Ed., *Sol-Gel Optics: Processing and Applications*, Kluwer, Dordrecht, the Netherlands, 1994.

19. G. Frenzer, W. F. Maier, Amorphous porous mixed oxides: Sol-gel ways to a highly versatile class of materials and catalysts, *Annu. Rev. Mater. Res.* 36 (2006) 281–331.

20. T. Boström, private communication, 2009.

21. C. E. Morosanu, *Thin Films by Chemical Vapour Deposition*, Elsevier Science, Amsterdam, the Netherlands, 1990.

22. H. O. Pierson, *Handbook of Chemical Vapor Deposition: Principles, Technology, and Applications*, 2nd edition, Noyes, Park Ridge, NJ, 1999.

23. F. A. Lowenheim, Deposition of inorganic films from solution, in *Thin Film Processes*, edited by J. L. Vossen, W. Kern, Academic, New York, 1978, chap. III-1, pp. 209–256.

24. S. Wernick, R. Pinner, *The Surface Treatment and Finishing of Aluminium and Its Alloys*, 4th edition, Draper, Teddington, U.K., Vols. 1 and 2, 1972.

25. W. Lee, K. Schwirn, M. Steinhart, E. Pippel, R. Scholz, U. Gösele, Structural engineering of nanoporous anodic aluminium oxide by pulse anodization of aluminium, *Nature Nanotechol.* 3 (2008) 234–239.

26. S. F. Meyer, *In situ* deposition monitoring for solar film production by roll coating, *J. Vac. Sci. Technol. A* 7 (1989) 1432–1435.

27. J. J. Steele, M. J. Brett, Nanostructure engineering in porous columnar thin films: Recent advances, *J. Mater. Sci: Mater. Electron.* 18 (2007) 367–379.

28. T. Motohiro, H. Yamadera, Y. Taga, Angular-resolved ion-beam sputtering for large-area deposition, *Rev. Sci. Instrum.* 60 (1989) 2657–2665.

29. J. A. Thornton, Influence of apparatus geometry and deposition conditions on structure and topography of thick sputtered coatings, *J. Vacuum Sci. Technol.* 11 (1974) 666–670.

30. S. Berg, T. Nyberg, Fundamental understanding and modeling of reactive sputtering processes, *Thin Solid Films* 476 (2005) 215–230.

31. M. B. Cortie, A. Maaroof, G. B. Smith, P. Ngoepe, Nanoscale coatings of $AuAl_x$ and $PtAl_x$ and their mesoporous elemental derivatives, *Current Appl. Phys.* 6 (2006) 440–443.

32. S. Zhao, Spectrally Selective Solar Absorbing Coatings Prepared by dc Magnetron Sputtering, Ph.D. thesis, Uppsala University, Sweden, 2007.

33. L. Harris, R. T. McGinnies, B. M. Siegel, The preparation and optical properties of gold blacks, *J. Opt. Soc. Am.* 38 (1948) 582–589.
34. C. G. Granqvist, R. A. Buhrman, Ultrafine metal particles, *J. Appl. Phys.* 47 (1976) 2200–2219.
35. L. F. Reyes, S. Saukko, A. Hoel, V. Lantto, C. G. Granqvist, Structure engineering of WO_3 nanoparticles for porous film application by advanced reactive gas deposition, *J. Eur. Ceram. Soc.* 24 (2004) 1415–1419.
36. C. G. Granqvist, L. B. Kish, W. H. Marlow, Eds., *Gas Phase Nanoparticle Synthesis*, Kluwer, Dordrecht, the Netherlands, 2004.
37. C. Cushing, V. L. Kolesnichenko, C. J. O'Connor, Recent advances in liquid-phase synthesis of inorganic nanoparticles, *Chem. Rev.* 104 (2004) 3893–3946.
38. T. Klaus, R. Joerger, E. Olsson, C. G. Granqvist, Silver-based nanoparticles, microbially fabricated, *Proc. Natl. Acad. Sci. USA* 23 (1999) 13611–13614.
39. T. Klaus-Joerger, R. Joerger, E. Olsson, C. G. Granqvist, Bacteria as workers in the living factory: Metal-accumulating bacteria and their potential for materials science, *Trends Biotechnol.* 19 (2001) 15–20.
40. R. Joerger, T. Klaus-Joerger, E. Olsson, C. G. Granqvist, Optical properties of biomimetically produced spectrally selective coatings, *Solar Energy* 69(Suppl.) (2000) 27–33.
41. D. Mandal, M. E. Bolander, D. Mukhopadhyay, G. Sarkar, P. Mukherjee, *Appl. Microbiol. Biotechnol.* 69 (2006) 485–492.

Appendix 2: Abbreviations, Acronyms, and Symbols

Abbreviations, acronyms, and symbols used only in the immediate vicinity of where they are defined are not given here.

A

A	Absorptance
A	Surface area
AGD	Advanced gas deposition
A_H	Hemispherical absorptance
A_{lum}	Luminous absorptance
AM	Air mass
A_q	Oscillator strength
A_{sol}	Solar absorptance
ac	Alternating current
α	Absorption coefficient
α	Deposition angle
α_{pol}	Polarizability

B

BAU	Business as usual
BIPV	Building integrated photovoltaics
β	Column inclination angle

C

C	Capacitance
C	Cost
C_{abs}	Absorption cross section
CB	Conduction band
CFL	Compact fluorescent light

COP	Coefficient of performance
C_{scatt}	Scattering cross section
CVD	Chemical vapor deposition
c	Speed of light

D

D	Dimension
DSSC	Dye-sensitized solar cell
d	Film thickness
d_c	Critical thickness (for percolation)
dc	Direct current
d_{dc}	Thickness of double layer
d_{eq}	Equivalent film thickness
d_i	Insulator thickness
Δt	Time correction
$\Delta \tau$	Temperature difference

E

E	Electric field strength
E	Emittance
E_a	Atmospheric emittance
EAA	Ethylene acrylic acid
E_{appl}	Applied electric field strength
$E_{a,sw}$	Atmospheric emittance in the sky window
EBA	Ethylene butyl acrylate
EC	Electrochromic
E_{cb}	Conduction band energy
E_g	Semiconductor band gap
E_H	Hemispherical emittance
E_{in}	Input energy
E_m	Emittance of heat mirror
EMA	Effective medium approximation
EMA	Ethyl methacrylate
EMMA	Ethylene methacrylic acid
E_s	Surface (substrate) emittance
ESL	Electron-stimulated luminescence
ETFE	Ethylene tetra fluoro ethylene
EVA	Ethylene vinyl acetate
e	Electron charge
e-beam	Electron-beam (deposition)

ε	Dielectric function ($= \varepsilon_1 + i\varepsilon_2$)
ε^*	Effective dielectric function
ε_h	Dielectric function of host material
ε_p	Dielectric function of particle
ε_0	Permittivity of free space
ε_∞	High-frequency dielectric constant

F

f	Filling factor
f	Frequency
f_c	Critical filling factor
Φ	Energy (or light) flux

G

GBO	Giant birefringent optics
g	Acceleration due to gravity
Γ_L	Luminous intensity
Γ_R	Radiant intensity

H

H	Height
H	Magnetic field strength
HOMO	Highest occupied molecular orbital
h	Planck's constant
hcp	Hexagonal close packed
$h\nu$	Photon energy
η	Efficiency
η_{CA}	Curzon–Ahlborn efficiency
η_{Carnot}	Carnot efficiency

I

I	Intensity
I_D	Diffuse intensity
I_{pb}	Parallel beam intensity
IPCE	Incident photon to current conversion efficiency
I_{ref}	Reflected intensity
IR	Infrared

I_{sol}	Solar intensity
ITO	Indium tin oxide (In_2O_3:Sn)

J

J	(Julian) day

K

K	Luminous efficacy
k	Extinction coefficient
k^*	Effective extinction coefficient
k_B	Boltzmann's constant
k_τ	Thermal conductivity
k_0	Wave vector

L

L	Depolarization factor
L	Lifetime
L	Longitude
LCD	Liquid crystal display
LD	Lorentz–Drude
LED	Light-emitting diode
L_L	Luminance
L_{loc}	Local longitude
LOR	Light output ratio (for luminaires)
L_R	Radiance
LSC	Luminescent solar concentrator
LSP	Localized surface plasmon
LUMO	Lowest unoccupied molecular orbital
L_w	Latent heat of water
L_x,L_y,L_z	Depolarization factors for axes x,y,z
λ	Wavelength
λ_c	Critical wavelength
λ_P	Plasma wavelength

M

M	Photon exitance
MG	Maxwell Garnett

MMA	Methyl methacrylate
MVOC	Microbial volatile organic compound
MW	Molecular weight
m^*	Effective electron mass
μ	Magnetic permeability
μ_0	Permeability of free space

N

N	Generalized refractive index
NA	Numerical aperture
N_e	Electron density
NIR	Near-infrared
n	Refractive index
n^*	Effective refractive index
n_h	Refractive index of host material
n_p	Refractive index of particle
ν	Particle volume

O

OLED	Organic light emitting diode
OSC	Organic solar cell
OTEC	Ocean thermal energy conversion
Ω	Solid angle
ω	Angular frequency
ω_L	Longitudinal phonon frequency
ω_P	Plasma frequency
ω_P^*	Shielded plasma frequency
ω_{PH}	Phonon frequency
ω_{SPR}	Surface plasmon resonance frequency
ω_T	Transverse phonon frequency
ω_τ	Damping frequency

P

P	Dipole moment
P	Power
$P(\lambda,\tau)$	Planck radiant exitance
PC	Polycarbonate
PCMB	[6,6]-phenyl-C_{61}-butyric acid methyl ester

P_d	Dipole moment
PEN	Poly ethylene naphthalate
PEO	Poly ethylene oxide
PET	Poly ethylene terephthalate
P_{in}	Input power
PMMA	Poly methyl methacrylate
P_{out}	Output power
PV	Photovoltaic
PVB	Polyvinyl buteral
PVD	Physical vapor deposition
P3HT	poly(3-hexylthiophene)
p	Aspect ration (of light pipe)
p	Polarization direction
$p(\omega)$	Oscillating dipole moment
pa	Per annum
p_{air}	Air pressure
Ψ	Rotation angle
ϕ	Angle (azimuthal)

Q

Q	Heat energy
QE	Quantum efficiency
Q_H	Heat removal

R

R	Radius of random unit cell
R	Reflectance
R	Thermal resistance
R_b	Reflectance from back side
R_d	Rate of condensation
R_f	Reflectance from front side
R_g	Universal gas constant
R_H	Hemispherical reflectance
R_{lum}	Luminous reflectance
R_{sol}	Solar reflectance
R_{therm}	Thermal reflectance
RUC	Random unit cell
$R_\square$	Sheet resistance
r	Radius

ρ	Resistivity
$\rho(\omega)$	Dynamic resistivity

S

S	Seebeck coefficient
$S(\lambda)$	Solar spectrum
SBS	Sick building syndrome
SFR	Skylight to floor area ratio
SPP	Surface plasmon polariton
SPR	Surface plasmon resonance
SSL	Solid-state lighting
SWNT	Single-walled carbon nanotube
s	Polarization direction
sw	Sky window (subscript)
Σ	Gas sensitivity
σ	Electrical conductivity
σ^*	Effective electrical conductivity
σ_{air}	Electrical conductivity in air
σ_{gas}	Electrical conductivity in gas
σ_{SB}	The Stefan–Boltzmann constant

T

T	Transmittance
T_a	Atmospheric transmittance
$T_{a,sw}$	Atmospheric transmittance in the sky window
TCO	Transparent conducting oxide
TIR	Total internal reflectance
T_{lum}	Luminous transmittance
TRIMM	Transparent index matched microparticle
T_{sol}	Solar transmittance
t	Time
τ	Temperature
τ_a	Ambient (atmospheric) temperature
τ_c	Critical temperature
τ_c	Output temperature (cold)
τ_{dp}	Dew point temperature
τ_e	Carrier lifetime
τ_h	Input temperature (hot)
τ_s	Surface (substrate) temperature

$\tau_{s,min}$	Radiative stagnation temperature
θ	Angle (polar)
θ_{crit}	Critical angle
θ_{wc}	Water contact angle
θ_z	Zenith angle
$\theta_{z,max}$	Acceptance value for zenith angle

U

U	Heat transfer coefficient (U-value)
U	Voltage
UHI	Urban heat island
U_{nr}	Nonradiative heat transfer coefficient
UV	Ultraviolet

V

V	Volume
VB	Valence band
VIP	Vacuum insulation panel
VOC	Volatile organic compound
υ_F	Fermi velocity

W

WOLED	White organic light emitting diode

Y

Y	Admittance
$Y(\lambda)$	Spectral response of the eye
Y_0	Admittance of free space

Z

Z	Thermoelectric performance factor
z	Distance
ζ	Number of reflections

Index